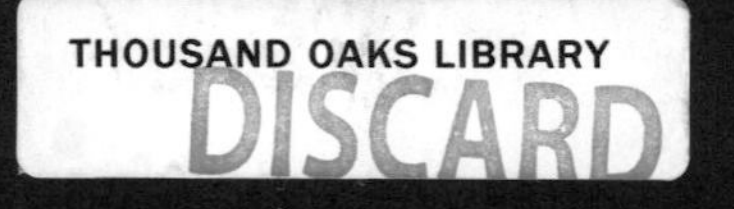

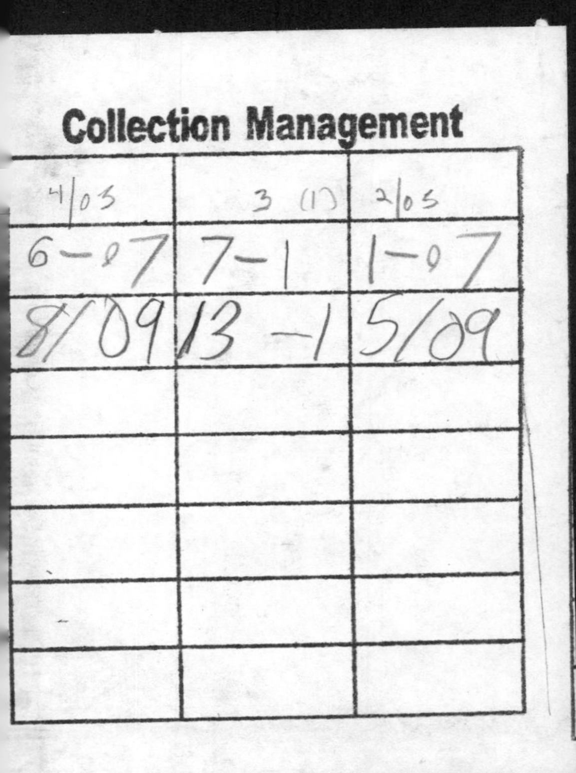

Owls

OF THE WORLD

Their Lives, Behavior and Survival

DR. JAMES R. DUNCAN

FIREFLY BOOKS

A Firefly Book

Published by Firefly Books (U.S.) Inc. 2003

First Printing

Publisher Cataloging-in-Publication Data
(Library of Congress standards)

Duncan, James R.
Owls of the world : their lives, behavior and survival / James R. Duncan. 1st ed. [320] p. : col. ill. , photos. , maps ; cm.

Includes bibliographical references and index.

Summary: Illustrated reference to 250 owl species worldwide, including common and scientific names, range maps and conservation status, and their mythology and legends.

ISBN 1-55297-845-1

1. Owls. I. Title.
598.9/ 7 21 QL696.S8D86 2003

Published in the United States in 2003 by
Firefly Books (U.S.) Inc.
P.O. Box 1338, Ellicott Station
Buffalo, New York, USA
14205

Published in Canada in 2003 by Key Porter Books Limited.

Electronic formatting: Jean Lightfoot Peters
Interior Design: Peter Maher
Jacket Design: Jacqueline Hope Raynor
Front cover: Stephen Dalton/Photo Researchers Inc.
Back cover: (left to right) Greg Haywood, Gerry Jones, S. Robinson, Gerry Jones, James Duncan, (middle) Jim Reimer
Spine: Gerry Jones

Printed and bound in Canada

To
Patsy, Connor and Brooke Duncan
For sharing my love and passion for owls and all life on earth and
in memory of Wayne C. Harris of Raymore, Saskatchewan, for his
contagious enthusiasm for owls and for generously sharing
his wealth of knowledge over the years.

Contents

Foreword

The Earth is home to over 200 species of owls. Like human beings, owls can be found almost all over the globe. Compared to most other creatures, owls have been poorly studied by scientists. This is due in part to their secretive and predominantly nocturnal nature. Yet owls are so familiar to people around the world that they have become icons, representing terrifying legends or omens of great fortune that persist in many cultures even today. Owl images often invoke strong emotions and memories in people.

Another influential and symbolic image of the 20th century was taken in December 1972. On a trip to the moon, at about 33,800 miles (54,400 km) away from our planet, Apollo astronaut Harrison Schmitt captured an image of the entire Earth shining against the infinite blackness of space. This image has transformed humanity's comprehension of life's tenuous grip on the planet. The biosphere—the thin layer of water, soil and air that supports an estimated 10 million species on Earth—suddenly appeared shockingly fragile and finite. Over the last 10,000 years, human populations have swelled to over 6 billion. Humans have emerged as a significant ecological force that increasingly influences the very support system that sustains all life in the biosphere.

This book on owls is a visual and descriptive celebration of a fascinating group of birds that share the planet with us. Owls are part of our collective experience—they are intertwined with our cultures and our ecology.

It is often forgotten that each generation sees much farther by standing on the shoulders of those who sought answers before them. Such is the rich heritage of owl study that I continue to cherish and enjoy, and I think that you will too. I hope this book encourages you to one night turn off your television or computer, put away your newspaper or book, slip into sturdy boots and clothes, and leave the comfort of your urban or rural home. No matter where you live, outside the owls are calling, inviting you to join them in a unique and personal natural adventure.

James R. Duncan
September 2002

Preface

Imagine, for a moment, the following scenarios, and that you are witness to some of the relationships that owls and people have experienced.

It is some 1000 years ago, and you are looking over the shoulder of an elderly aboriginal man in north-central Australia. You and he are in the cooling shade of a large rockshelter, and his paintbox of red ochre, white calcite, yellow limonite, black charcoal and grass brushes is nearby. All around you in the rockshelter are other paintings and engravings—this sacred rockshelter has been visited by people for over 10,000 years. As he works, a large striped owl, the *creation ancestor*, is beginning to appear on the rockshelter wall. The spiritual power of the owl is very strong; the elder believes he is not actually painting the owl onto the wall—but that the spirit of the owl is painting itself onto the wall through him.

Stepping back a little further in time, imagine yourself in the Pacific Northwest of the United States, along the shoreline of the mighty Columbia River. It is 1500 years ago. At sunset on this late autumn day, you are following a shaman to a sacred cave in the boulder-strewn grasslands of the expansive river gorge. Once at the cave, you squeeze through the small entrance hole and pass through the portal into the realm of the supernatural underworld. With a torch in your hand, you follow the shaman down the narrow, steeply descending shaft. As you descend to the main chamber, you pass other petroglyphs, and the images of these animals fairly jump to life in the flickering light of the flame. Once in the main chamber, the chanting shaman begins to chip the outline of an owl spirit onto the rock wall. As he carefully works his tools, he is summoning the spirit owl's visual powers to help him "see" the location of a misplaced bundle of amulets—so he can locate and return the treasured family heirloom to its rightful owner.

It is now the 20th century, and you are in Malawi, Africa. It is early evening and you are walking around a village, where the smell and haze of cooking fires fill the air. During your walk, you climb a small hill overlooking the village. From here, you notice that the cumulative gathering of firewood has restricted large trees to those that reside (and are protected from harvest) within the neighborhood cemetery. Here, as in other places in the world, owls are seen as bad omens, their calls portending sickness or death. This myth is strongly reinforced here, as trees in the cemetery are the only trees that grow large enough to contain nesting cavities for the resident barn owls (*Tyto alba*). At night, as the barn owls begin their evening hunt for small mammals, their white shapes are seen leaving the cemetery—thus reinforcing the perception that owls have an affiliation with death, dying, or sickness.

It is April 2, 1993, and you are sitting in an auditorium in Portland, Oregon, with a crowd of onlookers, intently listening as U.S. President Bill Clinton leads a public forum at the "Forest Conference." The focus of this forum is to address the human and environmental needs served by some 24.4 million acres (9.6 million ha) of federal forestlands in Oregon, Washington and northern California. At particular issue is the dispute over the management of the old-growth coniferous forests remaining on these public lands. In 1990, the northern spotted owl (*Strix occidentalis*) was listed as a species threatened with extinction; immediately, the management of its old-growth forest habitat was thrust to the forefront of scientific and political discussion. Coming out of this forum is a management strategy with substantial social and conservation implications. You then recognize that while the northern spotted owl was perhaps a catalyst, on a larger scale, the forum also reflected how a democratic society deals with forest management and species' conservation on a regional scale.

The strong images and beliefs that owls invoke have helped shape our world. Paintings and engravings from sacred tribal rock art sites reflect encoded meaning about tribal religion, law and traditions. Likewise, masks, stone carvings, coins and other media incorporating owls were developed to represent, at least in part, the spirits of the owls themselves. In more recent times, owls and their habitats have played important roles in wildlife conservation issues. A number of unique, and continually evolving, relationships are illuminated through the links between owls and humans through the millennia.

Owls are inherently interesting (if not outright fascinating!), and personal connections to science and nature are fundamentally important to conservation efforts in today's global society. In this book, biologists Jim and Patsy Duncan, and an array of contributing authors, have done an impeccable job of offering a tremendous wealth of personal experiences and an important synthesis of scientific information on owls and their habitats. I am confident it will help make a better world for owls.

David H. Johnson
September 30, 2002

Introduction

I am often asked: "Why are you so interested in owls?" This is difficult to answer. My emotional reaction to owls is nothing short of an aptitude, a passion, and a love affair of sorts. Not the kind of expression people expect from a scientist. After all, the widely held and incorrect stereotype is that scientists are always objective about their research subjects. And I am not alone. Consider Paul Johnsgard's personal dedication in his wonderful book, *North American Owls*:

To those who know owls to be something more
than ordinary birds
if something less than gods,
deserving our respect and love

I have had years to search for a more objective or rational reason why owls fascinate me, reasons acceptable to more orthodox fellow scientists. When needed, I trot out the somewhat sterile: "Owls are excellent research animals with which to examine the natural history of vertebrate predator-prey relationships." I also sometimes state, correctly, that: "As top predators with large home ranges, many owls are good indicators of healthy ecosystems and sustainable development." These are reasons consistent with my personal conservation ethic. But what really excites me about owls is that even in our modern times there remains so much to learn about them.

Owls are popular objects for children's art projects, especially after owls are studied at school. Drawn by Brooke Duncan, age 6.

Many young eyes watch out for danger at the entrance of a burrowing owl nest. An alarm call from a nearby parent will send the whole brood scurrying underground.

Owls are among the first birds that children readily identify and distinguish from other birds. When people see or interact with owls they express themselves differently than, say, with a crow, heron or chicken. Perhaps this is because an owl's head and face appears somewhat similar to ours. Like us, owls have large, forward-facing eyes. This gives them excellent binocular vision, an important aid for rapid depth perception while hunting visible prey or gauging distances to perches while flying. Their relatively large eyes relate to their nocturnal habits. Large eyes gather more light at night than small eyes. The staring, wide-eyed aspect of owls makes them look innocent or bewildered, like infant humans. Relatively large eyes are also known to stimulate the emotions of caring and affection in humans and perhaps in other mammals. Cartoon animators and advertisers, in recognition of this human trait, draw characters with oversized eyes. The good and lovable characters (e.g., Mickey Mouse) typically have big eyes, the villains, narrow and shifty eyes.

The eyes of an owl are surrounded by a feathered disk that gives it a primate-like head—big and round with no apparent neck. Like us, owls blink with their upper eyelids (most other birds use their lower eyelids). The exposed part of their short, curved bill appears remarkably nose-like,

Owl-like eye spots on this butterfly are suddenly exposed when it is disturbed by insect-eating birds.

completing the human-like image. These characteristics have resulted in owls becoming icons in human culture in diverse and sometimes bizarre ways.

Other animals besides humans have also taken notice of owls in subconscious and immediate ways. Many prey species react quickly to the presence of an owl by fleeing, giving alarm calls or even gathering in groups that drive away the potential predator. Behavioral biologists have elicited similar responses by using exaggerated large yellow eyes painted on flat or round objects, a phenomenon known as supernormal stimuli. Over the millennia, individual prey that did not flee or react quickly enough to the sudden appearance of owls were selectively weeded out by predation.

Many species have capitalized on the evolved effectiveness of the owl-eye stimuli. This signal has independently evolved in manids, butterflies and moths, beetles, flies, katydids, fish, frogs and a non-raptorial bird called the sunbittern. The northern pygmy owl has a set of owl eye-like spots on the back of its head to either ward off mobbing birds or to thwart potential ambushes from predators, including other owls. Not willing to wait for evolution to run a parallel course on humans, golfers, gardeners and others sometimes paint owl-like eyes on the back of baseball caps to ward off attacking songbirds while working or playing outside.

Nocturnal and retiring, hence often mysterious, relatively little is known about the 205 species of owls alive today. Indeed, new owl species have been discovered as recently as 2001 in remote and exotic corners of the globe. *Owls of the World* celebrates the lives and biology of these fascinating creatures that influence us and other species that share our planet.

CHAPTER 1

THE NATURE OF OWLS

Suspended on silent wings, this great gray owl pauses to reconfirm the location of its hidden but noisy prey.

OPPOSITE: **Some wide-ranging owl species vary in color across their range. Owls from the northern parts of their range, like this great horned owl, tend to be lighter in color.**

THE BASIC SHAPE OF OWLS IS typical of most birds because of the constraints imposed by their wondrous ability to fly. Birds have a reduced number of digits (fingers) and those remaining have fused or immobile joints, which reduces the weight of the outer wing limb (making them easier to flap) and provides a rigid structure for muscle attachment. Thus, birds' forelimbs are highly modified structures designed to support large and specialized flight feathers called primaries and secondaries. The tips of the outer large flight feathers (primaries) act like the

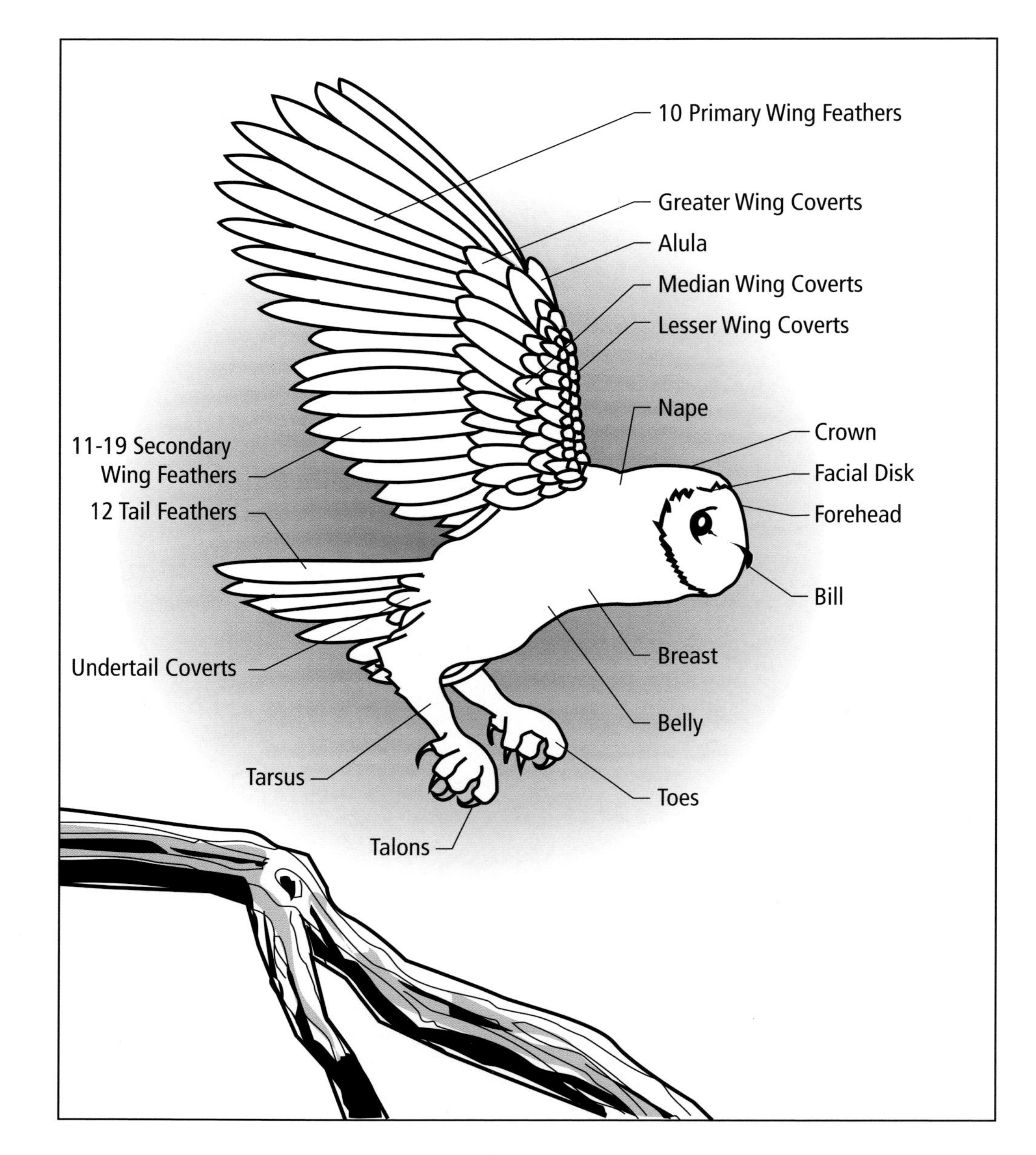

propellers of an airplane and thrust the bird forward as it flaps its wings. The inner large flight feathers (secondaries) are curved along their length (called camber), which, in combination with the whole wing, provides lift whenever air flows over them, thus keeping the bird aloft during flight. The capacity to generate lift must be greater than the mass, or weight, of the bird in order for it to fly.

Birds have special adaptations that reduce or redistribute their body mass, thereby facilitating flight. Hollow bones reduce their weight to the extent that the skeletons of some birds actually weigh less than their feathers! The large flight muscles and internal organs are arranged

This great gray owl dances nimbly on twigs revealing that despite its large size this 2 to 3 pound (1 to 1.5 kilogram) owl is seemingly as light as a feather.

to keep a bird's center of gravity low to increase flight stability. Thus, apart from a distinctive bill, neck and toes, an owl's general body shape, minus its feathers, looks remarkably similar to most other plucked birds.

Nevertheless, birds are quite variable in outward appearance, such as color, behavior and bill shape. Consider size variation, for example. The largest flying birds include the wandering albatross, California condor and Marabou stork, which have wingspans of 10 to 12 feet (3 to 4 m) and weigh 20 to 27 pounds (9 to 12 kg), as much as 7000 times heavier than the tiny 0.06 ounce (1.6 g) bee hummingbird. If we expand our comparison to include larger flightless birds, which presumably evolved from smaller flying ancestors, we find that the 330 pound (150 kg) ostrich is about 88,000 times heavier than the bee hummingbird.

Mammals, except bats, are not constrained by the limitations of flight. Hence, they show much greater variability in mass and body form: for example, the blue whale (264,552 pounds or 120 t) is over 70 million times the weight of the bumblebee bat of Thailand (0.06 ounces or 1.6 g).

The variation in body mass among owls alive today is somewhat smaller; the smallest owl in the world is the elf owl and the largest owls include the great gray owl, the Eurasian eagle owl and the great horned owl. Comparative measurements show that the great gray owl is longer than, but not as heavy as, the other large species.

Smallest and Largest Owls:

Species	Length	Wingspan	Weight
Elf owl	6.1 inches (16 cm)	15 inches (38 cm)	1.3 ounces (40 g)
Great gray owl	33 inches (84 cm)	5 feet (152 cm)	3 pounds (1450 g)
Eurasian eagle owl	28 inches (71 cm)	5.2 feet (160 cm)	9 pounds (4200 g)
Great horned owl	25 inches (63 cm)	5 feet (152 cm)	4 pounds (1800 g)

Feathers—Structure and Function

Except for the snowy owl and the northern hawk owl, species that frequent open habitat in northern climates in winter where they encounter brisk freezing winds and driving snow storms, owl body feathers are typically soft and frequently long, especially on their margins. The wing and other flight feathers are superbly adapted for silent flight. The upper surface of the flight feathers has a soft velvety pile. As the wing beats up and down, the flight feathers slide back and forth, rubbing over one another. In most birds this, in part, contributes to an audible "swishing" noise during flight. However, the soft upper surface of owl feathers produces little, if any, sound when the owl flies toward its intended prey or away from its enemies. The fringed leading edges of the owl's outermost large flight feathers also dampen the sound of flight by helping maintain smooth and relatively silent airflow over the wing surface.

The soft velvety upper surface of an owl's flight feathers reduces the "swishing" noise typical of other birds in flight when these large feathers rub against each other.

The serrated leading edge of an owl's outermost flight feathers helps reduce noise when the edge slices through the air during flight.

The Facial Disk

A special arrangement of feathers on the face also helps owls detect prey by collecting sound energy and focusing it on their ears. Owls have large ear openings that are covered with an anterior flap. On an individual owl, the ear openings differ in shape and sometimes size. The combined feathers on the flap and head make up what is called the "facial disk."

An earflap controlled by muscles and a filamentous ligament change the shape of the facial disk.

The facial disk of this great gray owl gathers sound energy and focuses it to the ear openings.

These feathers protect the face and ears when owls plunge into snow, water or vegetation after prey.

The distinctive feathers of the facial disk vary dramatically in shape, color, and density. Many are soft and almost skeletal in appearance, allowing sound to readily pass through them to the denser feathers underneath. The dense, flat and stiff feathers that form the ruff of the facial disk provide structural support and reflect sound toward the owl's ear openings. You likely have experienced this effect by cupping your hand behind your ear to hear quiet or distant sounds better.

In some species, like the great gray owl, certain concealed feathers at the top of the facial disk have relatively bright colors, suggesting that owl plumage coloration was not always so conservative. Along the edges of the fleshy earflaps are tightly packed bristle-like feathers that increase their sound-gathering effect. To some extent, owls can control the shape of the facial disk. When relaxed in sleep, the upper half of the disk appears to fold down somewhat. The ability to change the shape of the facial disk presumably helps fine-scale adjustments that assist the owl in locating hidden prey. In some owls the fleshy earflaps are connected to the skull by a membrane or ligament that undoubtedly functions to control the shape of the facial disk. Interestingly, the

The feathers that make up the facial disk of the great gray owl are highly specialized. From left to right; a—the surface feathers are barred in color, creating a pattern of concentric rings characteristic of this species and their filamentous barbs allow sounds to travel through; b—stiff feathers at the rear edge of the ear opening provide structural support to the disk; c—curved, stiff and densely webbed feathers line the back of the facial disk and focus sounds to the ears; d—orange-buffy feathers hidden from view within the facial disk suggest that this species had a more colorful ancestor; e—filamentous feathers at the top of the ear flap allow sounds to penetrate to the ear; f—smaller feathers along the front edge of the ear flap extend the surface of the structure.

The northern hawk owl, which hunts mainly by its keen eyesight, has a relatively small facial disk.

In open country with few perches, there can be stiff competition between owl species. Such conflict can result in owls killing other owls. Near our farm in Manitoba, Canada, great horned owls sometimes kill and eat snowy owls that invade their territories. Not only does this behavior provide the more aggressive individual with a meal, it also serves to reduce competition for local prey and hunting perches. Under low light conditions, ear tufts may be important in species recognition.

facial disk has evolved independently in some other birds of prey such as the northern harrier, which has been dubbed "The hawk that hunts like an owl."

Feathered "Ear" Tufts

Many owls have feathers on the top of the head that resemble a pair of "horns" or "ears." As noted above, owls' real ears, hidden from view by feathers, are located on each side of the head (see "Ears and Hearing"). The function of the "ear" feathers (tufts) remains unresolved, but some have suggested that they provide each species a unique look or silhouette that allows other owls to distinguish between species under low light conditions. Another explanation offers that the tufts make owls look like fierce mammalian predators, such as the lynx or bobcat, which may deter predators. Feathered tufts more likely contribute to the owl's camouflage by disrupting their outline, mimicking broken dead tree limbs. Tufts appear to also play a role in communication. A captive unreleasable great horned owl we looked after for many years would lay its tufts flat when irritated or frightened but would raise them when alert and watching distant animals.

Preening and Grooming Behavior

Owls frequently groom and clean their feathers to rid them of dirt and parasites. They use their bills and talons to help with this task. Owls have a special edge on their middle talon—a comb-like feature in barn owls—that is used during grooming and cleaning.

Like all birds, owls preen their feathers by gently grabbing them near the base and drawing the feather through the bill. This action rejoins feather vanes that have separated by reconnecting special hook- or Velcro-like structures. A uropygial gland, found at the base of the tail, secretes an oily substance that protects feathers and repels water. Owls spread this oil over their feathers using their bills. When irradiated by sunlight, some of this oil changes to vitamin D.

It is impossible for an owl to preen its own face and head, which is perhaps why owls seem to derive such pleasure from preening and grooming each other. Tame owls will even invite such preening sessions from their human caretakers. Mutual preening is an important behavior often observed between mated pairs and between adults and their offspring.

Molting

Like well-worn clothes, bird feathers wear out from everyday use. Sometimes the tips of feathers simply break off from repeated collisions with trees or other objects, or

Mutual preening is an important social behavior among owls. These eleven-week-old great gray owls are clearly enjoying each other's attention to detail.

Owls have a basic pattern of limited molt of major flight feathers, such that their wings have feathers of different ages that show varying degrees of wear.

The light-colored tips of the large wing and tail feathers of this great gray owl indicate that it hatched the previous summer.

may even be bitten off by would-be predators. Birds replace old and damaged feathers through a process known as molting. Many owls exhibit incomplete molt, in that they do not replace each feather every year, so they are never left flightless, a fatal condition for a winged hunter. Furthermore, the energy needed to grow new feathers is significant, and owls that are reproducing do not molt as many feathers as nonbreeding birds. In winter, we have recaptured male great gray owls known to have nested the summer before, and their shabby, worn feathers have, for the most part, not been replaced. An area of interesting research, the molt pattern of many owl species, remains completely unknown.

Plumage Color

Apart from the brilliant yellow eyes and bills of some owl species, owls are relatively drab compared to most other birds.

The ability to perceive color requires specialized cells in the retina of the eye called "cones," which require higher light intensity than the retinal cells (rods) used to detect black and white images (see "Retina and the Pecten"). As the intensity of light falls in the evening, the ability of organisms to perceive color dwindles. Since the majority of owls are active at night (nocturnal) or at dawn or dusk (crepuscular), bright colors would not be useful for courtship or territorial displays. Furthermore, owls typically roost during the day well hidden from predators and other birds that might harass them until the owl would be forced to relocate. Mottled patterns on the body feathers, and heavily barred wing and tail feathers, provide effective and protective camouflage.

Color Variation in Owls

Owls, such as the great horned owl, with large geographic ranges that span dramatically different habitat types, vary considerably in color from pale-whitish to dark browns and burnt-orange tones that seemingly relate to habitat type. This color variation appears to be a gradient, changing gradually over distance. Comparisons of different owl species in relation to geography suggest that northern species that survive in snow-bound winter conditions tend to be light and gray whereas more southern owls are darker and brown.

Some owl species occur in two or more distinctive color phases that are independent of age, sex or season. One such well-studied species, the eastern screech owl, essentially comes in two colors—gray or bright cinnamon-rufous (reddish)—a condition referred to as "dimorphic plumage." Genes inherited from the parents determine owl feather coloration, just like human eye color. For example, in humans brown-eye genes are dominant over blue-eye genes. At first, it was thought that in the eastern screech owl the rufous-color feather gene was dominant. But later, some intermediate brownish-colored eastern screech owls were discovered, suggesting that more than one pair of dominant-recessive gene

Many owls such as the crested owl occur in dark and light color phases.

The northern range of the eastern screech owl in Manitoba has been extended by their use of barns in winter for shelter. This frost-covered owl has fluffed out its plumage, trapping air warmed by its body to help it thaw a frozen mouse it is sitting on.

alleles likely controls feather color in this species.

The relationship of different color phases in owls to ecological factors such as temperature and humidity was noted as early as 1893. Detailed studies of color phases in owls have been conducted on the tawny owl, great horned owl, and the eastern screech owl. Over its entire range, the majority (about 60%) of individual eastern screech owls are gray. Biologists noticed, however, that there were few rufous-colored owls at the western edge of its range in drier habitats. Two not mutually exclusive reasons have been suggested to explain this. Owls with rufous feathers seem to be better camouflaged in the dim light and/or vegetation associated with cloudy weather (more frequent in humid areas). Secondly, reddish feathers wear out from abrasion faster than gray feathers, and are therefore not as durable in the relatively dry and dusty west. On a more local scale, one researcher in central Texas noted this relationship between feather color and the environment: rufous eastern screech owls were more common in relatively warm and wet urban habitat than in the surrounding drier and cooler rural areas.

Rufous-colored eastern screech owls are also relatively scarce in northern areas, because it appears that they die more often during severe winter cold snaps. To help understand why this was happening, biologists studied this owl under highly controlled laboratory conditions. They measured how much energy the owls used to keep warm as air temperature was lowered. They found that gray eastern screech owls used less energy to keep warm in cold temperatures (below 23°F/ –5°C) than rufous owls. At temperatures above 41°F (+5°C) however, gray eastern screech owls used more energy to keep warm than rufous owls. Rufous-colored feathers may offer less insulation against cold harsh winters, or plumage color may be linked genetically to physiological differences. Interestingly, such color-related mortality may affect males more than females. Because females are larger than males—in this species and in most owls—their smaller heat-radiating body surface relative to body mass suggests they radiate less body heat in cold weather than smaller males. Therefore, rufous-colored females likely survive better than rufous-colored males in cold weather. Indeed, biologists have reported that more females than males are rufous-colored.

Reversed Sexual Size Dimorphism

In most bird and mammal species the male is larger than the female. Humans incorrectly described this condition as *Normal* Sexual Size Dimorphism—incorrect because, in fact, females are normally the larger sex for the vast majority of all living animals described to date. Nevertheless, female owls are said to exhibit *Reversed* Sexual Size Dimorphism (RSD) because they are typically bigger and can be up to 40% heavier than male owls. Over the last 100 years biologists have put forth more than 20 theories to explain why owls and relatively few other birds (less than 10% of birds overall) are sexually "reversed" with respect to size.

Charles Darwin described Normal Sexual Size Dimorphism as the outcome of sexual selection in which larger males were more successful when battling for and acquiring mates, and therefore more frequently passed on their genes. Since then, other natural selective forces have been shown to play important roles in shaping the body forms of living creatures over evolutionary time. For example, researchers largely ignored the role of female mate choice in the evolution of RSD in owls until the 1970s, perhaps because prior to that time science had traditionally been a male-dominated discipline!

One explanation for RSD in owls relates to the extent to which the sexes partition their roles in breeding—males typically provide the female with prey during courtship, egg-laying and incubation, and female owls defend their eggs and young from predators while the male is away foraging. This role partitioning is perhaps facilitated by the owls' predatory adaptations (talons), enabling the female alone to vigorously defend the nest while the male is hunting. Another important factor favoring these distinct breeding roles is that many owls hunt prey that are alert to predators. A single owl hunting knows where it recently failed to catch prey, and therefore can choose to move to a new area of the

Patsy Duncan, a zoologist and my research partner, holds a breeding pair of great gray owls for banding. The male (right) is distinctly smaller than the female, a phenomenon known as reversed sexual size dimorphism.

A study in aerodynamic perfection, a common barn owl glides silently across a grassy meadow. Aggressive owls defending their young occasionally strike people without warning when they unknowingly wander close to their nests.

breeding territory where it is more likely to surprise its quarry. In contrast, a pair of owls that are taking turns hunting cannot communicate where they were hunting—one may hunt in an area where prey were previously alerted to the other owl's presence. A tag-teaming pair of hunting owls would therefore be less efficient at providing food for their brood than if just one hunted.

Factors that may favor female owls as the larger sex include egg production, incubation and nest defense. A larger female can produce more eggs, generate heat energy for incubation more efficiently, and defend her eggs and brood against larger predators. Conversely, smaller males are more agile and faster fliers—useful traits for aerial courtship displays and increased hunting efficiency. For some owls, large females might be better suited to starving themselves during periods when their young are growing rapidly and prey are scarce. However, starvation by nesting females may be a regular breeding strategy of some owls. My owl research partner, Patsy, and I regularly catch the same female great gray owls a few times over a breeding season. We routinely see females that weigh

Some prey species, such as this ruffed grouse, are almost constantly on the alert for predators such as owls.

3 pounds (1.5 kg) while incubating lose a third of their mass over one to two months. By partly starving themselves, attentive females ensure that their insatiable chicks eat more voles brought to the nest by their mates.

Diet and overall body size have been proposed as major factors influencing the degree to which the sexes differ in size. Larger owls tend to prey more exclusively on vertebrates and are more sexually size dimorphic than smaller owls. The great gray, snowy, eagle and great horned owls are among the most size dimorphic of the owls, while some, such as the diminutive insect-eating elf owl, show the least size dimorphism, with males only slightly smaller than females. But alas, there is no clean, linear relationship between owl size, diet and the degree of sexual size dimorphism. The relatively small (7 ounce or 200 g) boreal owl is as size dimorphic as owls 20 times its weight. The insectivorous mottled owl has the highest reported RSD among owls, and it probably never has to endure food shortages in its tropical habitat. Furthermore, the burrowing owl, which eats both vertebrates and insects, and other owls such as the rufous, powerful, barking and papuan hawk owls exhibit Normal Sexual Size Dimorphism!

Some authors have proposed that within an owl species, a larger female would eat larger prey than the smaller male, and thereby reduce competition for food between mated pairs. While this makes sense intuitively, there is currently insufficient information to test this idea. Confounding this subject is the

Recording the weight of breeding female great gray owls led to the discovery that females starve themselves during the nesting period in order to maximize the growth of their young.

observation that in the northern hemisphere there is an unexpectedly high degree of diet overlap between owl species, perhaps due to the lack of competition between owl species associated with the periodic superabundance of rodents and other prey.

Vision

The incredible ability of owls to see in dim light is well known but is also overestimated. Many still believe that owls can see in complete darkness, and that it is this ability that renders them almost blind in daylight. This belief, surprisingly widespread among many cultures, may arise from the behavior of some owls when they are discovered during daylight, or if found starving and injured. You can sometimes walk right up to an owl that seems not to see you. Sound unlikely? I have done this at least a dozen times, and many of my fellow owl enthusiasts have had similar experiences. There are at least two reasons that result in such outwardly strange behavior; the owls are simply young and/or naïve. As a result, biologists can sometimes reach out and gently grasp wild owls that are fully capable of flying. Young owls do not appear to perceive humans as a threat.

Patsy loves to recall one event during our years studying great gray owls that illustrates this point quite nicely. That summer we were trying to locate and band all of the young at active nests we had been studying. One pair had raised two healthy active chicks. We visited their nest one day only to discover it empty. Young owls typically leave their nests before they are able to fly, so we proceeded to search for them in the surrounding boggy forest. We found the first owlet about 15 feet (3 m) above the ground in a tree. It stood still while we found a long dead branch. The owlet remained "frozen" while I raised the long branch up to it and

The bold nature of some owls makes them vulnerable to being captured or shot. This northern hawk owl was fearless when approached to within arm's length.

then gently nudged the back of its legs. It then stepped backward onto the branch, and we lowered it to the ground. The young owl was capable of flight, but it seemed determined to take its chances on its "freeze and trust your camouflage" tactic. But that's not all. We wanted to find its nest mate, so I decided to imitate the adult male's "here's food" call. This sometimes stimulates young owls to respond with their food demand call, a sound like steam escaping from a pipe. After a few of my low-pitched "WHOOOOO's," it not only answered but also flew in, carefully navigating between dense trees. It promptly landed right on my head, peered over the peak of my baseball cap and looked down at me wondering, perhaps, what I had done with its supper!

Sick and injured owls are occasionally found in a starved condition that often results in owls being described as blind, stupid or tame. Near death, they can be so weak that they are unable to move. One winter, Patsy and I encountered a great gray owl in a tree on the side of the road. We unsuccessfully tried to lure the owl in to catch and band it. Thinking it was not that hungry, I approached it to see if it would respond to a live mouse in a bal-chatri trap, essentially a wire cage with nylon nooses (the mouse safe inside). The owl eventually flew in, landed beside the trap and seemed to look longingly at the mouse. After a long five minutes, I

A great gray owl with a severely damaged eye can survive in the wild because it relies primarily on sound to locate and capture prey. However, it may be more vulnerable to predation by eagles or other owls.

approached the owl, thinking to sneak up and net it, something we had never tried before. To our surprise, it allowed me to walk right up to it and gently place a net over it, after which it struggled only weakly. It was a juvenile, female bird, one that had lost its left eye, possibly accidentally when hunting. It was also extremely thin, weighing about half as much as a healthy individual. We brought the bird to a wildlife rehabilitation facility where eventually it regained its normal weight and was released.

Owls have a daytime visual keenness similar to that of most birds, and their sight is better than other night-roaming animals such as cats and rats. However, humans, hawks, eagles and falcons have daytime vision that is about five times as sharp as that of most owls. In spite of this, we have lured northern hawk and great gray owls from as far we can see them (over a half a mile) using only a mouse-sized lure dragged over the snow, suggesting that some owl species have keener daytime eyesight than humans.

Another common misunderstanding is that owls can see in complete darkness. Joel Peters immortalized this misconception in the following lines from his poem "The Birds of Wisdom":

Embodied silence, velvet soft, the owl slips through the night,
With Wisdom's eyes, Athena's bird turns darkness into light.

In winter, the Blakiston's eagle owl can been seen standing near open water waiting to catch a fish. The destruction of riparian habitat and the depletion of fish populations by humans have significantly reduced some local populations to dangerously low levels.

Of course, this is not true, for no animal can see in complete darkness because eyes, after all, are light receptors. Owls can, however, see well at light levels so low that humans perceive it to be completely dark. Some researchers have estimated that on a dark night owls can see as well as we can during an overcast day. The truth is that it is never completely dark outside at night, except perhaps in deep cave systems where a few species (bats, oilbirds, some fish and invertebrates) find suitable homes. Sources of light energy at night include sunlight reflected off the moon, auroral luminosity (northern lights) generated by electrical disturbances in the upper atmosphere, zodiacal light or the cone of faint light in the west after sunset and in the east before sunrise, starlight, electric and other forms of artificial light in areas where people live and work. Owls and many other nocturnal creatures have adaptations that allow them to use these sources of limited light to see. But even these creatures have their limits.

Thanks to the efforts of many patient and clever researchers, we now know just how well owls can see in near complete darkness. Barred, long-eared, barn and burrowing owls were used in some classic experiments started in the late 1930s to determine how well owls could actually see in the dark. Hungry owls were presented with dead mice placed in a room in which the level of light could be controlled. Dead mice were used to eliminate the possibility that the owls were using sound to find food. It was discovered that the owls could see the mice at a mere fraction of the light needed for human vision. It was also apparent that owls that hunt in open country and that were partly diurnal had poorer vision at low light levels than more nocturnal species that hunt in forested habitats.

The absolute visual threshold refers to the minimum light level needed to see

Reported Visual Light Level Thresholds (in millilamberts)

Barn owl	0.000,000,002	no avoidance of obstacles
Barn owl	0.000,000,003	avoidance of some obstacles
Barn owl	0.000,000,01	avoidance of most obstacles
Barn owl	0.000,000,02	avoidance of all obstacles
Tawny owl	0.000,000,17	lowest illumination to find dead prey
Ural owl	0.000,000,26	lowest illumination to find dead prey
Long-eared owl	0.000,000,27	lowest illumination to find dead prey
Night	0.000,000,43	forest understory on a moonless cloudy night
Human	0.000,015	lowest reported human threshold
Boreal owl	0.000,016	lowest illumination to find dead prey
Human	0.000,08	one researcher’s threshold
Eurasian pygmy owl	0.000,16	lowest illumination to find dead prey
Night	0.0043	open habitat on a moonless cloudy night

things. One researcher found that the tawny owl has an average threshold about two times better than ours. Most surprising was the discovery that the individual variation among the people and owls he examined was large, and that some humans with excellent night vision had the same or better light-gathering capacity than some of the tawny owls examined! Other researchers have reported that the barn owl can see up to 50 times better than humans at night.

Owls, however, can't hold a candle to some nocturnal mammals. The common house cat can see more than twice as well as the tawny owl at night thanks to a special reflective layer in the retina called the *tapetum lucidum*. The tapetum reflects light and results in the intense eyeshine visible when nocturnal animals are seen in light at night. While many nocturnal mammals have this specialized structure, the only birds thought by some to possess it are the nightjars and their allies. Yet biologists have reported eyeshine in barred, boreal, long-eared and great gray owls, while, apparently, some more diurnal owls such as the burrowing owl do not appear to have this structure. More detailed study is needed to better understand the extent to which owl eyes are adapted to low light levels.

Presumably, creatures like the tawny owl and the domestic cat have evolved the maximum light sensitivity and spatial resolution at low light levels possible for the vertebrate eye. While owls can see well enough to avoid hitting most branches and other dangerous objects at night, they cannot detect all dangerous obstacles in their flight path in dense forest understories on nights without moonlight. We have captured a few owls with severely injured eyes that may have had accidents while flying about at night. Perhaps the limits of their visual capacity dictate their 24-hour activity patterns. Many owls concentrate their hunting and other nocturnal activities during periods with relatively greater light levels, such as after sunset and before sunrise and during periods when moonlight is brighter (although sunset and sunrise are when small mammal prey are most active). Snow cover or water surfaces can also dramatically increase ambient light levels, reflecting light into the darkest corners of even dense spruce forests, allowing some species to hunt by day or night. Our studies have documented that the great gray owl actually gains weight over cold winter months, while living in forested areas with thick and widespread snow cover!

Eye and Skull Structure

Eyes, one of nature's truly marvelous designs, are important for the survival of the organisms that possess them. Birds typically have eyes that are relatively large and have the best long distance visual capacities among vertebrate animals. They can see color, which in some species can be important for finding food and mates, and in detecting threats. Some species have exceptional depth perception and motion detection abilities, while others, like owls, see well at night.

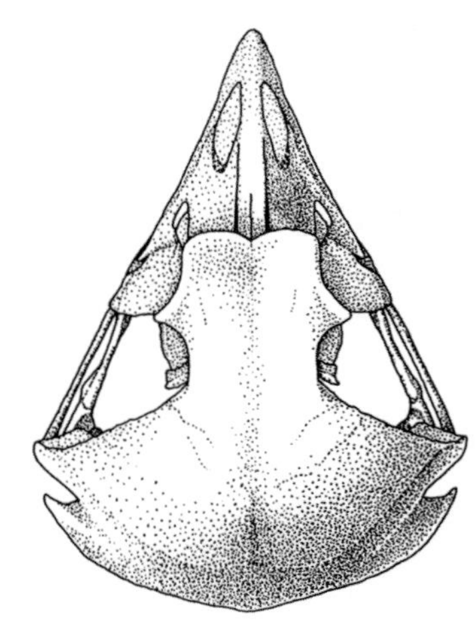

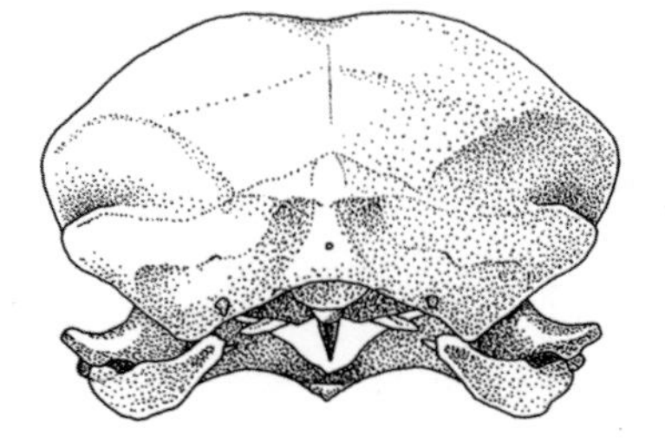

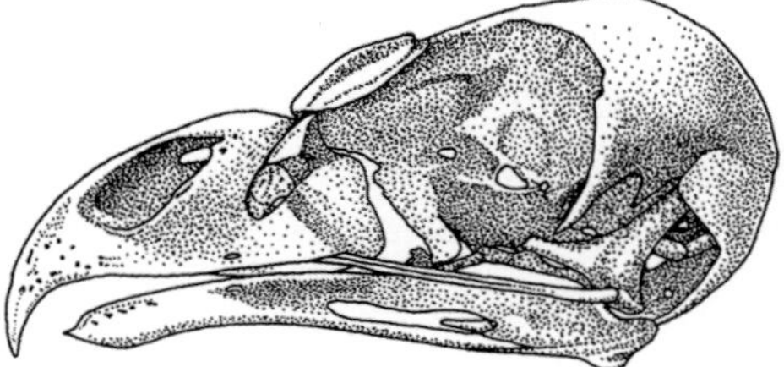

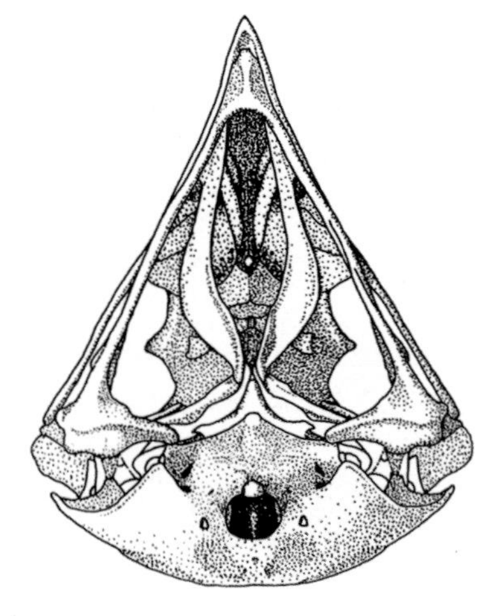

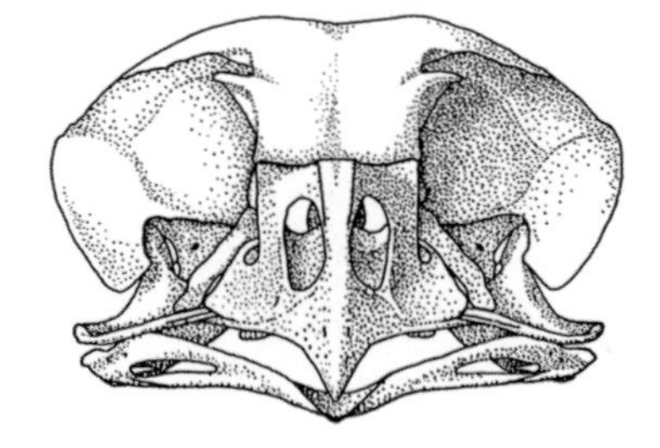

The highly modified skull of an owl. The need for large eyes to gather light and a strong bill to tear apart prey leave little room for the brain case, which has been pushed back and upwards. Five aspects of a great horned owl skull, drawn by Tom Reaume, November 2000.

The skull of an owl, it seems, is made up of the largest possible eyes, given the need to retain a foundation for a strong predatory bill, large ears and some brain capacity. Some owls have eyes larger than human eyes, yet these are contained within a dramatically smaller skull! To accommodate large eyes in a relatively small head, the owl's brain has been pushed upward and backward into a swollen cranium such that the ventral plane of the brain within the skull is about 45° from the horizontal line of the bill. For comparison, this angle is about 15° in smaller-eyed cormorants.

The variation of eye mass relative to body mass among owl species suggests that the degree to which they depend on eyesight versus hearing at night for hunting also varies considerably. The eastern screech owl has proportionately small and normal external ears but has extremely large eyes relative to its body (reported to be a 5.4:100 ratio). It typically hunts in forested understory at night. In contrast, the barn and short-eared owls have external ears that are extremely specialized for hearing, and an eye to body size ratio of about 1:100. These owls rely heavily on hearing to find prey in more open habitat with relatively greater average night light levels.

The eyes of most birds are relatively immovable, but the extremely large eyes of owls means that there is even less room for the muscles required to rotate the eyeball in its socket. Consequently, owls and other birds have particularly flexible necks to compensate not only for their large immovable eyes, but also in response to their specialized forelimbs. Wings are great for flight and other forms of locomotion, but a bird must use its bill and flexible neck to handle food or nesting material. The heads of birds and reptiles are joined to the neck with a single pivoting ball joint, whereas mammals have a more restrictive twin ball joint system. Therefore, mammal heads do not have the same freedom of movement as those of birds.

Owl necks, with 14 vertebrae or neck bones, are relatively short compared to those of other birds (swans have 25, for example). Even so, owls can readily turn their heads from a forward-facing position to look directly over their backs (180° from forward) and sometimes even further, to almost over 300° from forward. Therefore, an owl can gaze anywhere, without moving its body, by turning its head more than a full circle or more than a 360° arc! To ensure that the blood supply to and from their heads is not cut off during these contortions, owls have evolved an arrangement of the jugular veins with associated bypass connector blood vessels.

The Owl Eye in More Detail

Regular Eyelids and a Bonus Lid

Like all birds, owls have a third movable eyelid, the nictitating membrane, found in the inner corner of the eye. This thin and highly elastic membrane, semitransparent in owls and a few other birds, flicks across the eye, spreading tear secretions produced by its own gland and that of a lacrimal gland. The inner lining of the nictitating membrane has cells with brush-like structures that actually sweep the cornea with tears with each blink. This keeps the eye surface moist, free of dirt, and protects against infection from microscopic organisms. A unique mechanical arrangement of an elongated tendon, which circles the eye, pulls the nictitating membrane closed by contracting the attached muscle. This tendon passes through a sling-like groove

Human-like, owls wink by lowering their upper eyelid. A translucent third eyelid, the nictitating membrane, functions to clean the eye.

in a different muscle that functions to keep the tendon away from the fragile optic nerve.

In most birds, including owls, the regular eyelids are used to close the eyes only in sleep, with the nictitating membrane used for blinking. Photographs of owls about to catch prey reveal that they close the nictitating membrane just before impact, presumably to protect the cornea from scratches or other damage. Humans close their eyes mainly by lowering the upper eyelid, whereas most birds do so by raising the lower lid. Owls (and a few other birds such as parrots, toucans, wrens and ostriches) are more human-like in that their upper lids are usually lowered to close their eyes. Other times when owls usually close their eyes (partly or entirely) are when transferring prey, scratching their face, preening another owl, and copulating. I will leave it to the reader to draw any further eyelid action comparisons with parallel human behaviors!

The Hunting Life—The Importance of Binocular Vision

The life habits of an animal greatly influence the position of its eyes on its head. Prey species, like rodents, woodcock and curlews, typically have eyes set on the sides of the head, resulting in better peripheral vision, enabling them to see their enemies approaching from any angle

The visual field of an owl.

(up to a maximum 360° visual field). Conversely, predatory species often depend on the rapid assessment of the distance to prey in order to surprise and catch them. Binocular or stereoscopic vision enhances depth perception and results when the visual fields of two eyes overlap. Humans and most other primates have forward-facing eyes and can readily move their eyes synchronously within their sockets. As a result, almost 80% of the human visual field (180°) is binocular vision. It is thought that binocular vision in primates evolved primarily due to facilitate manual dexterity and less so for hunting.

Binocular vision is not the only way to determine distances or depth perception. Many birds estimate or assess distance by bobbing or turning the head. The advantage of binocular vision is the ability to gauge distance without needing to move the head. Head-bob movements can alert potential prey to an owl's presence, possibly thwarting a successful capture, or conversely could draw unwelcome attention from the owl's many predators.

Some predatory animals, such as chameleons, have evolved the remarkable ability to easily move their eyes within their sockets independently, enabling them to use binocular vision to pinpoint prey in almost any direction, while their bodies and necks remain rigid. In other words, chameleons have eyes that give them a full visual field of 360°. Conversely, the large eyes of owls are virtually immobile in their sockets. To maximize their capacity for binocular depth perception, owls have dramatically shifted the physical axis of their eyes within the skull, thus reducing their field of view to less than 110°. As mentioned earlier, owls have extremely flexible necks that compensate for this reduced visual field. Not a bad price to pay, perhaps, as an estimated 50 to 70% of their field of view represents binocular vision.

The shift in the physical axis of the eyes of owls has resulted in a pivotal shift

This young short-eared owl demonstrates just how flexible owl necks are. Viewing objects with an upside-down head helps raptors place the image on a part of the retina with a dense concentration of light receptor cells, resulting in a sharper image.

of the visual axis, and a lateral shift in the location of the fovea, the location of maximum resolution or sharpest vision on the retina. At the fovea, the overlying tissue is thinner and light receptor cells are packed more tightly in a funnel-shaped foveal pit. The result is that an owl perched on the ground or on a low perch sometimes has to rotate its head to see objects with greater clarity—it has to line up the visual image falling on the retina with the fovea.

The Lens and Accommodation

Light entering the eye is focused on the retina by the cornea and a highly refractive lens. Most birds require the ability to quickly focus the lens (accommodation) near and far in order to catch fast-moving prey or to fly quickly through dangerous obstacles. The ciliary muscles, located at the base of the iris, are very important for visual accommodation. They suspend the lens and alter its shape to focus on near or far objects. Birds typically have softer lenses than do mammals, an important factor needed to quickly focus the eye.

Accommodation is measured in diopters (recorded in meters, it is the reciprocal of the focal length of the lens). Cormorants have relatively soft lenses and coincidentally a high accommodation of 40 to 50 diopters; chickens and doves range from eight to 12 diopters, and most

Speed and agility are required to net the northern hawk owl for banding.

owls are at the low end with two to four diopters. For comparison, a human child has an accommodation of 13.5 diopters, which decreases to about six at the bifocal stage in life (40+ years). Owls are thus less able to focus on close objects, and often back away from food or prey that is too close before pouncing on it.

The ability of different owl species to focus on close objects was found to be greatest in small owls. Many small owls eat insects, and therefore must be able to focus and catch small and sometimes fast-moving prey at close range. The larger barn owl is a notable exception. It has a relatively high accommodation ability (over 10 diopters) among owls, and it can focus on objects as close as 4 inches (0.1 m) from its face. This ability is due in part to the barn owl's relatively small eyes and the ease with which smaller eyes can be thus focused. Conversely, the large-eyed great horned owl has an accommodation of around 2.2, and a near-point focus about 3 feet (0.85 m). Interestingly, Patsy and I published our first scientific paper on our observations of great horned owls catching and eating predacious water beetles in the air as the beetles flew up and out of the water. The owls were also seen walking up to beetles and adroitly picking them up at their feet from a gravel road! So, despite the great horned owl's reported poor accommodation and close-vision handicap, they—and smaller and less visually challenged owl species—appear fully able to catch and eat insects. As one fellow owl biologist who commented on these observations noted: "Maybe the great horned owls you saw hadn't read the research on owl-accommodation yet." Some researchers have also noted that the ring of ciliary muscles in the eyes of hawks and owls can significantly change the curvature of the cornea, which also plays an important role in enhancing the eye's ability to focus.

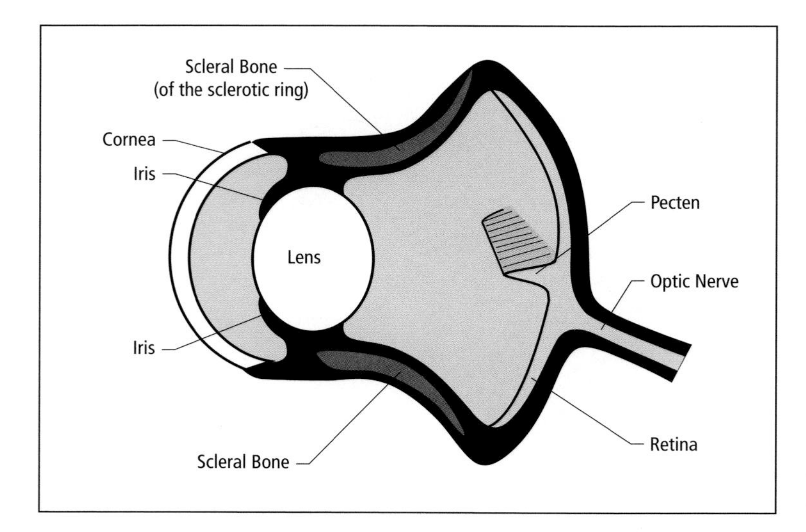

A cross-section of an owl's eye.

The fast-flying northern hawk owl has demonstrated an amazing ability to focus its eyes quickly at over 100 diopters per second, or over 10 times faster than humans. The nature of the ciliary muscles involved in focusing the eye, and perhaps their attachment to a stable and solid base in birds—the sclerotic ring—explains this difference. The ciliary body in the eyes of these birds comprises the faster-contracting striated versus smooth muscle-tissue type.

Eye Shape

Bird eyes vary in shape in relation to their natural history and may be flat, round or tubular. Birds that are active during the day have flat and occasionally round or globose eyes. Birds with narrow heads, like parrots or doves, have relatively flat eyes. Their eyes have a short focal length (the distance from the center point of the lens to the focal point or retina); consequently, a smaller image falls on the retina. Diurnal birds of prey and other birds with wider heads (finches, robins) have more rounded eyes and a longer focal length for increased visual acuity. Many nocturnal birds, including owls, have eyes that are relatively long and tubular in shape. This eye shape results in a long focal length; consequently, a relatively larger image is projected on the retina.

Bird eyes are anatomically similar to that of mammals (an outer tough scleroid coat, a middle vascular and pigmented choroid layer, and an inner light-sensitive retina, which sends signals to the brain via the optic nerve), but they share some eye structures with reptiles. One of these is a sclerotic ring of 10 to 18 shingle-like bony plates that support the shape of the eyeball. In some diurnal birds of prey, including the red-tailed tropicbird, and more notably in owls, the sclerotic ring forms a tube-like structure. The sclerotic ring is located in the sclera encircling the lens. These bones strengthen the eyes and provide attachment for the ciliary muscles, which help support and focus the lens.

Eye shape is also maintained by fluid pressure from two sources. The ciliary body produces a liquid called the aqueous humor, which is found between the cornea and the lens. There is a clear, viscous and jelly-like substance called the vitreous humor found in the space between the lens and the retina.

Iris, Pupil and the F-Number

The eye's vascular middle layer includes the pigmented or colored iris, a thin sheet of circular and radiated muscle fibers and connective tissue that controls the size of the pupil of the eye. The pupil is the visible dark circular opening at the center of the iris. Operating like a diaphragm that controls the aperture of a camera lens, the iris controls the amount of light entering the eye. Birds have striated muscle in the iris, which contracts faster than smooth muscle. Birds can adjust their pupil size in response to changes in light level much faster than mammals, whose irises have smooth muscle in them.

Owls have a large maximum pupil size that, combined with a large cornea and lens, enhances the light-gathering capacity of their eyes. One way to measure the light-gathering capability of eyes is the *f*-number used extensively to rate camera lenses. The *f*-number is calculated as the focal length divided by the pupil's (or camera lens) maximum aperture. The *f*-number 7 is expressed as *f*7 and it indicates that the focal length of the eye's lens is 7 times the maximum size of the eye's pupil or aperture. The smaller the *f*-number, the better the eye (or camera lens) can gather light. The average focal length of human eyes is about 1 inch (2.5 cm), and humans have an *f*-number about 2.8. A value useful for comparison purposes is the relative retinal illumination calculated as the reciprocal of the *f*-number squared. Comparisons of these values have been reported for a few species:

By these estimates the mainly nocturnal tawny owl has eyes that are from three to nine times better at gathering light than human eyes.

Retina and the Pecten

The inner third layer of the eye is the retina. Birds have a retina that is different from mammals in that it is relatively thick and has no blood vessels. Bird retinas have two characteristics in common with reptiles: the aforementioned bony sclerotic ring and another structure called the pecten. The pecten is a thinly folded structure that projects outward from the retina. It contains many small blood vessels that likely help supply the retina with nutrients and oxygen while removing carbon dioxide. The area of sharpest vision in the retina is called the macular disk. A tiny pit, the fovea, is located in the center of the macular disk and is the area of maximum optical resolution.

Birds have a retina that contains two

Focal Length (cm)	*f*-number	Relative Retinal Illumination	Species
0.85	1.38		Domestic cat
0.92	1.18		Tawny owl*
2.26	1.16	0.74	Tawny owl*
1.7	1.7	0.35	Little owl
	1.98	0.26	Pigeon
	2.13	0.22	Humans*
2.5	2.8	0.13	Humans*

* Note that the *f*-number can vary considerably between individuals, hence studies report different values for the same species.

kinds of photoreceptors: rods and cones. Rods are more sensitive to light, and cones are important for sharp color vision. Cone receptors far outnumber rod cells in most birds, and the fovea of a typical bird contains only cone cells. Owls appear to have evolved a relatively rod-rich retina and a fovea containing rod cells. The rod-to-cone ratio in the little owl is about 13:1, 10:1 for the barred owl. These are adaptations to help owls see better at low light levels. The price they have paid includes a reduced capacity for color vision, and a generally lower visual acuity.

Ears and Hearing

Some owls have prominent tufts of feathers ("ear tufts") on top of their heads that look like ears, although their real ears are found on either side of the head just behind the eyes.

Owls have evolved a highly developed hearing system primarily because the prey they catch are frequently concealed from sight by vegetation, soil, water or snow. Four features help owls hear better: large ear openings, specialized loose feathers around the ear openings, a movable flap of skin around the ear, and a facial disk of stiff feathers that collects and focuses sounds to the external ear opening.

Under extreme conditions owls can catch prey using only sound. In nature this occurs on dark moonless nights in forested areas as well as during broad daylight! Prey hidden by dense grass or thick snow cover are routinely snatched by owls.

To rule out the possibility that owls use odor or infrared (heat) radiation from prey to guide them, one biologist trained barn owls to catch mice in complete darkness in a soundproof room. He then tied a piece of paper to the tail of each mouse with a thread. As the mice moved about the dark room, the owls regularly struck the pieces of rustling paper behind them. The biologist further demonstrated that barn owls, and presumably other owls, need their facial disks of densely packed feathers to locate mice by sound; when he trimmed away the owls' facial disk feathers they were no longer able to consistently catch mice.

The peculiar structure of the feathers of the facial disk (see "Facial Disk") helps owls hear better. In addition, the shape of their facial disks can be controlled by muscles to assist with hearing. This control results in what appears to us as a remarkable variety of facial expressions, perhaps providing owls another means by which to communicate emotions to mates, parents, offspring and competitors. That owl heads are particularly large among birds of equal or larger size suggests the critical importance of large facial disks in gathering sound in order to detect prey by sound.

The actual ear opening also varies in shape and size in different owl species. Barn owls (Family Tytonidae) have round ear openings that are covered with a flap called the operculum. In typical owls (Family Strigidae) the ear openings are more varied, from small and round to large oblong slits.

Although humans can hear about the

A plunge hole in the snow marks where a great gray owl dove after an unseen vole—the blood suggests the vole did not escape. Thick snow cover forces many avian predators, like the short-eared owl and vole-eating hawks, to migrate to snow-free areas. In winter, great gray owls expand their territories into new habitats when the competition from other raptors is reduced.

same range of sound frequencies as owls, owls' hearing is up to 10 times more acute than that of humans for frequencies ranging from 0.5–9 kHz. Thus owls are able to hear even the faintest sounds (5–8.5 kHz) made by their prey scuttling under leaves and other cover.

To get a sense of how owls use their acute sense of hearing to find hidden prey, try closing your eyes and thinking about how you find something that makes noise. If your sense of hearing is not impaired, each of your ears can hear sounds equally well. If the source of sound, say a ticking clock, is louder in your right ear, you correctly assume it is to the right. By turning your head to the right until the sounds seems equally loud in both ears, you end up facing the sound.

Owls also use another auditory cue to pinpoint the location of concealed prey. When the source of sound is not directly in front of an owl, there is a very small difference in the time it takes for the sound to reach each ear. Sound reaches the closest ear soonest. By turning its head so that the sound arrives simultaneously in each ear, the owl can focus its attention on its next meal. Some studies have demonstrated that owls can distinguish such a time difference of about 30 millionths of a second! Owls can determine the location of a sound source within a 60° arc without having to turn their heads.

Some strictly nocturnal species like the barn owl or the boreal owl have evolved the most extreme skull ear specialization among vertebrates—asymmetrical ear openings in which one ear is higher than the other. This trait is also found, to a lesser degree, in some diurnal hunting species, like the great gray owl, which has perfected the capture of prey concealed under thick snow cover by sound alone. Other owl species, e.g., the tawny and long-eared owls, have evolved external ear asymmetry to achieve the same end. An owl's ability to locate sounds in the vertical plane appears to be more acute at higher frequencies (12–15 kHz). Species with asymmetrical ears can therefore locate prey in vertical and horizontal directions as both high and low frequency sounds are received simultaneously.

On numerous occasions I have watched hunting owls. Those that had not yet detected me (as far as I could tell) and were facing away from me reacted instantly when I uttered a single mouse-like squeak, even when thousands of feet (hundreds of meters) away. It appeared that the one squeak was enough for them to swivel their heads to locate the source (me) immediately and precisely. It seems that owl ears and brains have a calibrated calculation integrating inter-ear sound time and sound differences that allows instantaneous location of prey-like sounds. Makes me glad I am not a mouse!

Research has shown that the integration and interpretation of sound signals (time and intensity differences from left to right and up to down) is accomplished by a relatively complex part of the owl's brain, the medulla, which is associated with hearing. The owl medulla has about 95,000 neurons, more than three times as many as other birds'. The medulla appears to create a three-dimensional mental map image of the location of the sound source.

Once the owl detects the location of its potential prey, it may decide to fly at it. If its prey moves prior to capture, the owl's remarkable hearing ability allows it to adjust its flight path accordingly. The concentration of an owl listening to prey is a study in discipline. Some unfortunate owls succumb to predation by other owls or predators while intently listening for prey. This can occur even while an owl is flying at prey, its ears and eyes focused on the source of sound from its intended prey. Fellow owl enthusiast Gordon Court once observed a great gray owl flying toward prey suddenly get struck in flight by a northern goshawk, a large bird-eating hawk.

Feet, Toes and Talons

While owls share many features with other birds, they possess some definitive traits such as strong feet and sharp talons, used to capture mammals, birds, reptiles, amphibians, fish, insects, crabs and other invertebrates. The main tools of the bird predation trade include eight strong, sharp claws called talons. Although owls can bite hard, anyone who has handled owls will tell you that their talons deserve much more respect as they can inflict far greater damage! Talons grasp and squeeze prey, and often puncture vital organs. Conversely, talons serve as an effective defense, albeit not always successfully, against those animals that are foolhardy enough to attack an owl. Insectivorous owls tend to have unfeathered and delicate toes, whereas carnivorous species have stronger, feathered toes with more robust talons. Like most birds, owls also use their claws to tightly grip perches.

A black-banded owl demonstrates the reversable outer toe characteristic of owls.

Most bird feet have three toes pointed forward and one (the hallux) pointed backward. Among all the birds of prey (owls, hawks, eagles and falcons) only owls and the osprey can reverse the direction of their outer toes, resulting in two toes pointing forward and two backwards (see figure). This zygodactyl arrangement must be advantageous for securing a grip on branches, but its true value lies in helping catch concealed prey. The reversible outer-toe arrangement enables owls to create an eight-point box-like pattern with their talons when striking at prey, maximizing the chance of capture, especially at night with low light levels or when prey is concealed by vegetation, snow or water. This arrangement also helps fishing owls and osprey to grip slippery fish. These very different fishing raptors share another trait to help them hold on to fish—the scales on the bottom of their feet are modified as pointed spicules. The spicules are an extreme development of the rough and bumpy surface found on the soles of all owl feet.

The feet and toes of some owls, especially those living in colder higher latitudes (e.g., the snowy and northern hawk owls), are densely feathered, whereas owls living in warmer climes have more sparsely feathered feet and toes. Some tropical owls such as the Colombian screech owl have nearly featherless feet. This variation, known as Kelso's rule, can also be found within some species whose distribution spans many degrees of latitude, including the barred and stygian owls.

Because the forelimbs of birds are highly specialized for flight, their feet and bill are used to manipulate their food before eating it. Owls often swallow their prey whole, providing it is small enough.

Even owls as young as two to three weeks old routinely do this, urged on by the hungry cries of their siblings who will gladly steal their vole, grasshopper or small bird. I often wondered how young owlets manage to breathe while taking 10 minutes or more to swallow prey barely smaller than them.

Owls, young and old, usually seem to be in a rush to get food inside their stomach. Perhaps this is because they are vulnerable to predation while eating—an owl leisurely nibbling at a prey item may not be aware of an approaching hawk or lynx. Another reason to rush your meal is to prevent other predators from stealing your hard-won food. Birds such as ravens, eagles and some hawks are notorious for this behavior, which is called kleptoparasitism. I have watched ravens and northern shrikes try, albeit unsuccessfully, to steal recently caught voles from great gray and northern hawk owls. Owls, therefore, have many reasons to gobble their food down as quickly as possible.

There are, however, some occasions when owls do take their time to eat even small meals. Using a strongly hooked bill and sharp talons, owls can tear at and separate the most delicate morsels from a dead mouse or other prey. This was demonstrated to us one winter by a cooperative northern hawk owl that had just caught a meadow vole. The owl was perched with its prey on a dead tree only 16 feet (5 m) from us. Watching it through binoculars, I thought how it seemed to take great pleasure in nibbling gently around the dead vole's head, eventually tearing off and releasing bits of skin and fur, which floated down swirling in the breeze. After a few minutes, the owl unceremoniously bit off and swallowed the vole's head (owls often eat the heads first). Next, the owl delicately but deliberately opened the vole's abdomen and removed its intestines and stomach and dropped them down onto the snow below its perch. It proceeded to eat the rest of the vole in bits. After the owl had flown off to another perch, we walked up to inspect the vole's gut pile on the snow. Only the intestine and stomach were there, still moving in rhythmic peristalsis contractions. With surgeon-like precision, the owl had selectively removed the gut and eaten the kidneys, liver and other organs.

Owls more commonly prepare their prey with such delicate precision when feeding very small, newly hatched young. This behavior is dramatically different from their normal swallowing of whole prey. Newly hatched owls are somewhat naked (psilopaedic), blind and helpless (altricial), and can hold their heads up for only a few seconds at a time. The female owl therefore must feed her newly hatched young tiny pieces of prey that are relatively free of hair or other debris. Once torn from the carcass, the female holds small morsels of food over the mouth of the young begging chick. Mucous from the female's mouth coats the morsel of food and perhaps aids in the digestion of the food (mucous contains enzymes) or at least makes it easier for the chick to swallow. Female owls will eat the head and other prey parts too large for their young. Apparently, very small young owls are not fed the gastrointestinal organs of prey.

Both of the above examples demonstrate that with only a sharp beak and strong feet and talons, owls are capable of processing food with surprising dexterity. Despite this ability, however, owls typically eat prey whole or in big chunks as fast as they can. This creates some interesting problems that must be overcome when it comes to digesting their meals. Imagine what would happen if you ate chicken wings, bones, feathers and all! In owls, the selection of what is digestible and the rejection of those indigestible parts mostly takes place after the food is swallowed.

Digestion—Owl Pellets

Owls and many other birds, such as kingfishers and crows, have evolved a mechanism for dealing with the unpalatable bony and other hard parts of the prey they eat. Like most animals, owl stomachs contain acidic gastric juices—the basal pH is about 2.35—to assist in breaking down food. After the soft tissues are digested and passed into the intestine, the remaining mass of indigestible material (teeth, bones, exoskeletons, hair, etc.) is regurgitated or "coughed up" as a compressed mass called a pellet. The amount of fur on a meadow vole, for example, is significant; regurgitating pellets prevents the fur and other indigestible material from interfering with the

The indigestible bones and teeth within an owl pellet are used to identify what owls eat.

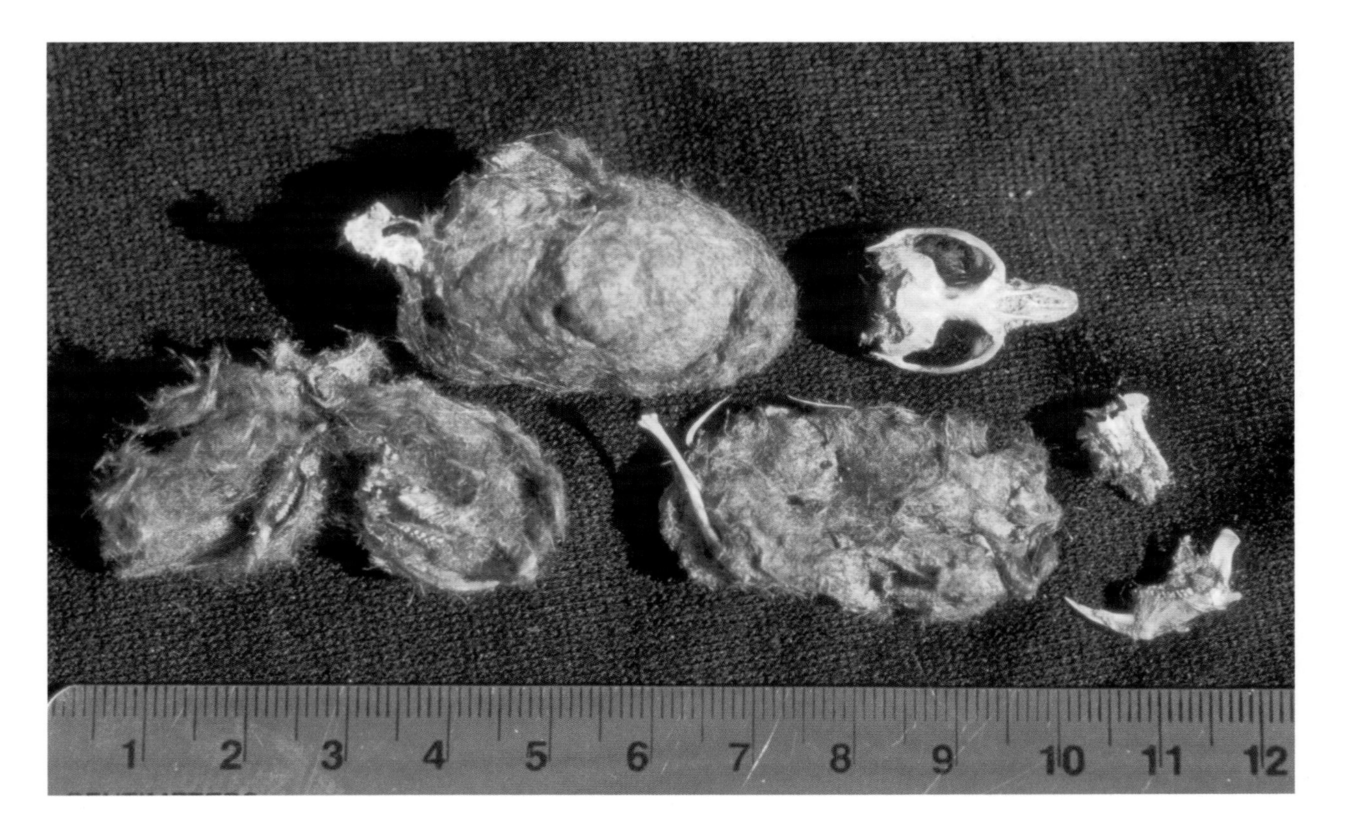

absorption of nutrients in the intestine and decreases the risk of damaging the digestive system or blocking the cloaca or vent.

How do owls regurgitate a mass of highly acidic waste without burning the delicate lining of their throat and mouth? In April 1987, I learned how firsthand. I was helping my then-supervisor Robert Nero at a shopping mall wildlife display that included a live unreleasable great gray owl. The live owl, tethered safely on its perch behind the display table, had grown up with people and was quite undisturbed by the large and noisy crowds that gathered in front of her. That day, while I was standing beside her, the owl leaned forward, lowered and shook her head, and coughed up a pellet.

I reacted before I could think and caught the warm and slimy 2.5 inch (6 cm) pellet in my hand. Recalling Fran Hamerstrom's tale about tasting a pellet from a golden eagle, Dr. Nero challenged in a voice that the crowd could hear: "I dare you to taste it!" Hesitating only briefly, I gently extended my tongue to taste the gray slick mass. The crowd was shocked. "It tastes sweet," I said. Curious, Bob took the pellet from me and bit into it! His face twisted into a grimace. He quickly spat out the pellet, took a slurp of water and stated in a muted voice that it was not sweet but extremely bitter.

An owl pellet is coated with a pH-neutral film of mucous as it moves up from the stomach. This protects the owl's throat and mouth from the low-pH acid-soaked mass of fur and bones. The lubricating mucous coating on a pellet must also make it easier to cough up.

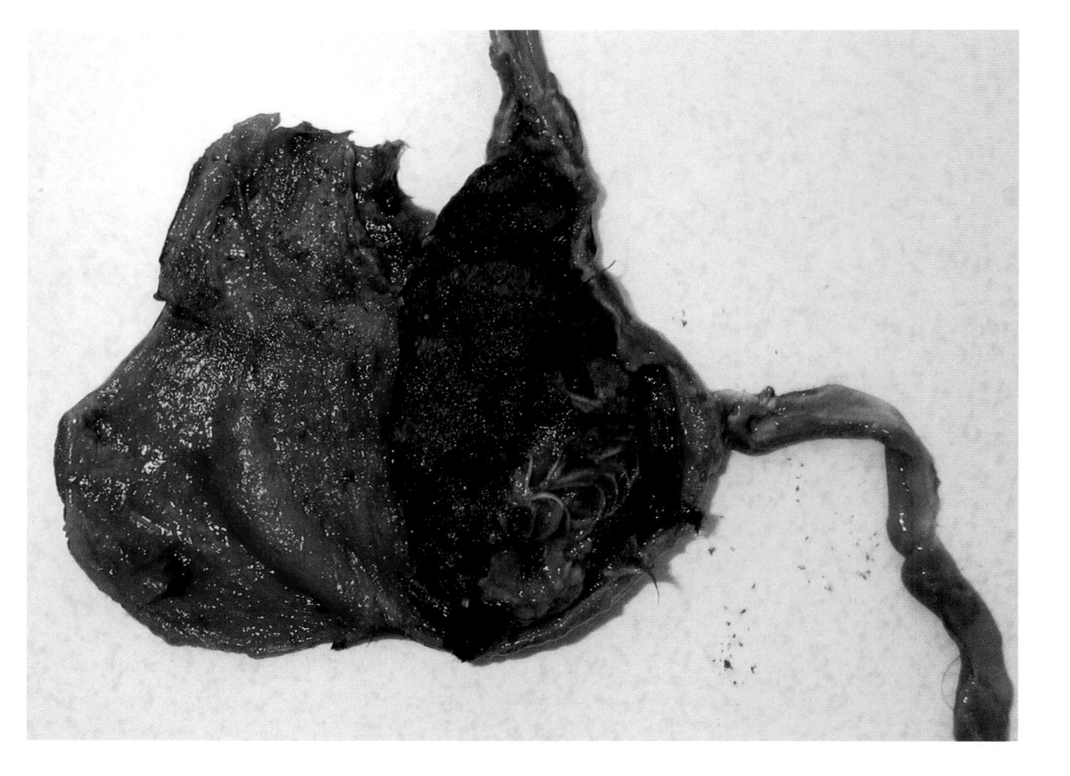

The stomach of a dead owl opened up shows how digestion of a rodent takes place. Bones of the rib cage are visible at the bottom near the entrance to the intestine, while fur (the dark mass) has been segregated to the upper parts of the chamber.

Therefore, whereas I had only licked the outside of the pellet, I experienced only the sweet-tasting mucous coating. Poor Bob had bitten energetically through the coating of mucous and had thus encountered the bitter-tasting contents. Needless to say, that was one experiment we did not repeat. After all, one's devotion to owl research can only go so far!

Owl Pellets and Owl Diets

Because the gastric juices of owls are only one-sixth as acidic as those of hawks, owl pellets contain about 10 times the bones found in those of hawks. Owl pellets therefore provide a tremendous opportunity for the study of owl diets because contained within is at least a partial record of the owl's last meal. Comparative studies, examining what owls actually ate to what was subsequently detected in their pellets, have shown that this is a reliable method for many owls. However, it is recognized that pellets do not contain the remains of soft-bodied prey such as earthworms or some amphibians.

An owl usually produces one pellet for each meal. But one meal might consist of many prey items—one snowy owl pellet I examined had the remains of at least 27 meadow voles! Research has shown that owls can be fairly regular in the interval between eating a meal and coughing up a pellet, but that larger meals take longer than smaller meals to digest. The time of day when the meal was eaten and the presence of other owls also affect digestion time.

Some biological supply companies sell owl pellets and dissecting kits complete with keys to identifying general skeletal parts as an educational resource for teachers. These are usually expensive, but the pellets are sterilized and odorless. Some teachers and parents collect their own owl pellets for their students or children, but care must be taken as the droppings of mammals can sometimes look like an owl pellet and contain tapeworm and other parasites dangerous to humans. Those handling owl pellets should use proper sanitation—schoolchildren have become seriously ill with salmonella from owl pellets.

Owl Hunting and Caching Behavior

Throughout this book you will find many references to the variety of prey species that owls catch and eat. Owls typically spend very little time hunting, usually perching only while waiting to see or hear prey. Only a few, such as the long-eared owl, hunt on the wing low to the ground with long glides alternating with flapping flight. Great horned owls have been described as uttering a frightful scream while hunting from forest perches, reportedly trying to flush nervous rabbits and other potential prey from their hiding places. Fishing owls sometimes wade through shallow water searching for prey. Insectivorous owls can yield great dividends by foraging on insects attracted to streetlights at night, providing that they can avoid being killed by speeding cars and trucks. Owls are opportunistic hunters by nature, and many species take a great variety of prey. Some, however, are diet specialists like the great gray owl, which will slowly starve to death in the presence of non-target prey, such as snowshoe hares or grouse, that they are physically capable of taking.

Watching owls catch their prey is fascinating. The intensity with which they hunt is remarkable. Their bodies tense as they decide whether or not to leap off their perch and swoop or pounce on their intended prey. In hunting flight they are streamlined, with legs held back. When closer to their prey, they will swing their legs forward and spread their talons in a broad pattern. Immediately before making contact, the feet are thrust out in front of the face, and the eyes close before making the grab.

Unlike many birds, such as pigeons or hawks, owls do not have a crop to store food. They either have to catch smaller prey items at regular intervals or hide the uneaten remains of larger prey to eat once the food in their stomach has digested. This was demonstrated one winter on our farm when an unusually pale great horned owl killed one of our large roosters at dusk just 18 feet (5.5 m) from our kitchen window. We watched the owl try to drag the rooster away to hide it in the woods, but it was too heavy. It fed on the fresh warm carcass for about 20 minutes, with frequent pauses to look around. Its

The ability to hover allows owls to reconsider or adjust an attack on prey. This barred owl was uncertain about attacking a shrew.

Barely skimming the snow surface, a northern hawk owl is about to grab a small mouse. In winter, many people feed wild owls live rodents and some owls become very tame.

nervousness was perhaps attributable to the possibility of being killed or robbed of its prey by a fox or other predator living in the wild areas of our farm. After feeding, it flew to a nearby perch where it could watch over its barely eaten carcass. There it stayed until the next morning, when just before sunrise it flew down and sat on the now frozen rooster to thaw it out. After about 10 minutes it started eating the partially thawed body, and alternated between thawing and eating for another 20 minutes. It again roosted nearby, where it remained until dusk when the whole thaw-eat process was repeated. This pattern continued for about six days until the owl was able to carry the rooster carcass deeper in the woods to finish it off away from our prying eyes.

Owls can also take advantage of unusual opportunities when small prey such as voles become available in unusually high concentrations. When small mammals fleeing wildfires or rising floodwaters concentrate in sheltered areas or on high ground, many predators avail themselves of such opportunities. Patsy and I watched a northern hawk owl in early spring hunting along a snow-covered ditch that was rapidly filling with melt-water. It caught a vole, which dashed across a road after being forced out of the ditch by the rising water. The owl had just finished swallowing its prey when a

A tawny owl demonstrates the typical capture method used by most owls to grasp their prey. After seizing the rodent in its feet, the owl often reaches down to bite its head to kill it before it can bite back.

second vole ran across the road. The owl swooped down and caught the second vole. Too full to eat, the owl landed in a nearby spruce tree and hid the vole in the branches. A few minutes later a third vole emerged from the ditch and it too was captured. This time the owl flew to the ground and placed the vole in a small depression in the snow at the base of a fence post. To our amazement, the owl actually used its bill to push snow on top of the vole, completely hiding it from view. Many other owls are known to hide or cache prey, retrieving it when they are hungry. Some species stockpile prey in nests with their young. Hungry young owls consume these food surpluses, especially when adults have difficulty catching prey during extended periods of bad weather.

Behavior

Concealment Displays

Despite being intimidating predators in their own right, owls are also prey for many species. Owls are cryptically colored to help them hide by day in obscure locations such as against the trunk of a densely branched or leafy tree. Undisturbed, they assume a resting, relaxed posture. But should a human or another terrestrial predator approach the owl, it typically reacts by slowly stretching upward while compressing its plumage, assuming a “tall-thin” posture. Some owls augment this concealment display by erecting their ear tufts and closing their eyes to mere slits, all the while keeping a peeping eye on the threat. If eye contact is made with

A thick blanket of snow provides a relatively warm humid environment at the "pukak" layer adjacent the ground. The pukak layer harbors an active community of animals such as spiders, shrews and voles. Voles can tunnel up into the snow column to feed on grass seeds not available to them in late summer. Only a few owls have adaptations that permit them to feed on small mammals hidden by snow.

A great gray owl's slightly spread wings typically keep it near the snow surface after a deep snow plunge after a vole. Afterward, it nervously glances around for possible predators.

Although primarily nocturnal, in times when food is scarce some owls are forced to hunt during daylight hours. This boreal owl was hunting in the afternoon beside a trail where many voles had been running across the snow.

the owl, it often completely opens its eyes and flees.

Some owls dramatically switch from a concealment display to a revealing display, seemingly based on their perception of risk. A distant eagle or other avian predator can cause owls, even those as large as the Eurasian eagle owl, to sleek or compress their plumage, perhaps minimizing the chance of being detected. However, when approached too closely by the winged threat, the owl reverses its strategy and appears to almost explode, its head, body and wing feathers held erect with the effect of making the owl look relatively massive. This latter display may be preferable to flight, given that some owls are indeed killed by eagles and hawks.

Fight or Flight—a Fright Response Display?

Some small owls, such as the northern saw-whet, boreal, northern pygmy, Eurasian pygmy and young northern hawk owls, react differently to an approaching threat. Dramatic and sudden movements are seemingly intended to let the predator know it has been spotted. The eyes are flashed open, exposing erect, whitish feathers between the eyes. One of the wings is jerked up such that the wrist lies in line with the bill, and the feathers of

A hunting great gray owl blends into a background of snow-covered branches. The plumage and markings of owls helps to break up their shape and camouflage them.

In response to a distant passing bald eagle, a known predator of owls, this great gray owl stands fully erect and compresses its plumage as if to avoid detection.

However, when an eagle approaches too close, the owl reacts differently, enlarging itself.

the crown are selectively raised, creating an erect ear tufts-like appearance. The true function of this display is puzzling, and I wonder if it may be the result of a conflict—a combination of parts of a concealment display come undone by an overwhelming urge to flee the approaching threat.

Threat Displays

Threatened owls have effective ways to tell other species and us to back away. The threat or defensive display makes the bird seem much larger than it normally appears. The owl may first react to a threat by raising or ruffling its body plumage. This is similar to a cat's raising the hair on its back when confronted by an angry dog. The owl, further stimulated, may then lower its head and spread its wings wide, turning them such that their upper surface faces the threat. The tail is often spread and sometimes raised. The head may sway from side to side while the owl hisses or snaps its bill. A barn owl can take this display even further by bowing deeply and shaking its head vigorously.

The Oriental bay owl has a remarkable display when it is frightened. Rocking its body to and fro, it suddenly tucks its face between its legs to more or less face backward. In this seemingly awkward position, it shakes its head and then suddenly flips its face forward, revealing its large and strikingly dark eyes and open bill. It may then escape by flying through remarkably dense foliage and vines. The common barn owl and some scops or screech owls are also known to display a less dramatic version of this bluff-and-flee behavior.

As mentioned earlier, desperate owls can also produce a fast series of loud clacking sounds when they repeatedly squeeze their extended lower bill against their upper bill and it snaps back in place. This expression of a nervous owl's fear or anger in a threatening situation is like a dog's growling or a cat's hissing. It also alerts other owls of danger. Young owls in

A recently fledged northern hawk owl reacts to an approaching threat (me!) with a fright response display. Note the raised right wing drawn up to the bill.

A fledged long-eared owl chick performs an inverted wing display, making it appear larger and more formidable to would-be assailants.

the nest or on the ground are quick to bill-snap when approached too closely, and some do it so intensely that they bite their tongues, causing profuse bleeding. This sound alerts nearby parents, who swoop in hooting and bill-snapping themselves to defend their young. Sometimes, albeit rarely, attacking adults fly in silently and strike without a sound! Likewise, bill-snapping adults, alert to an approaching threat, can invoke bill-snapping in their young. Less frequently, and seemingly under more stressful situations (e.g., actual handling of adults or young), bill-snapping is accompanied by hissing, caused by the rapid release of air from the lungs.

Breeding and Territorial Displays

The breeding behavior of many birds is partly controlled by hormones released by glands in response to signals from the environment. One environmental signal that reliably corresponds to generally good breeding conditions is changing photoperiod—hours of daylight relative to darkness. In the northern hemisphere, the increasing day length associated with spring causes egg- or sperm-producing organs to grow, releasing hormones that cause birds to behave differently.

Most owls live independently on their own until the breeding season. Typically, affected owls become more vocal, calling or singing to advertise their presence, locate potential mates, and establish or re-affirm breeding territories. Those species that remain on breeding territories with their mates year-round increase their interactions in preparation for nesting. Some owls, especially those that eat a wide variety of prey and depend on nest structures that are rare, such as tree cavities, benefit from guarding a territory year-round. It is not uncommon to hear the territorial call of such owls, e.g., the tawny owl, after the breeding season. Some believe that in this situation they are aggressively excluding the young of the year from their territory as well as re-establishing boundaries with neighboring adult owls. However, owl species that do not maintain year-round territories also call more frequently after the breeding season. The pre- and post-breeding season photoperiod are similar, therefore, despite their different functions; the increased calling behavior at these times may simply be a function of hormone release.

Courtship Displays

Courtship displays are intriguing to watch, but are unfortunately among the least-studied behaviors of wild owls. About a month before nesting, male owls start to vocalize to either attract a new mate or renew the bonds with an existing

female. Ritualized hunting and chase displays appear to be multifunctional, involving both territorialism and courtship, two closely related breeding phenomenon. The most dramatic of these are the dual-purpose (aggression and courtship) wing-clapping display flights of the long-eared, short-eared and marsh owls. It is unlikely that female owls are as confused as biologists in determining when the chasing behavior of males near their nesting territories subtly changes from aggression to affection regarding approaching members of their own species. For owls where both male and female remain together on a shared territory, other behaviors are indicative that breeding is imminent. Simply perching together permits an exchange of quiet vocalizations and tender mutual preening, behaviors thought to reaffirm pair bonds.

The investment of individual owls in their future mates has escalated significantly by the time that courtship feeding occurs. The male must prove his worth by delivering prey to his lady in waiting, and this event is often a prelude to copulation. However, the roles of the sexes in this regard are not simply hardwired or stereotypic. I once observed a seemingly desperate or love-struck female eastern screech owl deliver prey to a male of her interest, and then wait impatiently for the male to courtship-feed her the prey item, which he did. Science has clearly long ignored the importance of female mate-choice in the behavior and

An over the back wing stretch limbers up a great gray owl before it flies out to chase another owl from its winter territory.

evolution of owls, and perhaps other creatures!

The proximate goal of courtship behavior is to ready the pair for an intimate exchange—the transfer of sperm from the male to the female. However, even the most functional act of copulation has become a ritualized part of courtship. The common barn owl will copulate with a mate hundreds of times during the peak of the female's fertile period, namely about two weeks prior to egg laying. Sperm competition (overwhelming another male's sperm sometimes obtained by the female from extra-pair copulations) alone cannot explain why this behavior occurs more frequently than required to merely fertilize a mate's eggs, because copulations start before the female is fertile and continue for many weeks after the young have hatched! Furthermore, extra-pair copulations are known to occur in only the burrowing owl and the flammulated owl.

Courtship feeding also serves an important role for successful reproduction. Female barn owls apparently need to be fed by the male while resting near the nest in order to gain the weight and obtain the energy needed for egg production. Yet female great gray owls, and perhaps other species, acquire the weight needed for egg laying independently of their future mates over winter months prior to pair-bonding with a mate.

It takes many insects to raise a brood of rapidly growing African wood owls. After the young owls leave the nest their parents locate them by exchanging contact calls. Biologists sometimes imitate these calls to locate young owls to band them.

Breeding Strategies

The breeding biology of birds has been subject to careful recorded study since as early as the 1700s. Owls, like most birds, exhibit a range of breeding strategies. The common barn owl represents one extreme. It regularly breeds when it is only one year old, lays large clutches (up to 14 eggs), can breed more than one time per breeding season, but alas enjoys only a brief life. Species with this kind of breeding strategy usually live in extreme and unpredictable habitats. Populations of these owls can usually recover quickly from population crashes and can quickly exploit newly created habitats.

Other species, such as the eagle owls, mature more slowly over several years, have lower breeding rates and live longer. Strongly territorial, mated pairs will often use the same nest for their entire lives, and may forego breeding in years when prey is scarce. These species are more seriously impacted by habitat loss and slower to recover from associated population declines.

Nesting

While the female incubates her eggs, the male provides her with food as often as 10 times a day or more, depending on food availability. This sometimes exceeds the needs of the female and can lead to a stockpiling of food in or around the nest. However, once the young start to hatch, the male accelerates his prey capture rate in an attempt to keep up with the demands of hungry growing chicks. Each young owlet is fed many prey items each day,

Safe for now, this young great horned owl will face many dangers within its first year of life. Great horned owls tend to live longer and have smaller broods than most other North American owls.

more so for insectivorous owls, given their smaller prey size.

Owls in the nest are of different ages and sizes because female owls typically start to incubate after the first or second egg is laid. The resulting hatch asynchrony is an insurance policy for breeding owls against times when not enough prey is available to feed all the young in the nest. In such desperate situations the older, larger chicks are dominant and get fed first, while the

A barred owl nesting in a diseased tree.

Two surplus dead meadow voles lie in waiting for hungry mouths. When the chicks started hatching at this nest, the male great gray owl delivered more prey than the tiny hatchlings could consume.

youngest chicks will weaken until starved. These weak or dead chicks are occasionally eaten by their siblings, and therefore not wasted. While seemingly cruel, this strategy ensures that at least some strong young will fledge.

Distraction Displays

Breeding owls, especially females, commonly exhibit distraction displays in response to a predator, including humans, threatening their young. Typically, this behavior intensifies as the owlets get larger and leave the nest. Long-eared owls are remarkable for the intensity with which they thrash shrubs and tree branches, seemingly to mimic an injured animal. This display lured potential predators away from vulnerable eggs or young and has also fooled owl researchers into following what they thought was an injured owl. Some female owls may give a strange kissing-squealing noise (that can be mimicked by sucking on the back of your hand) to lure wolves, coyotes and other mammalian predators away from their young.

The nesting biology of a few owls has been very well studied, but this research has mainly occurred in temperate latitudes. This limits our ability to compare and contrast the variety of reproductive strategies of owls around the world. The individual owl species accounts presented in the second part of this book indicate that the breeding ecology for most of the tropical and southern hemisphere owls remains completely unknown. Support for such studies is sorely needed if we are to expand our knowledge of these diverse species, if for no other reason than to ensure their conservation.

The Classification of Owls

Organizing the amazing diversity of life on earth into useful categories has been a challenge to humans through the ages. Perhaps at some point in our distant past it sufficed to categorize things as dangerous, of little consequence or useful—such as food. Perhaps the amazing adaptability of humans stems in part from our aptitude and ability to name and organize things into categories. Much time and energy has been invested during the last two centuries in the scientific naming of living things, including owls. Many people are so familiar with the concept of naming species that they assume that once named, the organisms are known, stable and somehow familiar. Unfortunately, this assumption is only partly true. A species name simply gives us a common language in which to express our state of knowledge (or lack thereof) about the complex biological object named.

The list of 205 owl species used in this

book (see Part II and the appended checklist) should therefore be regarded with suspicion! I state this not because I question the credibility of the scientists whose cumulative efforts created it, but rather because any list of species actually misleads the reader. A species is an artificial unit created by people to categorize life, whereas life on earth is most accurately thought of as a continuum of variation and change.

Perhaps to the dismay of many scientists, the subtle differences between what we call "species" are, in the end, greatly influenced by opinion. In the end, even the most rigorous numerical methods, using measurements of physical structures such as bones, proteins or genetic material, are a subjective classification of life. This is the reason the list of owls in this book may vary from those found in other owl references. Alas, all species lists are artificial by nature's standard. Nonetheless, a list of species provides a handy reference point for discussion and a means by which we can study the relationships between species.

Scientists agreed to the use of "dead" languages like ancient Greek and Latin to name species because they are static languages, unlike living and evolving languages in use today. Scientific names provide a common language in which to express and exchange knowledge about a species within the scientific community.

Owls that nest on the stick nests of other birds risk losing eggs or young if the structure is too old or flimsy. Even in the best of nests, starvation can also take its toll on young owls. Of six eggs laid in this old broad-winged hawk nest, only two long-eared owl chicks survived to fledge.

They also provide a common language in which to express our confusion about how many species there really are!

Owls share some common features with hawks, eagles and falcons—a group known as diurnal birds of prey. These features include a strongly hooked bill for tearing meat, a fleshy cere at the base of the bill and robust hooked talons for grasping and killing prey. These similarities led famed taxonomist Linnaeus in 1758 to lump owls and diurnal raptors as closely related species. Over 130 years later, others documented evidence that owls were actually more closely related to a group of birds called the nightjars and similar species, a suggestion more recently supported by DNA evidence. The strong similarities between owls and diurnal raptors is therefore a consequence of evolution—the work of convergent selective pressures molding two relatively unrelated groups of birds over millions of years.

A Taxonomic Summary of the Owls

- Animals—**Kingdom Animalia**
- Animals with Backbones—**Phylum Vertebrata**
- Birds—**Class Aves**
- Owls and Allies—**Order Strigiformes** (9 families)
  - Owlet-Nightjars—Family Aegothelidae
  - Australian Frogmouths—Family Podargidae
  - Asian Frogmouths—Family Batrachostomidae
  - Oilbird—Family Steatornithidae
  - Potoos—Family Nyctibiidae
  - Eared Nightjars—Family Eurostopodidae
  - Nightjars and Allies—Family Caprimulgidae

- **Barn Owls—Family Tytonidae** (2 genera, 16 species)—possess a heart-shaped facial disk
  - **Subfamily Tytoninae**—tyto owls—14 species (*Tyto*)
  - **Subfamily Phodilinae**—bay owls—2 species (*Phodilus*)

- **Typical Owls—Family Strigidae** (25 genera, 189 species)—have a rounded facial disk
  - **Subfamily Striginae** (117 species)
    - **Tribe Otini**—scops owls and screech owls—68 species (*Otus, Pyrroglaux, Gymnoglaux, Ptilopsis, Mimizuku*)
    - **Tribe Bubonini**—eagle owls and allies—25 species (*Bubo, Ketupa, Nyctea, Scotopelia*)
    - **Tribe Strigini**—wood owls—24 species (*Strix, Jubula, Lophostrix, Pulsatrix*)
  - **Subfamily Surniinae** (63 species)
    - **Tribe—Surniini**—pygmy owls and owlets—38 species (*Surnia, Glaucidium, Xenoglaux, Micrathene, Athene*)
    - **Tribe—Aegoliini**—saw-whet owls—4 species (*Aegolius*)
    - **Tribe—Ninoxini**—hawk owls—21 species (*Ninox, Uroglaux, Sceloglaux*)
  - **Subfamily Asioninae**—eared owls—9 species (*Pseudoscops, Asio, Nesasio*)

An enlarged view of a vestigial wing claw from a great gray owl.

The Evolution of Owls

The oldest known bird discovered in the fossil record is *Proto avis*—a chicken-sized animal that lived during the upper Triassic about 225 million years ago.

It likely had both dinosaurial and avian features, including three curved claws on the forepart of each wing. Wing claws were also a feature on that famous fossil bird *Archaeopteryx lithographica* from the late Jurassic about 150 million years ago. This pigeon-sized bird had a toothed jaw, and its relatively complete fossilized skeleton has provided the best evidence that at least one kind of bird evolved from dinosaurs.

Remarkably, some birds alive today still retain certain features characteristic of these early ancestors. The hoatzin of South America has a featherless light-blue cheek patch perhaps similar to the featherless head of *Archaeopteryx*. The

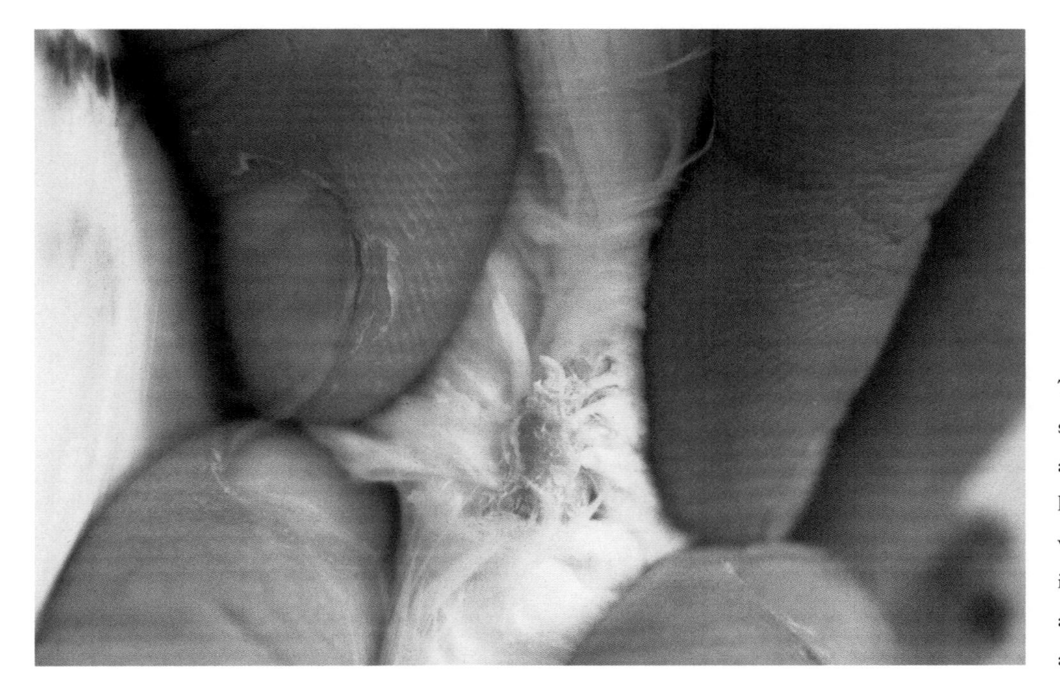

The wing claw of a snowy owl is too small and weak to serve any known function—this vestigial structure invokes images of an ancient bird-reptile ancestor.

hoatzin's young have functional hook-like claws on their wings, enabling them to swiftly climb up trees well before they are capable of flight. *Archaeopteryx* is thought to have been a weak flyer and therefore may have used its similar claws in this manner.

Owls, and many other modern bird species, have retained a vestige of this clawed wing adaptation. One can sometimes find nonfunctional claws up to 0.65 inches (16.5 mm) long on the second and/or third wing digits of great gray owls. These are most often found on young owls, and are seemingly lost as the owls mature. While the wing claws of owls may be nonfunctional, equally interesting is the perhaps-remnant behavior of young owls after they leave their nests. Unable to fly, they use their wings, especially their wrists, for balance and support as they climb steep leaning trees (an ancestral behavior pattern still useful without functional wing claws).

Fossils

The search for fossils must be a frustrating task. When one is found, it often consists only of small fragments. While owls are well represented in the fossil record, our knowledge of the sequence of their emergence and evolutionary relationships are nonetheless based on limited information. New fossil discoveries are rare, and only over time will they either corroborate or refute the ever-changing proposed evolutionary relationships.

For some time the earliest record of an owl fossil found in Romania was thought to be a species in the family Bradycnemidae from the Upper Cretaceous almost 100 million years ago. Some questioned the early identification, and the specimen was later confirmed to be a small dinosaur. Based on DNA evidence from species alive today, it is estimated that owls separated from a mutual ancestor of the closely related Caprimulgiformes over 65 million years ago. The oldest known owl fossils are from the Paleocene, which occurred 56 to 65 million years ago. Based on fossil and DNA evidence, these ancient owls evolved over 50 million years ago into two families of modern owls, the Tyto and Strigid owls. The first to separate was *Ogygoptynx wetmorei* (Ogygoptyngidae), found in the early Paleocene (Paleogene) deposits (58 million years ago) in Colorado, U.S.A. The exact origin and timing of emergence of either main group is uncertain. Early evidence suggested that the Strigidae emerged about the Eocene-Oligocene interface followed by the Tytonidae in Miocene. Later, this evidence was re-examined, and misidentified species were correctly assigned to Tytonidae. Authorities now conclude that Tytonidae preceded Strigidae in the evolution of owls, with Strigidae being of Neogene origin.

Perhaps the most dramatic owl fossils were those of huge barn owls that must have terrorized the night sky of Cuba about 10,000 years ago. Evidence suggests that giant barn owls existed in the Pleistocene in Caribbean and Mediterranean areas between 10,000 and 30,000 years ago. These large owls, known as *Ornimegalonyx*, stood over 3 feet (1 m) high, two to three times the size of modern barn owls and perhaps twice the size of the eagle owl. Using the most powerful talons of any known owl, *Ornimegalonyx* likely preyed on animals as big as sloths and giant rodents such as the capybara, which are over 4 feet (1.2 m) long.

There was a major radiation of owl species by the Eocene or late Paleocene (more than 50 million years ago) associated with a Tertiary radiation of mammalian prey species. Tytonidae were notably diverse in the Paleogene and especially in the Eocene epoch. Strigid owls were first recorded only in the lower Miocene, with *Bubo poirrieri* (France) and *Strix brevis* (North America) reportedly the earliest confirmed Strigidae about 22 to 24 million years ago. The genus *Tyto*, including the extant barn owl, also dates back to the middle Miocene.

The Range Limits of Owls

Climate, vegetation, parasites, pathogens, predators, competitors and prey availability are some of the factors governing the distribution of owls around the world. Cold winter temperatures, snow cover and food availability play an important role in determining the northern reaches of numerous species. A year-round study of the eastern screech owl in Manitoba, Canada, revealed that northernmost populations persisted only with the help of humans! Owls survived cold, snowy winters by eating mice and sparrows attracted to cattle barns, which also provided shelter for the owls. When spring arrived, the owls apparently deserted their winter farm accommodations and returned to natural breeding habitat in adjacent forested riparian areas and parklands. Global warming may result in range shifts, perhaps even an expansion of the eastern screech owl to the north. It will be interesting to observe just how closely the fate of this small human-adapted species is linked to climate-related habitat changes and an ever-expanding human population.

The western and southern range limits of the eastern screech owl appear to coincide with that of another closely related species, the western screech owl; however, range overlap between these two species has been documented in some areas. This creates an exciting opportunity for new studies on the evolutionary dynamics of how such similar species compete for or partition the resources in their environment.

Overall, the range limits of owls are poorly known, and new efforts to research and implement effective auditory surveys will greatly assist in their documentation. The distribution maps presented for each species in Part II of this book are therefore often a best guess based on available habitat and known owl occurrences.

Biogeography

Owls can be found in all parts of the world except on some oceanic islands. The study of species distribution, and the factors that

Zoogeographic regions of the world.

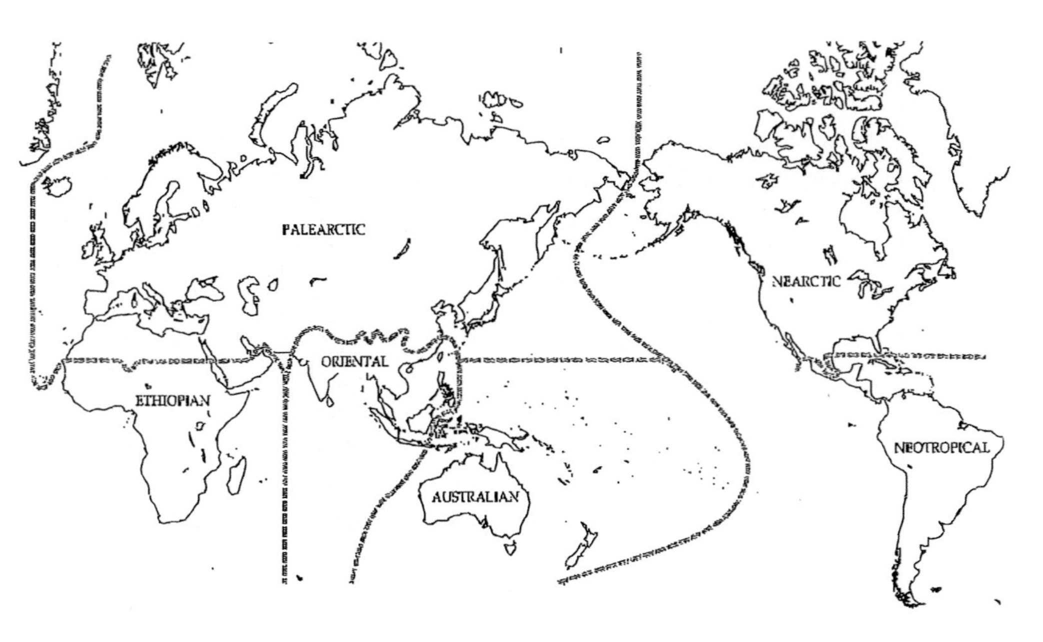

influence it, is called biogeography. Just like species concept, using an arbitrary boundary to separate natural entities, geographic regions and species' ranges in this case, is like trying to determine exactly where one color of a rainbow begins and ends. But try we must if we are to approximate and discuss the cause and effect of owl distribution on earth.

The world can be divided up in infinite ways, and we most frequently think of it in terms of our current political borders. Owls and other life forms are distributed according to their habitat needs (where to roost, hunt and nest) and are restrained by physical and ecological barriers to dispersal. Biogeographic regions are therefore based on the distribution of life that reflects the cumulative effect of a group of barriers. Given the mobility of birds, some biogeographic regions based on their distribution correspond exactly to great land masses. For the discussion of owls in this book I have selected one particular biogeographic system because of its historical importance, its long use as an international standard and its relative simplicity (only six regions).

In the first half of the 19th century a need for a geographical grouping of animals arose. Armed with sufficient geographical knowledge, at least for inhabited parts of the earth, and a reasonable number of described and classified species, some biologists attempted such projects as early as 1815. But it was not until 1857 that Philip Luthley Sclater devised the first widely recognized regional zoogeographical system based on the distribution of avian families. Sclater's system is still widely used today among students, researchers and natural historians.

What has made Sclater's zoogeographic system so enduring is that it's based on certain major barriers that have separated the animals of five

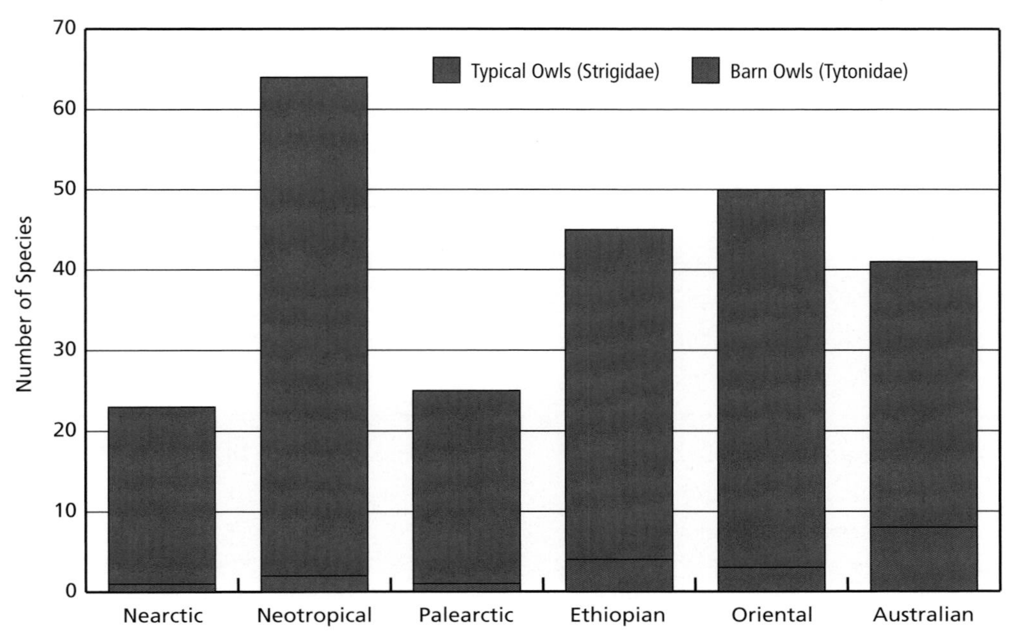

The distribution of owl species by zoogeographic regions.

continents for a geological time frame long enough to permit substantial evolution and faunal separation. Interestingly, Sclater's system was not influenced by the Darwin-Wallace theory of evolution. Furthermore, Sclater's system is often erroneously credited to Alfred R. Wallace based on his 1876 published classic work, *The Geographical Distribution of Animals*. In reality, Wallace modified Sclater's bird-based biogeographical regions to fit all animals generally.

I chose to summarize the pattern of owl species distribution according to these six traditional zoogeographic zones (see Appendix 1, A Checklist of the Owls of the World). Yet, in doing so I must point out that we still have much to learn about owl distribution, and therefore this review is based on limited knowledge.

Effective owl conservation efforts should consider the implications of the major barriers to animal movement as reflected in Sclater's zoogeographic regions. The extirpation of a fringe or naturally rare species or population from a faunal region is significant, even if the species remains common in an adjacent faunal region. Each region hosts unique co-adapted genetic varieties of species suited to the region's ecological conditions. Many scientists consider populations of species that occur at the edge of their range as a type of genetic engine of evolution and ultimately a source of new species.

Owls are truly cosmopolitan creatures, occurring worldwide, including many island land masses such as Madagascar, New Zealand and the West Indies. No one species, or genus, is completely cosmopolitan, although four main kinds of owls are nearly so. Among the typical owls (subfamily Striginae), screech owls (Otus), horned owls (Bubo) and pygmy owls

(Glaucidium) occur in all the main regions of the world except the Australian region, and barn owls (subfamily Tytoninae) are absent only in temperate Asia and in other cold places where typical owls occur.

Some species, such as the Manus hawk owl, endemic to the tiny forested Manus island in the Admiralty Islands, have extremely restricted distributions, and are known from only one or a very few locations. In contrast, the common barn owl is one of the world's most widely distributed land-dwelling birds. Owls reach their greatest diversity in the tropics, and species diversity dwindles with increasing latitude. A review of information about the 205 owl species found on earth is presented in Chapter 6 of this book where the relationships between their distribution, natural history and conservation can be explored.

Invited Contributions

Continuing with the subjects of species, evolution and genetics, the following invited contributions from renowned scientists serve as personal accounts of recent studies on owls. John Penhallurick reveals how genetic tools are changing our understanding of how owls are related to each other. Jerry Olsen and Susan Trost share the excitement and sense of adventure about their recent discovery of a new owl species—the Sumba hawk owl. Lastly, Heimo Mikkola examines how owls can surprise us, behaving in ways that challenge accepted scientific concepts. When biological rules are broken, such as when different species of owls pair together in what we might consider a misguided love affair, it causes us to re-examine the concept of a species. After all, as we too often forget, simply naming something as complex as a "species" does not in itself tell us much about it. Canadian natural history author Louise de Kiriline Lawrence asked noted animal behaviorist Konrad Lorenz why birds sometimes pair with the wrong species (hybridize), to which he replied: "My dear, it happens in the best of families."

What's in a Name? DNA Sequencing and the Classification of Owls

by John Penhallurick

Taxonomy is the science of classification. Historically, judgments about owl taxonomy, like those of other bird families, have largely relied on morphology, comparing details of plumage, or anatomy, or skeletons. More recently, ornithologists have become aware that convergence, the process whereby birds of different origins, but occupying similar environmental niches, become more similar to each other, can be very misleading.

Of all bird families, the adaptation to nocturnal life characteristic of most owls has imposed strong convergent pressures, such that birds that look similar may turn out to be only distantly related. A prime example involves the Eurasian pygmy owl of the Old World (Eurasia) and the mountain pygmy owl of the New World (North and South America). As recently as 1990, Sibley and Monroe placed the mountain pygmy owl in the Eurasian pygmy owl superspecies, and stated: "Sometimes considered conspecific [an organism belonging to the same species as another] with *G. passerinum* [Eurasian pygmy owl]." As pointed out below, the relationship between these two species is in fact relatively remote within the Strigidae. Their nocturnal adaptations mean that all owls share such features as relatively dull colors (black, browns, grays and reds); large eyes adapted for night vision, which, in turn, profoundly influence the structure of the skull and neck; and, often, feathers of the wings adapted for silent flight.

A primary tool in the re-analysis of relations between different groups of birds has been the analysis of DNA. In the simplest terms, DNA is made up of a long sequence of four unique molecules called nucleotides or "bases." A sequence of DNA that can be identified at a specific location on a chromosome is called a gene. A gene or a group of genes determine a particular characteristic in an organism. Genes undergo mutation when their DNA sequence changes, and hence serve as a measuring stick or molecular clock. By isolating a particular gene in different species or subspecies of owls, and then determining the sequence of bases in the same gene, locations at which the bases differ can be identified. The greater the number of such disagreements, the greater the time since the taxa (a group of species) in question diverged, and the more distant their relationship.

Scientists believe that the precise calibration of distances between taxa made possible by DNA sequencing of marker genes makes it possible to obtain nonarbitrary decisions as to the boundaries between genera, subgenera, species and subspecies. But this is a relative measurement, and not an absolute or universal clock; for example, we cannot say that all organisms with a 1.6% difference in bases should be called different species. Although it cannot be said that molecular data have solved the problem of establishing the boundaries between genera, subgenera, species and subspecies unambiguously, nonetheless, by examining the distances between well-recognized species and genera within a family, we can make soundly based decisions.

The most commonly used gene used in DNA sequencing of birds is cytochrome b in mitochondrial DNA (mtDNA) because it evolves faster and more predictably than other genes and DNA types. The figure on page 78 shows an owl relationship map called a Maximum Likelihood Tree that represented the percentage differences between every possible pairing of the cytochrome b sequences for the owl studied. Several radical taxonomic changes should be readily evident. The genus *Scotopelia* of *S. peli* Pel's fishing owl; the genus *Ketupa* as represented by *K. zeylonensis* brown fish owl; and the genus *Nyctea* of *N. scandiaca* snowy owl all have to be submerged in the genus *Bubo*, since each of them is more closely related to some members of *Bubo* than are other members of *Bubo*.

However, I wish to focus on owls

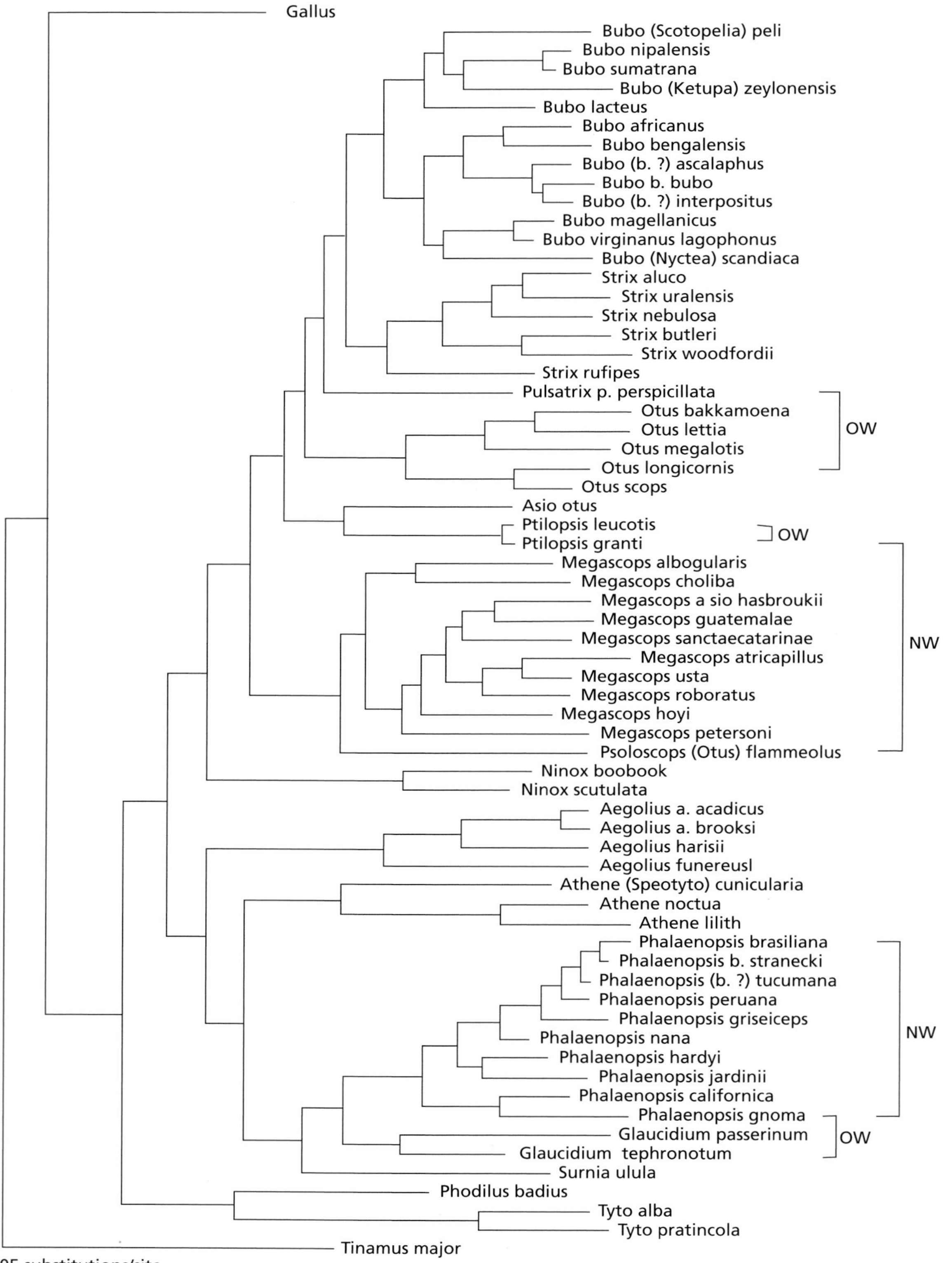
Gallus
Bubo (Scotopelia) peli
Bubo nipalensis
Bubo sumatrana
Bubo (Ketupa) zeylonensis
Bubo lacteus
Bubo africanus
Bubo bengalensis
Bubo (b. ?) ascalaphus
Bubo b. bubo
Bubo (b. ?) interpositus
Bubo magellanicus
Bubo virginanus lagophonus
Bubo (Nyctea) scandiaca
Strix aluco
Strix uralensis
Strix nebulosa
Strix butleri
Strix woodfordii
Strix rufipes
Pulsatrix p. perspicillata
Otus bakkamoena
Otus lettia
Otus megalotis
Otus longicornis
Otus scops
OW
Asio otus
Ptilopsis leucotis
Ptilopsis granti
OW
Megascops albogularis
Megascops choliba
Megascops a sio hasbroukii
Megascops guatemalae
Megascops sanctaecatarinae
Megascops atricapillus
Megascops usta
Megascops roboratus
Megascops hoyi
Megascops petersoni
Psoloscops (Otus) flammeolus
NW
Ninox boobook
Ninox scutulata
Aegolius a. acadicus
Aegolius a. brooksi
Aegolius harisii
Aegolius funereusl
Athene (Speotyto) cunicularia
Athene noctua
Athene lilith
Phalaenopsis brasiliana
Phalaenopsis b. stranecki
Phalaenopsis (b. ?) tucumana
Phalaenopsis peruana
Phalaenopsis griseiceps
Phalaenopsis nana
Phalaenopsis hardyi
Phalaenopsis jardinii
Phalaenopsis californica
Phalaenopsis gnoma
NW
Glaucidium passerinum
Glaucidium tephronotum
OW
Surnia ulula
Phodilus badius
Tyto alba
Tyto pratincola
Tinamus major
0.05 substitutions/site

hitherto assigned to the small-bodied genera *Otus* and *Glaucidium*. *Otus* had members in Eurasia, Africa and North and South America. But compare the position of the four Old World *Otus* owls (depicted as "OW" in the tree) with the 10 American New World *Megascops* owls (labeled "NW"), which have hitherto also been placed in *Otus*. It should be apparent that these two groups are only very distantly related.

As M. Wink and P. Heidrich, who performed the analysis on which the above tree is based, state: ". . . members of the genus *Otus* appear in at least three different monophyletic clades (groups), indicating that the genus is polyphyletic. It therefore needs a systematic revision. The screech owls of the New World represent a distinct group which is separated from Old World Otus by genetic distances of between 12 and 16%, equivalent to 6–8 million years."

The Old World *Otus* have as their closest relatives members of a clade containing *Bubo*, *Strix* and *Pulsatrix*. The sister clade for this whole assemblage is a clade containing *Asio* and *Ptilopsis*. In fact, the two species in *Ptilopsis*, *P. leucotis* northern white-faced owl and *P. granti* southern white-faced owl, were also previously assigned to *Otus*, but there is already widespread acceptance of the fact that they should be assigned to a separate genus. Only then do we have a sister clade containing the New World owls formerly assigned to *Otus*, which need to be assigned to a different genus, namely, *Megascops*.

It will also be noticed that *Otus flammeolus* flammulated owl is rather distant from other New World owls that we now assign to *Megascops*. In the figure, *flammeolus* is placed as a sister clade to a node representing the entire remainder of the New World *Otus* species, but with an equally distant relationship to the Old World *Otus* species. *Flammeolus* is indeed distant, in genetic terms, from other New World "*Otus*" owls examined: *asio* 14.8%; *atricapillus* 14.5%; *guatemalae* 14.0%; *choliba* 15.9%; *hoyi* 14.3%; *petersoni* 14.6%; *roboratus* 15.1%; *sactaecaterinae* 14.5%; and *usta* 14.3%. But it shows even greater percentage distances from all the Old World *Otus* species included in the sample: *bakkamoena* 15.8%; *lempiji* 15.9%; and *scops* 17.6%. By way of comparison, the distance between *Asio otus* and *Bubo bubo bubo*, two taxa in different genera, is 13%. In short, the evidence cited regarding *flammeolus* requires that it be assigned to a new genus distinct from both *Megascops* and *Otus*. The oldest (in fact apparently the only) available name appears to be *Psiloscops*.

Finally, we come to the genus *Glaucidium*, which hitherto has had both Old World and New World members. In the relationships between the Old World members, represented by *G. passerinum* Eurasian pygmy owl and *G. tephronotum* Red-chested Owlet (bracketed together as "OW" in the figure), and the 10 New World ("NW") members listed under *Phalaenopsis*, we do not find the polyphyletic relationships that compelled us to split *Otus* into different genera; the two groups form sister clades. But important here is the distance between the two clades.

Wink and Heidrich point out that "Old and New World species cluster in monophyletic clades which possess a common ancestry but diverged more than 7–8 million years ago." The genetic distances between the Old World *G. passerinum* and the New World taxa *peruanum* (14.8%), *bolivianum* (14.9%), *brasilianum* (14.9%), *tucumanum* (15.0%), *californicum* (9.0%), *gnoma* (10.6%), *griseiceps* (14.2%), *hardyi* (15.4%) and *jardinii* (13.8%), representing divergences of 6 to 8 million years ago, indicate that they should be assigned to a distinct genus. It would be inconsistent to have most genera within the Strigidae differing from each other by 10 to 12%, but to have a single genus containing members that differed from each other by up to 15%. The oldest available name appears to be Phalaenopsis, to which we reassign the New World taxa in the figure. I would like to thank Professor Michael Wink of the University of Heidelberg for providing this figure.

Discovery of a New Owl Species—the Sumba Hawk Owl

by Jerry Olsen and Susan Trost

In December 2001, for the second time in six months, we visited the island of Sumba in Indonesia. The coast of Sumba, and its main town Waingapu, was stifling and uncomfortable, but the island's high plateau was cooler. At dusk on our second night there, our Sumbanese guide (he asked not to be named) led us through mountain forest up a steep muddy trail to a grassed-over flat at the top of a hill. Fireflies floated in low bushes, and giant insects buzzed past our ears and through holes in the canopy. Native pigeons coming in to roost grunted like little pigs. The guide stopped at some jagged gray-white pieces of limestone laid against the large trunk of a deciduous tree and whispered: "Here people pray to the gods for good hunting in this forest." The bird hunters on this mountain would rest at the limestone altar and speak to their ancestors.

We waited for a moment, listening to the forest, and then pulled a portable cassette player from our pack. In the high grove we stood in a huddle and broadcast the voice of an unknown owl, a single repeated "hoot," like a child blowing the same note every three seconds on a flute.

The island of Sumba is located at the crosshairs of 10° South and 120° East along the Lesser Sunda chain in southeastern Indonesia. The island is 130 miles (210 km) long with a surface area of about 4632 square miles (12,000 square km); at 4019 feet (1225 m), the highest point is Gunung (Mount) Wangameti. The island experiences a dry winter and a wet summer season, and annual rainfall is between 20 inches (500 mm) on the south coast and 79 inches (2000 mm) on the inland hills. Closed-canopy forest, mostly deciduous with some evergreen, now covers less than 11% of Sumba and, due to the clearing and repeated burning of vegetation to provide land for grazing and cultivation, this forest is confined to relatively small and fragmented pockets.

Since the late 1980s, ornithologists have written of an unknown owl on

Sumba, apparently a scops owl, the large group of tiny owls with ear tufts, that occurs over most of the world except Australasia. Some ornithologists, such as Stephen Debus, had written that we should, with some urgency, find and describe this scops owl, and determine its conservation status, that is, what it was, and how safe it was after so many trees had been cleared from Sumba. Some writers dismissed the reports of the little owl as a misidentification of the relatively large, endemic Sumba boobook, a medium-sized hawk owl with dramatic rufous bars across its front. But recent writers said that the bird was more likely the Flores scops owl, Flores being the big island 28 miles (45 km) north of Sumba. The song of this unknown Sumba owl was unlike any known scops owl, and the song of the Flores scops owl was unknown, and still is unknown at the time of writing, so ornithologists thought that the two owls were the same. However, one ornithologist, Ben King, played the song of the unknown Sumba owl on a Flores mountainside near the place where the Flores scops owl had recently been rediscovered. He failed to elicit a response. Moreover, the Flores scops owl has ear tufts, a characteristic of scops owls, but the mystery owls seen on Sumba had no visible ear tufts.

For half an hour we played the tape in the grove where our guide said he had seen the mystery owl some weeks earlier. But no owls replied, and after standing for another half hour in the dark and listening to geckos and insects, we decided to walk back to the car and drive the twisting road down to Kilometer 49, west of the town of Waingapu on the Lewa Road, a place where birdwatchers had heard the mystery owl. We pulled the car to the edge, and our driver stayed with it. By moonlight and torch we walked a ridge covered with scorched grassland and scattered gray limestone pieces some 2625 feet (800 m) to an isolated arm of a forest, then found our way between high trees. Waist-high shards of limestone poked up through bushes, vines and around tree trunks, so we took care not to stumble. At 18:30 hours, from a grassy clearing, Sue broadcast the taped voice of the mystery owl. At first we waited and heard nothing and prepared to be disappointed once more, listening to croaky gecko calls, churring night insects, and the coos and grunts of roosting pigeons. After 10 minutes, two owls called back at us, and we tracked them as they drifted down the hillside from tree to tree to the forest edge, concealing themselves behind leaves and branches in the canopy. While Sue

A newly discovered owl species—the Sumba hawk owl.

played the recording down in the clearing, our Sumbanese guide and I entered the forest with the camera, pulling at vines between jagged tips of limestone, trying to glimpse the hidden owls and hopefully take some video. The two owls were shy of us, and each time we drew close enough to see them in the torch beam, they moved behind leaves or flew to another tree. But after some minutes the two owls settled, and one sat directly over our heads and called to the unfamiliar voice coming from Sue's recorder. The owl was reddish, the size of a quail, and had loose feathers and bright staring eyes, like flat yellow buttons, on the front of its face. The owl had no ear tufts, as scops owls are supposed to have, and we realized that the owl might be a different species, maybe a new species of hawk owl.

Over the next two weeks we returned with our guide and located two other pairs in adjacent forest. One was accompanied by a dependent juvenile that flew with them in the canopy and called for food. On December 30, 2001, our guide showed us a specimen of the little owl that he said a bird hunter had shot in a patch of forest 2.5 miles (4 km) east of the patch we had been watching. We studied, measured and photographed it, then returned that night to Kilometer 49 and found all three of our pairs still calling from their forest territories, some distance from the place where the bird hunter had killed the owl that turned out to provide the first description of its species.

When Professor Michael Wink at the University of Heidelberg in Germany received feathers from the specimen, he analyzed nuclear marker genes to determine which taxonomic group the owl belonged to. It was not a scops owl but a hawk owl like the southern boobook that we studied and knew well in Canberra, however, this little owl differed from southern boobooks by 8.2% nucleotide substitutions and from the brown hawk owl of east and southeast Asia by 9.1% nucleotide substitutions. The owl differed in other ways from all known hawk owls: it was smaller, and most hawk owls have a double-noted call like the European cuckoo, the call put into cuckoo clocks, but the song of the new owl was a single note, a monosyllabic "hoot" they repeated every three seconds. There is no similar song known for any hawk owl.

We proposed for the new owl a common name, the little Sumba hawk owl, and a scientific name—*Ninox sumbaensis* (the "hawk owl from Sumba"), but said in our research paper that its conservation status was unknown. We hope that with too little forest in its native place, the little Sumba hawk owl doesn't go the way of the laughing owl in New Zealand and the Norfolk Island boobook of Australia, but instead, with its flute-like call, is allowed to live and defend its high forest sanctuaries on the island of Sumba.

STRANGERS IN THE DARK: HYBRIDIZATION BETWEEN OWL SPECIES *by Heimo Mikkola*

The biological species concept states that when different species mate they do not normally produce fertile offspring, even if hybridization were to take place either in nature or in captivity. It also is a common belief that isolation mechanisms are relatively effective in explaining the infrequency of hybridization between different owl species. Owls reportedly have the lowest hybridization rate (about 1%) among birds. Instances of interspecific crossings between diurnal raptors are also uncommon except in captivity. Bird groups in which hybridization is particularly common include warblers, grouse and hummingbirds. Estimated rates for game birds are over 20% and are even higher (40%) for waterfowl (swans, geese and ducks). Habitat loss and climate change may bring species that are normally separate (allopatric) into contact, and this new sympatry could result in an increase in hybridization between closely related species. To examine this possibility, I have assembled a preliminary record of known owl hybridizations.

Eastern Screech Owl × Western Screech Owl

Fossil fragments of a kind of screech owl of the *Otus kennicottii/asio* type from the Upper Pliocene period in Kansas seem to confirm the American origin of these closely related species. In present times, local sympatry between the eastern and western screech owls takes place along the Big Bend of the Rio Grande area in Texas and adjacent Mexico and possibly also in the Arkansas River area in eastern Colorado as well as in the Cimarron River in western Kansas.

In July 1962, Joe T. Marshall, Jr., found a mixed pair tending young at Boquillas in the Rio Grande area. Apparently one juvenile had some feathers typical of the eastern screech owl. The pair reportedly exchanged vocalizations—the male's western screech owl bouncing-ball song and double trill stimulating the female to respond with an eastern screech owl long trill. This interaction culminated when the female and male indulged in billing and head-preening. The male was itself a hybrid, with a pale green bill, leading Marshall to conclude that this hybridization, while opportunistic, was not an isolated event. He concluded nonetheless that there was no regular zone of hybridization between the eastern and western screech owls. Hybridization occurred here due to low owl densities along marginal habitat (a fringe of small mesquites and willows along a river that crosses a desert) resulting in limited mate choice opportunities.

Whiskered Screech Owl × Western Screech Owl

The strong resemblance in coloration and size between the whiskered screech owl in the mountains of southern Arizona and adjacent Mexico and the western screech owl has raised important questions regarding the former's specific ecological position and geographical origin. Also, no difference has been found between the anatomy of the syrinx (the vocal organ of a bird found in the trachea) of these two species. There have been undocumented instances of hybridization between these two species in Arizona. Generally, however, the whiskered screech owl lives at higher elevations than the western screech owl and below that of the flammulated owl.

The tawny owl (left) is considerably smaller than the Ural owl (right).

Pharaoh Eagle Owl × Eurasian Eagle Owls

The pharaoh eagle owl is doubtless closely related to its Eurasian counterpart, but vocalizations and DNA evidence both indicate that they are distinct. The apparent sympatry of the *hispanus* race of the Eurasian eagle owl with the pharaoh eagle owl in the Algerian Atlas Mountains also argues in favor of them being different species. However, the pharaoh eagle owl intergrades with the *hispanus* race of the Eurasian eagle owl, and there is a zone of intermediates. Furthermore, the two species have interbred on several occasions in Israeli zoos, where their fertile hybrid offspring also interbreed freely with each other as well as with both "pure" species.

Tawny Owl × Ural Owl

The tawny owl and the Ural owl are thought to be well-separated species, although they obviously have a common ancestor. During the Middle Pleistocene period in the Czech Republic, Hungary, Austria and in southern France there was an owl called *Strix intermedia*. Based on the size and shape of fossil leg and wing bones, this owl could have been an intermediate form in the evolutionary process that led to the tawny and Ural owls we know today. The extensive overlap of the ranges of the tawny and Ural tawny owls in southern Finland provides favorable opportunities for hybridization to occur. However, despite the intensive study of these species, no field records of hybrids are known to exist.

Professor Wolfgang Scherzinger decided to test the risk of hybridization between tawny and Ural owls in captivity prior to reintroducing captive-raised Ural owls into German National Parks such as the Bayern Forest. The concern was that in Central Europe a wild Ural owl might choose the wrong mate (a tawny owl) if it was hard-pressed to find its own kind. Ural owls are rare or endangered, and hybridizations could destroy them as a species. The captive hybridization experiments showed that there was no genetic barrier preventing fertile cross-breeding of the two species.

First, Professor Scherzinger raised two hybrid young from a German tawny owl mother and a Swedish Ural owl father. The hybrids (a male and a female) showed maternal and paternal as well as intermediate characters. The "vocabulary" of the hybrids was more varied than that of either parental species, adding new inventions to the original repertoire of the parents.

In captivity, with no other mates of their own kind provided, tawny owls will cross-breed with Ural owls to produce fertile offspring.

A typical spotted owl (left) and barred owl (right) and a wild natural hybrid "sparred owl" (center). The more aggressive barred owl is thought to have invaded the range of the spotted owl within the last few decades.

Later he bred crossings between the "good" species, made backcrossings of hybrids with parent species, and also obtained offspring from this third generation. Professor Scherzinger wrote to me in the following terms: "I am being very careful that hybrid birds do not escape into the wild, as they would spoil my re-introduction project, which has resulted in five to 10 breeding Ural owl pairs in the area to date." He also noted, to his relief, that he had not detected hybrids in the wild.

Barred Owl × Spotted Owl

In the New World, the barred and spotted owls are the ecological equivalents of the European tawny and Ural owls. Moreover, these four *Strix* species seem to be very closely related. Although the barred owl is somewhat larger and stronger than the spotted owl, both species seem to have arisen from the same stem by periods of geographic isolation during subsequent ice ages in the Pleistocene period.

Hybridization between these two closely related species was prevented by geographic isolation, their original ranges being separated by two mountain ranges and a desert. However, human-altered habitat, including the fragmentation of old-growth forests, has facilitated the spread of barred owls from east to west. The barred owl seems to adapt easily to logged areas. As early as 1989, Professor Karel H. Voous predicted that the barred and spotted owls could not be expected to coexist for any length of time because the more aggressive barred owl is able to displace the spotted owl.

However, since 1990 several cases have been recorded in which spotted owls have interbred with the more common barred owls. Unofficially, the offspring were dubbed "sparred owls." They had markings on the back of their nape and head similar to those of the barred owl, but their breasts looked more like the spotted owl. Rectangular bars on the head and facial coloration were intermediate between the two species. The bars on the tail resembled a spotted owl but were farther apart.

Three adult spotted/barred owl hybrids, two in Washington and one in Oregon, were confirmed between 1989 and 1992, and one juvenile hybrid was produced by a female barred owl paired to a yearling male spotted owl in Oregon in 1992. Recently, Eric Forsman wrote: "We see more barred owls every year, the range continues to expand southward into California. They have become quite common on many of my study areas in Oregon and Washington." Thomas Hamer wrote further that: "We are beginning to find many more hybrids in Oregon and Washington and the situation may become acute if it starts to wash out the pure spotted owl gene pool." So, in addition to old-growth habitat loss, the spotted owl may also be threatened by increased habitat-induced interactions with the closely related barred owl.

This review of examples suggests that owl hybridization rates may be greater than previously thought. Habitat alteration may force closely related species to interact more frequently, increasing the opportunities for cross-breeding. In addition, hybridization is sometimes encouraged by keeping closely related species together in the same aviaries. When new species are discovered, the possibility that they may actually be hybrids needs to be considered. This is especially true when the ecology of so many owls remains poorly studied. Additional reports, comments and observations of owl hybridization from readers would be greatly appreciated.

CHAPTER 2

Owls in Mythology and Culture

Bruce G. Marcot and David H. Johnson

Throughout human history, owls have variously symbolized dread, knowledge, wisdom, death and religious beliefs in a spirit world. In most western cultures, views of owls have changed drastically over time. Owls can serve simultaneously as indicators of scarce native habitats and of local cultural and religious beliefs. Understanding historical and current ways in which owls are viewed, and not imposing western views on other cultures, is an important and necessary context for crafting owl conservation approaches palatable to local peoples.

Introduction

I am a brother to dragons, and a companion to owls.

—*Job 30:29*

Thou makest darkness, and it is night: wherein all the beasts of the forest do creep forth.

—*Psalms 104:20*

Long before there were ornithologists and graduate students, keen observers in other tribes and bands roamed the forests and plains. In their search for resources they encountered owls—winged denizens of the night—and incorporated such spectral figures into their mythology and culture.

The North American Cherokees call them *uguuk*, the Russians *sovah*, the Mexicans *tecelote*, the Ecuadorians *huhua* or *lechusas*, and aboriginal peoples of the Kaurna area of Australia *winta*. In Chinese, they are *mao tou ying* or, literally, "cat head eagle." For centuries, indeed millennia, owls have played diverse and fascinating roles in a wide array of myths and legends. In this era of rapidly shrinking habitats for many owls of the world, a first step toward garnering concern for their conservation is to better understand the role of owls in cultural stories, religions and lore. In this chapter, we hope to foster an appreciation for the extent to which owls have become part of the mythos of human societies. We offer this as a celebration of the remarkably

OPPOSITE: **Owl mask with articulated mandible. 32 cm × 27 cm × 22 cm. Collected by J.T. White in 1894 at Kasaan Bay, Alaska. It is superbly designed and carved, one of the finest examples of the region's sculpture.**

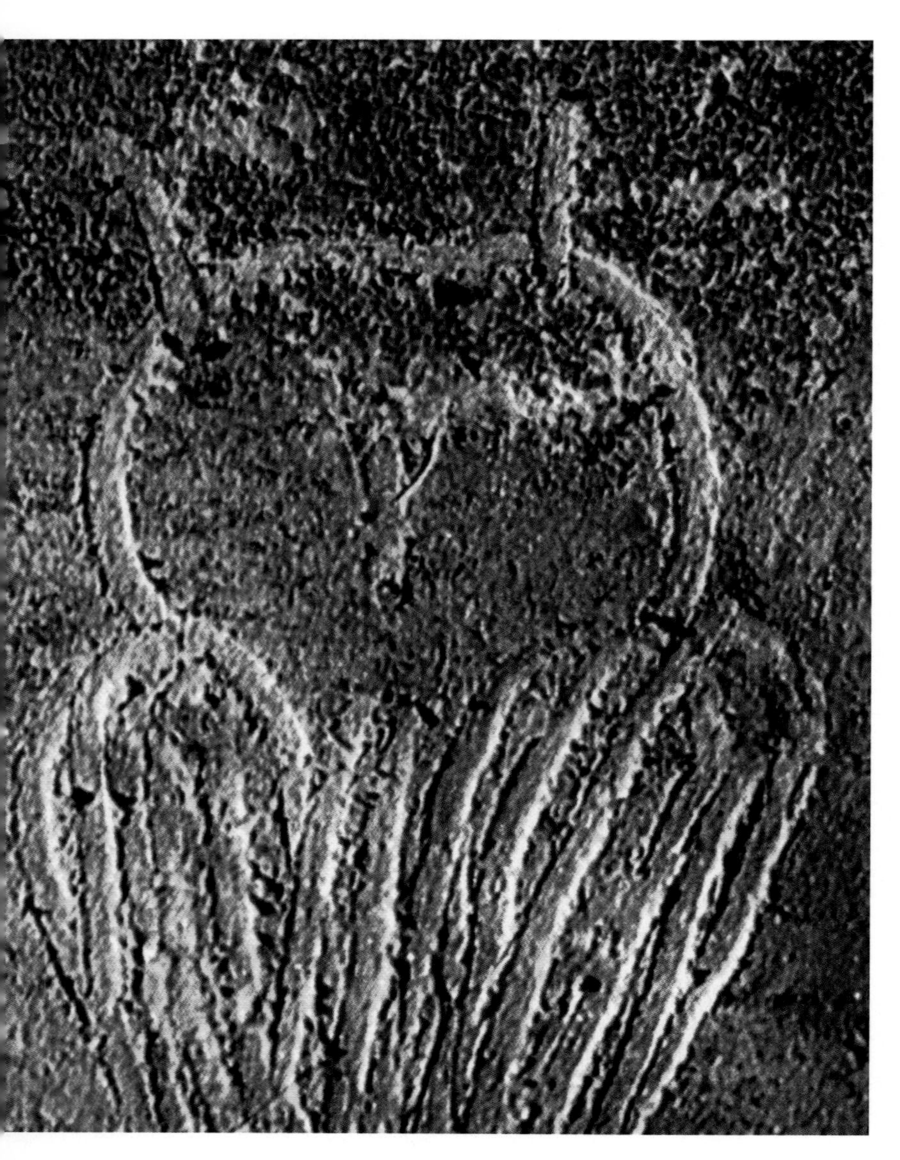

Owl petroglyph in the Hillaire Chamber of the Chauvet Cave in France. Radiocarbon dating has indicated humans used the cave an estimated 35,000 years ago.

diverse ways that cultures have responded to owls the world over. No other bird family has aroused more universal fascination and interest, and can better serve as a basis for conservation. For conservation must proceed from respect for diverse cultures and creeds, as much as for the organisms that share our sphere.

Symbols Old and New

But wild beasts of the desert shall lie there; and their houses shall be full of doleful creatures; and owls shall dwell there, and satyrs shall dance there.

—*Isaiah 13:21*

And thorns shall come up in her palaces, nettles and brambles in the fortresses thereof: and it shall be an habitation of dragons, and a court for owls.

—*Isaiah 34:13*

I see a likeness between the old, animist forest, where one could not be sure whether a screech owl's call came from a bird or an Omah[1]*, and the evolutionary forest, with its unclear distinctions between tree and fungus, flower and fir cone. The tree-fungus relationship is as mysterious in its origins and implications as the owl-Omah one. Both belong to a world that goes deeper than appearances, where a buried interconnectedness of phenomena renders behavior ambiguous, where one cannot walk a straight line.*

—*Wallace 1983:83*

Owls have always been part of the root metaphors of how humans relate to the land. One of the earliest human drawings, dating back to the Upper Paleolithic period at least 30,000 years ago, was of an owl—probably a long-eared owl—painted on the wall of Chauvet Cave in France. Rock paintings or petroglyphs of owls have also been found in other disparate locations including the Victoria River region of northern Australia and the lower Columbia River area of Washington state, U.S.A.

In the Victoria River region of Australia there is a group of prehistoric art sites at a rock outcrop known as Jigaigarn,

Prehistoric rock painting of an owl from the Weliyn rock-shelter, Victoria River region, Northern Territory, north-central Australia.

where human use has been dated to about 10,000 years BCE (before the Common Era). There sits an unusual formation of a large sandstone boulder balancing on a tiny base. This balancing rock was said by the local people to have been placed there by, and is still imbued with the presence of, the creation ancestor Gordol, the Owl. A large rockshelter below this formation contains many engravings and paintings, dominated by a huge one of a striped figure of the Owl. Also, many grooves were pounded into the rock, in the belief that making such marks conjures the power of the rock and the Owl ancestor.

Owls played roles in the ancient Mayan cultures of Mesoamerica. A carved bas-relief of the ancient Mayan Ruler 3 of Dos Pilas, in what is now Guatemala, following the death of Ruler 2 in 726 CE (Common Era), is shown adorned with a screech owl, apparently a symbol of ruling power or the resurrection of government. The ancient Mayans also believed that an owl hooting long and loud was a bad omen. They represented the screech owl as *Mo An*, the bird of death, and the Mayan death god *Yum Cimil* had attendants that included vulture, dog and owl. The Worcester Art Museum of Massachusetts has in its collection an artifact from the Guanacasta-Nicoya region of Costa Rica in Central America, a small 1.5 × 2.5 inches (4 × 6.4 cm) pre-Columbian "mace head" figure of what appears to be an *Otus* screech owl that has

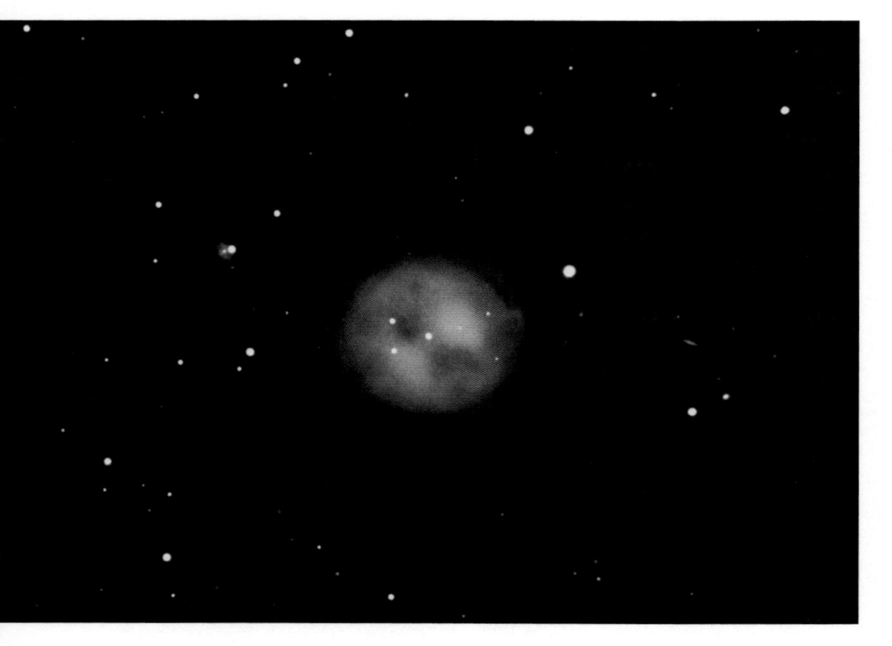

The Owl Nebula, known in astronomy as M97 or NGC3587, is a planetary nebula in the constellation Ursa Major.

been dated to 1 to 500 CE. Made of white stone, this owl head may have been a heraldic finial, an ornamental cap to a ceremonial staff such as is often found with offerings of jade and metates in elite, ancient Mesoamerican burials.

Owls also are very much a part of modern culture, in the sky as well as on the land. In the constellation Ursa Major, at a most dim magnitude of 11.20, is a planetary nebula designated by astronomers as the Owl Nebula (more formally called M97 or NGC3587) because of its resemblance to an owl. In a more terrestrial venue, a query of the U.S. Geological Survey database on place names revealed 576 features in the United States in some way named "owl," such as Owlshead Canyon, Owl Mine, Owl Creek and Owl Hollow. Records of the Canadian Permanent Committee on Geographical Names lists 88 current and 17 additional historic places in some way named "owl." Doubtless, many other countries have similar designations.

Etymologically, the word "owl" goes back to the Middle English word "oule," which may derive from the Old English "ille," which is cognate with the Low German "ule," in turn going back to the German "eule." The ultimate root of the modern word "owl" was presumed by Lockwood to be a proto-Germanic word "uwwalo" or possibly "uwwilo."

Another derivation of "owl" is the Icelandic "ugla," which is cognate with "uggligr," which gave rise to the Scandinavian "ugly," which led to the Middle English "ugly" and the Modern English word "ugly." The Icelandic "uggligr" does not mean "ugly" in modern connotations (that is, unpleasant to behold), but rather it means "fearful or dreadful." This is precisely the connotation of owl symbols and totems in many myths and legends. Thus, the very names that we use often speak of a deep history of traditional viewpoints and cultural perspectives.

In modern Korean, owl is "ol-bae-me." Bae-me is often used as a suffix to denote an animal, so the root Korean word "ol" is similar to the English "owl." Further, in Hindi, owl is "ul" (similar to the German "eule" or Low German "ule") or "ulu" if referring to one of the large owls (the Hindi or Urdu term for smaller owls is "coscoot"). The ancient Roman "bubo," the ancient Greek "buas," the modern Hindi "ulu," the Maori "ruru," and the modern Hebrew "o-ah" are each an obvious onomatopoeia, as is the modern Nepali "huhu." In Korea, the

Oriental scops owl is called "seo-juk-say," resembling the owls' call (and "say" or "sae" is Korean for "bird"). Thus, in many cultures, the sound has become the name.

An awareness and understanding of the deep, complex perceptions of owls in the past may help support efforts to protect those species today. For example, the ancient cultural importance of owls in Europe helps modern conservationists there. The same is true in America. The blend of traditions carried to the United States by white immigrants and black slaves from West Africa means that North American owl species have a strong cultural profile that may aid conservation measures. E. Ingersoll in 1923 traced the bird beliefs amongst African-American slave and ex-slave communities. Such beliefs seeped into the dominant European-American culture the way that African rhythms were given to the world through the blues and jazz music of black North America. Thus, current U.S. folklore about owls is an eclectic blend of European, African, Native American and Asian traditions.

Asymmetrical pair of owl masks collected by Sheldon Jackson on the lower Yukon River, western Alaska, in the 1890s. Similar masks were collected by H.M.W. Edmonds, which he identified as reflecting the "*inua* [spirit] of the Short-eared Owl." The Yup'ik carvers greatly valued the representational dimension and manufacture of their work, as masks were used in dances to elicit the goodwill of members of the spirit world.

This can be extended to an environmental principle for the West (meaning all areas occupied by those of European descent, and also by mixed-race societies such as South Africa, where a highly developed conservation tradition exists). Any animal or plant with a strong cultural profile, no matter how negative that cultural perception may once have been (such as with bats, wolves, sharks and owls), is at a major advantage, for conservation, over an animal with no cultural profile whatsoever (such as some rodents and sparrows). The advantage is that they are rooted and recognized in the social consciousness. In the case of owls, the deep fears and anxieties they generated and the prophetic status they once held (and still hold) present environmentalists with a handle with which to engage the interest and sympathies of a wider audience. But the critical element in these situations is the fact that most western cultures no longer perceive owls as omens of evil, or retain only the dimmest vestiges of these old beliefs.

However, for some or even many Africans, Native North Americans, Asians and South Americans, these perceptions of owls are living traditions with deep and powerful roots. For example, in Africa, owls are still genuinely believed to be evil. Surveys of attitudes toward owls in Malawi revealed that owls were regarded as bad birds by a very high percentage (more than 80%) of the people surveyed. In West Africa, most people do not like owls and regard them as evil. In fact, the standard pigeon-English name for owl in West Africa is "witchbird." Rather than garnering support for endangered species such as the rare Congo Bay owl, the ancient African mythic traditions relating to owls may present a barrier to their conservation unless conservationists understand and make use of such myths. A classic parallel case is the aye-aye of Madagascar where this endangered mammal, down to a last few dozen, has been ruthlessly persecuted because of its cultural profile as a witch-creature. The challenge for conservationists is to turn the barrier to an advantage by understanding the cultural society and helping to craft conservation actions while taking these into account.

Modern carving of a stylized owl from a city park in south Victoria, Australia.

Conservationists should understand the role a bird like an owl may play in some societies. Conservation policies for a species of conservation concern, such as the Congo Bay owl, should not be formulated without understanding local attitudes and any uses of that particular species. Conservationists too often inculcate their own positive view of the animal in question, and thus fail to change local cultural attitudes, that is, to replace deep fear or anxiety with admiration and respect.

Markers of Gods, Knowledge, Dread, Wisdom and Fertility

From ancient Athens, the silver four-drachma coin bore the image of the owl on the obverse side as a symbol of the city's patron, Athene Pronoia, the Greek goddess of wisdom who, in an earlier incarnation, was goddess of darkness. The owl—whose modern scientific name *Athene* carries this heritage—came to represent wisdom from its association with the dark. An ancient Greek saying was "to send owls to Athens," the modern equivalent being "to send coals to Newcastle" or to engage in something useless, since owls and Athens were, in a sense, considered synonymous. The role of owls in ancient Greece has been traced in detail.

Many children have grown up with nursery stories of wise old owls. From the ancient Greek legends to the wise owls in *Winnie-the-Pooh* and "The Owl and the Pussycat," we have all seen images in folk tales and fables of owls as the quintessential bearers of knowledge and sagacity.

In many other cultures, owls represent wisdom and knowledge because their

Image of an owl on the cork from a modern bottle of fine Italian wine, from Azienda Agricola (agricultural business) in Valle dell Asso, Galatina.

nocturnal vigilance is associated with that of the studious scholar or wise elder. According to Saunders, in the Christian tradition owls represent the wisdom of Christ, which appeared amid the darkness of the unconverted. To early Christian Gnostics, the owl is associated with Lilith, the first wife of Adam who refused his advances and control. The owl also had a place as a symbol in the King Arthurian legends, as the sorcerer Merlin was always depicted with an owl on his shoulder. In Japan, owl pictures and figurines have been placed in homes to ward off famine or epidemics.

The Blakiston's fish owl was called *Kotan Kor Kamuv* (God of the Village) by the Ainu, the native peoples of Hokkaido, Japan. The traditional Ainu people were hunter-gatherers and believed that all animals were divine; most admired were the bear and the fish owl. The owls were held in particular esteem and, like the people, were associated with fish (salmonids) and lived in many of the same riverside locations. The fish owl ceremony, which returned the spirit of fish owls to the god's world, was conducted until the 1930s.

Many different stories of owls pervade Chinese culture, where owls play both good and bad roles and the stories often are passed down among generations within families. In China, owlets have been believed to pluck out their mothers' eyes. N.J. Saunders in 1995 noted that "The owl's night excursions, staring eyes and strange call have led to a wide-spread association with occult powers. The bird's superb night vision may underlie its connection with prophecy, and the reputation for being all-seeing could arise from its ability to turn its head through almost 180 degrees."

In a similar vein, on Andros Island, Bahamas, an historically extinct species of flightless owl (*Tyto pollens*), scientifically known only from subfossils[2], stood one meter tall and may have been the source of old local legends of "chickcharnies" or aggressive leprechaun-like imps that wreak havoc, have three toes and red eyes, and can turn their heads all the way around. This owl likely inhabited the dense stands of old-growth Caribbean pine, so much of which had been clearcut on Andros during the latter 20th century by American companies.

Some Native American cultures link owls with supernatural knowledge and divination, possessing special powers not found in other animals. For example, in the Menominee myth of "The Origin of Night and Day," Wapus (rabbit) encounters Totoba (northern saw-whet owl) and the two battle for daylight (*wabon*) and darkness (*unitipaqkot*) by repeating those words. Totoba errs and repeats "*wabon*" and daylight wins, but Wapus permits that night should also have a chance for the benefit of the conquered, and thus day and night were born. In *The Night Chant of the Navaho*, one of the gods is represented by the burrowing owl, who befriends the story's hero and sets him free when he is taken captive by the Utes.[3]

The Pawnees view the owl as a symbol of protection; the Pueblo associate it with Skeleton Man, the god of death and spirit of fertility; and the Ojibwa see it as a symbol of evil and death, as well as a symbol of very high status of spiritual leaders of their religion. On a warm afternoon in August 1985, one of the authors (David H. Johnson), observed Ojibwa peoples at a weekend cultural celebration in Duluth, Minnesota, using the dried wings of the great horned owl to fan themselves after participating in native dances.

In his book *Mother Earth Spirituality*, McGaa described the four directions of the Sacred Hoop (the four quarters with the power of earth and sky and all related life) of Native Americans. In this description, the snowy owl represents the North and the north wind. The traditional Oglala Sioux Indians (from northern Great Plains of North America) admired the snowy owl, and warriors who had excelled in combat were allowed to wear a cap of owl feathers to signify their bravery. An old-time society of the Sioux was called The Owl Lodge. This society believed that nature forces would favor those who wore owl feathers and, as a result, their vision would become increased.

A. H. Miller in 1935 described how remains of owls worked into various artifacts were found in prehistoric Native American middens of the western United States. The partial ulna of a western screech owl was drilled, leading Miller to speculate it was used as a tiny whistle similar to whistles created from ulnae of long-winged birds such as cranes, as found in other North American Indian middens. Miller reported that one whistle, made from the ulna of a crane and drilled in a similar way, reportedly produced a tone closely resembling that of a northern pygmy owl. Miller also reported finding in the middens remains of an immature long-eared owl. He speculated that a young bird may have been kept by a child or medicine man, and even held in captivity for ceremonial rites, as "some Pacific Coast tribes consider the owls as incarnations of nocturnal spirits of mystic powers."

Some Native American nations have strong taboos against owls. For example, the Apaches view the owl as the most feared of all creatures. Historically, Apaches shared the widespread Athabascan fears of owls as the embodied spirit of Apache dead. John Bourke, in his *Apache Campaign in the Sierra Madre*, related a famous story of how Apache scouts tracking Geronimo became terrified when one of the U.S. soldiers found and brought along a great horned owl. The scouts told Bourke that it was a bird of ill omen and that they could not hope to capture the Chiricahua renegades if they took the bird with them. The soldier had to leave the owl behind.

In another example, the consortium of Yakama tribes in Washington State use the owl as a powerful totem. Such taboos or totems often guide which forests and natural resources are to be used and managed, even to this day and even with the proliferation of "scientific" forestry on Native American lands.

Owls have played various roles in Russian traditions. For example, in Slavonic cultures, owls were believed to announce deaths and disasters. Russians and Ukrainians sometimes call an unfriendly person a "sych," which is also the Russian common name of the little owl. Traditionally, the little owl has been disliked and feared by people believing that these birds announce deaths. However, Russian common names of other owls, such as the Eurasian scops owl—*Splyushka*, resembling its call, or *Zorka*, meaning dawn—do not carry this negative connotation. In old Armenian tales, owls were associated with the devil.

The Spirit Chasers

He discovereth deep things out of darkness, and bringeth out to light the shadow of death.

—*Job 12:22*

The screech-owl, screeching loud,
Puts the wretch that lies in woe
In remembrance of a shroud.

—*Puck, in* A Midsummer Night's Dream,

Shakespeare wrote of "The owl, night's herald" (*Venus and Adonis*, 1593, line 531) and recognized the role that owls have as the "fatal bellmen, which gives the stern'st good-night" (*Macbeth*, 1605–1606, act II, scene ii, line 4) to that final, deepest sleep. In this way, owls have been seen as harbingers of the end of the world and of the ultimate fate of humans.

The owl was the guardian of the Acropolis, and the Roman statesman Pliny the Elder wrote that owls foretell only evil and are to be dreaded more than all other birds. The archaic word "lich," from the Anglo-Saxon "lic" and the German "leiche," means a dead body, and the archaic term "Lich-owl" refers to the screech owl, which, in superstition, supposedly foretells death.

In many cultures, owls signal an underworld or serve to represent human spirits after death; in other cultures, owls represent supportive spirit helpers and allow humans (often shamans) to connect with or utilize their supernatural powers. Among some native groups in the United States Pacific Northwest, owls served to bring shamans in contact with the dead, provided power for seeing at night, or gave power that enabled a shaman to find lost objects.

As with the owls of the ancient Roman statesman Pliny the Elder, many forest owls have played key roles as signallers of death. The mountain tribes of Myanmar (Burma) know the plaintive song of the mountain scops owl in such legends. In one Navajo myth, after death the soul assumes the form of an owl.

In India, the brown wood owl, forest eagle owl and brown fish owl are found in dense riparian forests of *Ficus* near streams and ponds, sites often considered as sacred groves, or in cemeteries that bear the area's last large trees with cavities and hollows. Old-forest owls, particularly the forest eagle owl, play major roles in many Nepali and Hindu legends. As heard calling at night from cemeteries and sacred groves, such owls are thought to have captured the spirit of a person departed from this world. In one sense, then, many of these owl species can serve as indicators of the religious value of a forest; conserving the religious site equally conserves key roost or nest sites.

Members of the animistic Garo Hills Tribe of Meghalaya, northeast India, call owls "dopo" or "petcha." Along with nightjars, they also refer to owls as "doang," which means birds that are believed to call out at night when a person is going to die; an owl's cry denotes the death of a person. Such beliefs are common throughout the world. In Sicily, when a Eurasian scops owl is heard calling near the house of a sick person, it is believed the person will die in three days, and if no sick person is present then someone will become ill with a tonsil ailment. In the pueblo tribes of the United States Southwest, including the Isleta of New Mexico, owls are viewed as messengers or harbingers of ill health and ill fortune. An Isleta tribal member relayed to one of the authors (Bruce G. Marcot) a story of once hearing an owl by her house. Shortly thereafter, her younger brother fell ill, and she and her other siblings went outside and shouted at the owl to drive it away because they were taught that it brought the illness.

The aboriginal peoples of North Queensland, Australia, view owls in a similar way. In January 2000, a female aboriginal elder relayed that owls are special to her people. A little apologetically, she added that owls are also considered an ill omen, signifying a death in the family—but only if the owl stays around the home site for several days.

Throughout India, owls are construed as bad omens, messengers of ill luck, or servants of the dead. In general, owls often have been treated badly both in daily life and even in Indian and Pakistani literature and daily lexicon. For example, in India and Pakistan it is very common to call a foolish person an "ulu" or "an owl," or an "ulu ka patha" or "a pupil of owl."

But in Indian mythology the owl also has been treated at times reverently and given some place of prestige. For instance, Laxmi, the Hindu goddess of money and wealth, is depicted as riding on, or being accompanied by, an owl. Even in present times, some people of India, particularly Bengali, believe that if a white owl enters a home it is a good omen, indicating a possible flow of wealth or money into that home.

The call of the Ceylon forest eagle owl subspecies consists of "shrieks such as of a woman being strangled" but "the dreadful shrieks and strangulating noises are merely its 'mating love-song,' which would also account for their rare and periodic occurrence." In related accounts, S. Ali and S. D. Ripley in 1987 described its noises as "a variety of weird, eerie shrieks and chuckles" and a scream "like that of a demented person casting himself over a precipice." V. C. Holmgren in 1988 also noted that in history, eagle owls have been variously called bird of evil omen, death owl, ghost owl, mystery owl, knows-all owl and even rat owl.

In India, the devil bird or devil owl can be found in graveyards and is associated with big dead trees—and death. Graveyards often contain the last old-growth trees, and in India, Muslims, especially, revere everything in a cemetery, including the vegetation. Hindus, as well, keep sacred groves of ancient trees, especially banyans. Thus, the eerie cries of the devil owl are heard mostly in cemeteries, portending death. And here converge myth, culture and biology to a consistent whole, as they should for successful conservation of cultures, people and wildlife.

Some Native Americans, for instance, wore owl feathers as magic talismans. For example, along the northwest coast of Alaska, the Yup'ik peoples made masks for a final winter ceremony called the "Agayuyaraq" ("way, or process, of requesting"), also referred to as "Kelek" ("Inviting-in Feast") or the Masquerade. This complex ceremony involved singing songs of supplication to the animals' "yuit" ("their persons"), accompanied by the performance of masked dances, under the direction of the shaman. In preparation for the ceremony, the shaman directed the construction of the masks, through which the spirits revealed themselves as simultaneously dangerous and helpful. The helping spirits often took the form of an owl. The majority of masks contained feathers from snowy owls. Carvers strove to represent the helping spirits or animal "yuit" they had encountered in a vision, dream or experience. In all cases, the wearer was infused with the spirit of the creature represented. Together with other events, the ceremony embodied a cyclical view of the universe whereby right action in the past and present reproduced abundance in the future. Among the Yup'ik, seabirds, loons and owls were commonly seen as embodiments of helping spirits.

Another native North American culture, the Kwakiutl of northern Vancouver Island in British Columbia, Canada, used an owl mask made of wood, and believed that owls were associated with darkness and the souls of the dead. Other owl masks are made by the Haida people of Queen Charlotte Islands of British Columbia and Prince of Wales Island of Alaska, who use the owl as a crest emblem.

In Central Asia[4], feathers of the Eurasian eagle owl, particularly from its breast and belly, were valued as precious amulets protecting children and livestock

Laxmi, the Hindu goddess of money and wealth, is depicted as riding on, or accompanied by, an owl.

The snowy owl as an emblem on a cigar box, ca. 1960. Minnesota, U.S.A.

from evil spirits. Talons of the Eurasian eagle owl were said to ward off diseases and cure infertility in women.

Among the Maori of New Zealand, to hear the call of an owl near the junction of two trails meant that an enemy force was nearby. The Maori viewed owls as creatures of sagacity, but to hear one calling at night conferred a feeling of uneasiness, as hostile raiding parties were deemed to be approaching. The Maori rendered such warning calls of owls as "Kou! Kou! Whero! Whero! Whero!" Many other Polynesians also hold a similar belief of an owl calling at night, but Native Tongans believe that an owl calling in the afternoon signals that there soon will be a birth in the family.

The Maori also believed that the owl and bat, which they called "ruru" and "pekapeka," originally were denizens of the underworld of "Rarohenga." This is why, in Maori belief, owls and bats do not move around in daytime but only under darkness. "Ruru" was looked upon as a bird of evil omen, and the presence of an owl at one's house signalled dire misfortune. The Maori would capture "ruru" or morepork owls with throwing sticks and slip-nooses. The "ruru" was occasionally eaten by the Maori and was presented preserved in a vessel to an assembly of guests while chanting a song that commenced: "He ruru taku nei" (My gift is one of owls). Maori legend has it that "ruru" was the talisman image used in the posture-dance as the glaring eyes of the owl. In another Maori tribal legend, two owls were used to guard special inland pools and to warn the owner when anyone went near them. The owls' names were "Ruru mahara" or thoughtful owl, and "Ruru wareware" or stupid owl.

On Java and Borneo, scops owls have benefited by having been viewed in

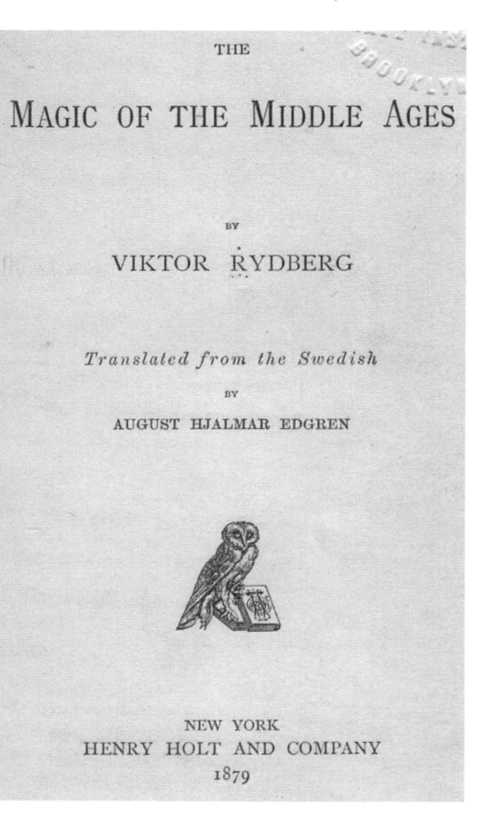
THE

MAGIC OF THE MIDDLE AGES

BY

VIKTOR RYDBERG

Translated from the Swedish

BY

AUGUST HJALMAR EDGREN

NEW YORK
HENRY HOLT AND COMPANY
1879

Title page from an 1879 volume on magic by Viktor Rydberg, showing an owl perched on books. In some cultures, owls have been viewed as bearers of knowledge and sagacity.

legends there with reverence or as an ill omen. However, these owls are killed in China and Korea for medicinal use and many have been lost there annually for such purposes.

H. Mikkola in 1997 reported how, for centuries, some cultures have used owls in traditional healing and folk medicine. In Europe, as in Africa, owl eyes were ingested to purportedly improve eyesight, and owl feet were eaten to ward off snakebite. In Malawi, Africa, owls are used by medicine men in ceremonies of magic to bewitch and kill people. In Togo and Ghana, hides and half-rotten owl carcasses are used to protect against evil spirits, and in South Africa owls are used for healing. In China, Korea and Thailand, the eggs and dried bodies of owls are used as medicines for many ailments, including rheumatism and loss of virility. Another old story is that if children eat owl's eggs they will never be addicted to strong drinks.

Informational sign in front of a bookstore in St. Petersburg, Russia, 2000. Text translation: "New Books, Store/Lounge."

Owls in Literature and Paintings

The Dutch of the 16th and 17th centuries often used owls in their paintings and viewed owls negatively because, owls being mostly nocturnal creatures, they avoided light and appeared to stagger about when caught by day. The Dutch of that time were strict Calvinists, and their painting and prints often were moral fables warning against a life enslaved by the pleasures of the senses. Owls were symbols of spiritual darkness, ignorance, drunkenness and debauchery, and thus used as admonitions against such vices. Indeed, for the Dutch, the term "uilskviken," meaning a "nest of owls," was the equivalent of calling someone a "nincompoop," similar to how Pakistanis or Indians call stupid people "an owl" ("ulu"). Another popular Dutch saying was "Wat baet er kaers en bril, alsden uyl niet zien en wil?" or "What need does the owl have for candles or spectacles if he cannot or will not see?"

In Aesop's fables, the owl, hated and mocked by other birds (indeed, many songbirds will "mob" real owls during daylight hours), was used as a decoy to lure songbirds to snares surrounding the owl's perch. Such use is displayed in the

Listed in a 1577 compendium on demons, Andras was named as the Grand Marquis of Hell and appeared with the body of an angel and the head of a wood owl. In medieval history, owls often were viewed as symbols of power and magic.

1665 oil painting *Rest on the Hunt* by Dutch painter Abraham Hondius. In the painting, a sly hunter, having just shot a Eurasian bittern, is plying a young lady with wine. In the background a live owl is shown tethered to a pole, and beneath the owl is a cage of live European bullfinches. The painting illustrates the Dutch phrase "Sweet talk has its poison," in that, just as the small and innocent birds would be lured to their doom by the owl, so too would the young lady by the hunter. In fact, the entire painting is one big smutty joke as the Dutch word for "bird hunting" was "vogelen," slang for sexual intercourse.

Owls also have appeared as demons. The owl demons, Andras and Stolas, were listed among 72 demons in Joannes Wierus's 1577 volume *Pseudo-Monarchia Demonorum*, and in J. Collin de Plancy's 1863 compendium *Dictionnaire Infernal*. Stolas was the Grand Prince of Hell, who appears in the shape of an owl. Stolas can assume the shape of a man and appears before exorcists; teaches astronomy and prophecy based on the study of plants, as well as the value of precious stones; and commands 26 demonic legions as their general.

Rydberg wrote that the magician can conjure the powers of the Creator by combining specific objects in the elemental world, including a piece of fine onyx, a dried cypress branch, the skin of a snake and the feather of an owl.

To the modern Garo Hills Tribe of western Meghalaya, India, the owl plays an essential role in preserving the mantras or divination spells of their largely secret animistic beliefs that constitute their magical religion called "Jadoreng." To record the spells and to protect their potency, mantras are to be written with a smoothened, sharp-pointed piece of red sandalwood as a pen, using ink made from a concoction of common red ink, the juice of the vilva plant, the juice of black basil, fresh blood of "dohka bolong" (a large variety of raven) and fresh blood of "dohpo skotottong," the spotted owlet.

Toward a Tolerant Conservation

Do you think I was born in a wood to be afraid of an owl?

—*Jonathan Swift*, Polite Conversation, *Dialogue i.*

Owl mythologies have come virtually full circle in Europe and America. From the worst bird in the world the owl has become almost the most popular. And old mythologies actually make owl conservation easier. The old bad news has become a way of making owls appealing for a contemporary audience. Whereas owls used to be persecuted, with governments imposing bounties on them along with other predators, the same governments are now protecting them with legislation.

This transformation of the owl's image is yet to be fully researched, but we can offer some initial thoughts. First, understanding European and American owl myths may help us better understand or interpret contemporary African and

Asian attitudes. It also may help us understand owl taboos among today's Native Americans and Canadians, tribal South Americans and other First Nations of both New and Old World cultures.

Second, if owl mythologies have evolved so dramatically in the West, then perhaps they offer an insight into the way owl mythologies could eventually metamorphose in other parts of the world, such as Africa and Asia.

Third, by understanding the patterns of owl mythology in modern-day Africa and possibly South America and Asia, we might be able to better understand our own cultural past.

Overall, the contemporary conservation community has not grasped the deeply negative image of owls in less developed parts of the world. In fact, western conservationists in general tend to think of birds or other wildlife mostly or only in terms of their own ecological-science-based and conservation-oriented system. By failing to appreciate other patterns of belief about birds, they are putting themselves at a disadvantage.

In the case of the owl, this is especially significant because the overwhelming nature of owl beliefs is that the beast is powerful even in western cultures. Until the 1950s owls were routinely nailed to barn doors in France and the United Kingdom to ward off lightning and the evil eye. There is even evidence to suggest that these practices continue today in parts of rural Britain. In Africa, deep taboos about owls are still powerful and living traditions. The beliefs in other old and modern cultures that owl body-parts transfer health and that owls attend gods, and imply knowledge and wisdom, also attest to the power of the owl.

H. Mikkola in 1997 suggested that in some cultures and locations such as southern Africa, owls could be bred for their cultural use as well as conservation. By ensuring a supply of owls for traditional healers and for local cultural beliefs and use, conservation of owls could be attained.

It is important to understand not just historical views of owls in myth and culture but also how such views have changed over time to the present and traveled geographically as human cultures have moved.

A club recovered from the Lake Ozette site in northwestern Washington state, U.S.A. This site reflects a Makah village that was buried in a mudslide 500 years ago. An owl's head was carved at both ends of the club. It is likely that the unworn club was used for ceremonial purposes.

A major problem with bird folklore is that some texts repeat the same ideas without any critical analysis of the truth of what they say, or whether the idea still applies in contemporary times. Most of these ideas about owls probably refer to the early modern period and were almost certainly gathered in the late 19th to early 20th centuries. But now many of them have completely died out, depending on the cultural changes of the people in question. For example, Batchelor's book, *The Ainu and Their Folklore*, analyzes this northern Japanese community's beliefs about animals, especially owls, at the end of the 19th century. These old cultural ideas sometimes are still retold as if they were living traditions, when, in fact, modern Ainu may consider their ancestors' owl beliefs as no more than old wives' tales, a part of their own colorful but quaint past. We should, therefore, attempt to understand how cultures and views have changed.

Conclusion

For in the end, we will save only what we love, we will love only what we understand, and will understand only what we are taught.

—Lao-Tzu

In this brief review, we have highlighted only a few of the many roles of owls in myth and culture. We encourage readers to pursue additional sources, particularly those listed in the bibliography.

Owls have served as marvelous and fantastic symbols of recreation, esthetics, art, science, lore, political power, ethics, magic, religion and even death. In the case of owls, the deep fears and anxieties they have generated and the prophetic status they once held, and still hold in some cultures, present environmentalists with a handle with which to engage the interest and sympathies of a wider audience. By inviting owls into a full cultural circle, we can build a more tolerant understanding of all societies and ages, and incorporate wildlife conservation into the broader tapestry of human endeavors.

Our special thanks to our colleague Mark Cocker who provided edits and information on owl anthropology and mythology, and who helped strengthen the academic accuracy of our work. We also thank those who searched for or helped us trace stories of owl lore: Ajai Saxena, V. B. Sawarkar and Baban, from India; Muhammad Mahmood-Ul-Hassan, from Pakistan; Eric Hansen and Paula Shattuck, from the United States; Max Sova, from Russia, Armenia and Central Asia; Jiin-Wen Tsai ("Wren") and Helen Lau from Taiwan and China; T. Tekenaka, from Japan; Robin Tim Day, from Korea; and Frances Schmechel from New Zealand.

Mark Lynch and Janet Guerrin of Worcester Art Museum and Massachusetts Audubon Society graciously provided invaluable information on owl lore. We also thank two anonymous reviewers for their comments on an earlier version of the manuscript, and Paula L. Enriquez and Heimo Mikkola for providing their work on owl mythology.

Endnotes

[1] In their stories, the Klamath Mountain Indians of northwestern California, U.S.A., referred to "Bigfoot," an elusive bipedal hominid supposedly inhabiting the deep forests, as *Omah*.

[2] One such record comes from Smithsonian Institution Archives, Record Unit 7006, Collection Division 11, Alexander Wetmore Papers circa 1848–1979, fossil birds, Box 162 of 237, Folder 60, *Tyto pollens*, Figures 10–16, "Bird Remains from Cave Deposits on Great Exuma Island in the Bahamas," 1937.

[3] The Cambridge History of English and American Literature in 18 Volumes (1907–21).

[4] In Russian literature, Central Asia is the region to the east of the Caspian Sea including Turkmenistan, Uzbekistan, Tajikistan, and Kyrgyzstan.

CHAPTER 3

The Study of Owls

Owls fascinate people. To some this is a passing fancy or focus, perhaps based on personal experience, or simply a matter of seeing an interesting story or news report on owls.

In order to find or see owls, people often play commercial tapes of owl calls, imitate owl calls, pound or rub on suspected nesting trees or shine lights on nests. While the effects of this behavior are largely unknown and need more study, it is likely negative if repeated, especially during the sensitive period when owls are incubating or brooding.

OPPOSITE: **Tracking down flightless young great gray owls to band them can leave scent trails through the forest. Other bird studies have shown that predators can follow these trails and locate vulnerable young birds. Luckily, young owls have plumage the color of bark, and are good at the game of hide and seek!**

A nervous northern saw-whet owl exits a nest box at the slightest noise. Early in the nesting season some owls are quick to abandon their eggs when danger approaches. With only one entrance, unwary owls are easily trapped by predators such as the pine marten.

ABOVE: **The northern saw-whet owl nests in natural tree cavities or in those built by woodpeckers.**

ABOVE RIGHT: **A quick nest inspection confirms that the nest is active. Eggs are more susceptible to heat than cold so such visits are confined to cooler times of the day.**

The repeated playback of an owl's call can be especially disturbing to nearby owls. This has been documented for some species such as the eastern screech owl. In general, the impact of research methods can be tested and then redesigned to minimize its effect on owls. This is important to ensure the owls' welfare, but equally so because the results of a study are influenced by how it is carried out.

Many birds build their own nests, and people who study them locate nest sites by observing this conspicuous behavior. Owls, however, do not build their own nests. To help find nesting owls to study, biologists either record potential nests during the year, and check these in the breeding season for nesting owls or install artificial nests, and monitor them for nesting owls. Artificial nests are particularly useful in the conservation of rare owls at risk of becoming extinct, especially to mitigate the loss or damage of breeding habitat.

Basic information such as clutch size, egg or owlet survival and nest use over time should be interpreted with caution if artificial nest sites are used. They may vary considerably from natural nests, and might therefore cause erroneous conclusions for population growth and survival estimates. Owl studies that rely on

RIGHT: **A newly banded and radio-marked barred owl is released after getting caught in a mist net. The lure in this case was a mounted barred owl and audio playback of the species' territorial song. The resident breeding owl reacted aggressively to the lure, flying into the unseen net when it attacked the perceived intruder.**

artificial nests should include data on nest height, orientation, density and nest use by competitors or predators.

The effects of radio transmitters and other markers, and their attachment to owls, also need to be tested by comparing the flight capabilities and behavior of marked versus unmarked owls in flight cages. This is especially important for smaller owl species such as the diminutive elf owl.

This chapter, and indeed this whole book, can barely scratch the surface of the variety of owl studies over the years. Owl research most commonly investigates behavior, population dynamics, diet, habitat use, nesting biology, distribution, anatomy, physiology and genetics. Artists also study owls in valuable ways that science cannot, yet both approaches require creativity and dedication.

Sometimes owl biologists turn their attention to people, searching for reasons why they are so compelled to collect owls. Self confessed "owlaholics" seek out rare items at garage sales, stores and even on web sites such as eBay. An eBay search based on the keyword "owl" turned up 4196 auction items, some of which sell for thousands of dollars.

Artists like Ed Brown of Balmoral, Manitoba, Canada, study and share their knowledge of owls in rigorous and creative ways. Ed's international award–winning carving of an eastern screech owl with a hoary bat is proof that both scientists and artists use the human mind as an instrument for understanding. Oil and acrylic on tupelo and basswood.

This owl set is but a fraction of the thousands of items in owl enthusiast Bryan Mitchell's collection.

Owl Behavior

Sometimes we learn about owl behavior when we least expect it. Until 1996 I had no idea an earthly creature was capable of uttering such demonic sounds. I was blazing a trail through the boreal forest in a remote area of southern Manitoba, Canada, accompanied by my two husky dogs. By 9:30 p.m. I was almost out of daylight and was several miles from camp, but I had reached a strikingly beautiful patch of ancient black spruce trees. I settled down on a moss-covered log to rest and to share my last bit of food with the dogs. The late spring sun's last rays cast an ethereal mood in the old forest. Advancing dark storm clouds threatened from the northwest. I was worried that it would get dark faster than I had anticipated, making it hard to follow my blaze marks back to camp. After packing up my gear, the dogs and I started along our trail.

After taking no more than a few steps, a violent and unearthly scream exploded from what sounded like inches behind my head. My first reaction was swift and automatic—I screamed and dropped to the ground in a state of sheer panic. The dogs yelped and cringed too, their tails tucked between their legs. We spun to see what evil thing was about to destroy us. There on a branch about 10 feet (3 m) high was a dark-eyed owl with distinct barring across its chest—a barred owl. This small animal (a mere 1% of my body mass) had orchestrated an intense and visceral reaction in my dogs and me. It was likely reacting, effectively I might add, to what it may have perceived as threats to a nearby mate and young. This owl encounter remains the most frightening experience I have ever had. What a bird!

Owls are also good indicators of the health of the environment because of their role as predators, and because some species depend on large areas of intact natural habitat. However, little is known about the population status of most species because owls are wide-ranging, shy during the breeding season and often occur at low densities. Consequently, most owl species, especially those in forested and tropical habitats, are not well followed by existing bird monitoring programs such as Breeding Bird Surveys or Migration or Christmas Bird Counts. Owl hoot surveys are popular events that recruit hundreds of citizen scientists to listen for owls. Using owl playback calls to elicit a response from owls on breeding

Although rare, even tree-nesting species like the great gray owl will nest on the ground when no suitable tree nests are to be found. We relocated this bird 500 miles (800 km) north of where we had initially radio-marked it 2 years earlier.

territories has greatly increased the known distribution of many species. But working with owls at night can be dangerous, as Jamie Acker of Bainbridge Island, Washington, U.S.A., described: "In early March, prior to barred owls going on eggs, while calling for northern saw-whet owls at 5 a.m. in a forest, I observed a large owl fly overhead. I had a report of a pair of great horned owls here the week before from a reliable source. I did not receive a response back from either a great horned or a barred owl tape, so I turned and started to walk out. I had gone 200 feet (61 m), crossed an overflowing stream and changed direction with the trail when I got hit hard on the back of the head by what I later saw and heard as a male barred owl. Later, on several occasions during the nesting season, I unsuccessfully searched the area for a barred owl nest. There were three potential trees with cavities in the area, but none of them produced a bird upon being rapped, when the female should have been on eggs or with small young. In mid-May, when young barred owls are close to branching, I was again out owling at 4 a.m. and played the barred owl tape in the same vicinity. The male barred owl flew in high and silent. I turned to walk away,

The unique sonogram of a great gray owl's low-pitched "Whoo whoo whoo" call. Most owls have distinct calls that are easy to learn.

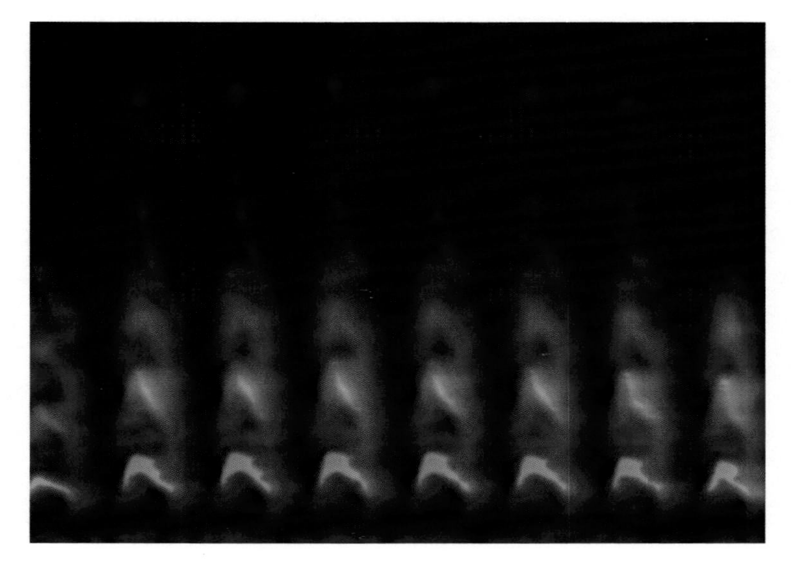

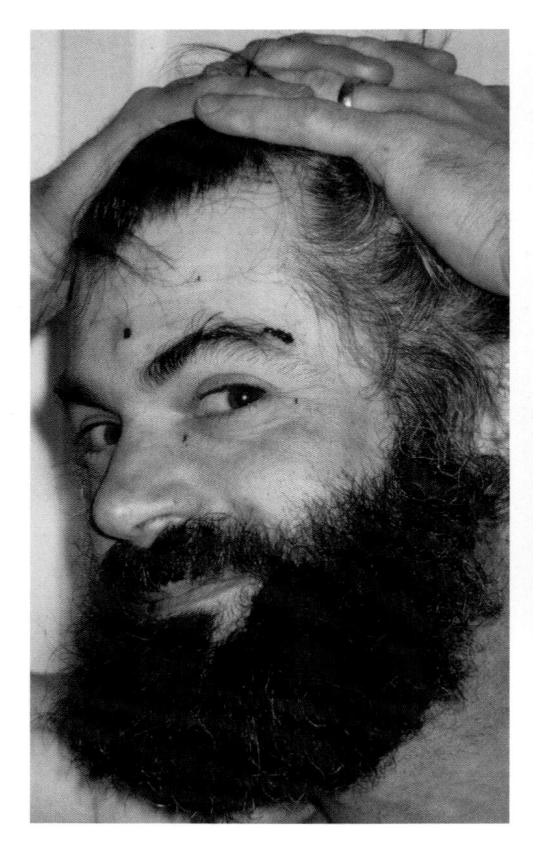

Jamie Acker shows what can happen when territorial owls react to a surveyor using playback calls.

The tiny meadow vole is an important link in the conversion of solar energy to owl biomass. The population cycles of many predators are driven by meadow vole numbers.

and then for some reason, felt uneasy and turned just in time to catch a face full of talons. As the photo shows, I was very lucky not to lose an eye. In hindsight, there must have been a nest there, and the female just stuck tight to her eggs/young when I rapped on the tree. The male was defending the nest site. Later that fall, over in north Seattle, there were several reports of joggers getting attacked by a barred owl. I was never billed for the tetanus shot I received—it was a 'first' for the doctor."

Sometimes owl hooters can mistake the identity of a calling owl, although most owl calls are distinct. One amateur naturalist in a suburban community of the United Kingdom wanted to initiate a simple study of a tawny owl living in his neighborhood. One night, he vocally mimicked its call, and to his delight one answered. Over the ensuing months, he frequently went out to elicit the owl's call. Months later, his wife went to the laundromat and started talking to another woman. They discovered that their husbands both studied owls in that area of town—the two men had unknowingly been recording each other's calling behavior!

Populations Dynamics—Predator-Prey Studies

Many noted biologists have stated that to really know a predator you must also study its prey. Prey availability is an important factor in governing the often dramatic fluctuations of owl populations. Large-scale irruptions of the great gray owl are a classic example of such noteworthy events, and they have been recorded since the late 1800s, both in Europe and North America. These occasional movements within and bordering their boreal forest breeding range are indeed characteristic of the species. In winter 2000/2001, unprecedented numbers of owls were recorded mainly along the southern margin of the boreal forest from Alberta and across Saskatchewan to Manitoba, Ontario and adjacent Minnesota. Owl numbers increased over winter, with some Manitoba birders recording more than 100 in a single day! Great numbers of large owls suddenly interacting with rural, suburban and even urban-dwelling people spawns numerous concerned inquiries from the general public. In Manitoba these inquiries usually land on my desk. Luckily, my biologist wife and I have been trapping voles (field mice) every October for the past 17 years as part of our long-term program monitoring great gray owls and small mammals in southeastern Manitoba. Our data indicated that high prey (meadow vole) populations accompanied above-average owl breeding success in 1999 and 2000, producing large numbers of great gray owls that subsequently made up part of the invasion. Furthermore, notwithstanding the presence of large numbers of owls in southeastern Manitoba in spring 2001, no nesting was observed that year. Our data for this region suggest that owl breeding fluctuations and invasions are natural events that are correlated with cyclical meadow vole populations. It seems likely that this locally demonstrated relationship can be applied across much of the range of the great gray owl as at least a partial explanation for their observed numbers and movements.

Sampling owl prey populations can be critical to understanding owl ecology.

Owls found dead are valuable for a variety of research projects. Tissue from this dead owl, found starved and frozen in a wood shed, will be used to help determine the genetic metapopulation structure of the boreal owl in North America.

Owl Pellets and Diet Studies

There are many ways to learn about what a predator eats. One way is to look for the digested remains of prey in the stomachs of owls found dead. I have done this on numerous occasions and can testify that it requires a strong constitution and a very tolerant family! My seven-year-old daughter is fascinated when Patsy and I periodically convert our kitchen table into an owlish morgue, but her older brother does not care for it as much.

A more time-consuming approach is to find and watch owls hunting. However, it is very difficult to accurately identify prey items caught at considerable distances, even with binoculars or a spotting scope, a task that becomes virtually impossible at night, when many owls hunt.

Many researchers have collected considerable data on prey captured by recording the items that breeding owls bring to the nest site to feed their rapidly growing young. Depending on the owl species, the prey taken and the number of young being fed, some breeding owls bring hundreds of prey items daily. Some clever devices have been designed and built to record these feeding trips by camera or video, even at night using infrared or other light sources. Usually these camera devices are triggered automatically every time an owl delivers a prey item to the nest and do not appear to cause any undue harm to the adults or young, owing to the quick-contracting striated muscles of the eye's iris. Of course, other precautions must be taken when working close to active owl nests, especially not to lure predators to the location by leaving human or food scent trails to and from the nest site.

Teachers wanting to give their students a hands-on biological learning experience sometimes use owl pellets. Students are asked to carefully pick owl pellets apart to identify prey items consumed, separating hair or feather fluff from bones, teeth and other indigestible bits. Washing hands after handling owl pellets is critical to avoid possible salmonella poisoning.

Nesting Biology

Studies on the nesting ecology of owls have shown some remarkable relationships between owls and other organisms.

A northern hawk owl quickly dispatches a recently caught small rodent to avoid being bitten on the toe or foot.

Mock attacks on inanimate objects are important in the development of hunting skills in young owls. Owlets that are well fed play more frequently and learn to hunt faster. The instinctive play behavior of this young captive eastern screech owl was used by the author to test color preferences.

Temporarily relieved from their cramped nesting cavity, four young boreal owls wait to be banded. When first discovered by western scientists, young boreal owls were thought to be a different species because they appear so different from adults.

Snake–owl Symbiosis—Perhaps unique to small-cavity nesting owls is an unlikely partnership between a predator and otherwise typical prey. In central Texas, Fred Gelbach and others documented that some eastern screech owls delivered live blind snakes to nests. The small, insect-eating snakes then burrowed into the dirt, old food, feces and other debris at the bottom of the owl's nest. By eating ant and fly larvae that normally thrive in this environment, sometimes to the detriment of the nestling owls, the snakes served to keep the nest pest-free. In nests without such reptilian live-in housekeepers, ant and fly larvae eat extra cached food stored by the adult owls for times when prey are scarce. Owl nests with live snakes raised more young than those without—evidence that the inhibition of the typical killing behavior in this case gives some owls an advantage over those that simply kill blind snakes to feed their young.

Ant–owl Interactions—Even very small animals like ants can wreak havoc on the lives of others much larger than themselves; ever try to fall asleep on the ground near an ant hill? Some ant species, like the yellow crazy ant, actively seek out bird nests to kill and eat the young. But at least one ant, the acrobat ant, inadvertently helps some cavity-nesting owls by defending the shared nest cavity space against intruders. The ants bite, spray repellant secretions and swarm non-owl intruders, including owl researchers! The ants apparently thrive on uneaten prey and fly larvae found within the debris at the bottom of the nest cavity.

Owl Banding

Besides being an exciting hobby, marking wild birds is a vital tool for studying the behavior, survival and movements of birds. In fact, bird banding has been recognized as the most useful activity in

the study of birds over the last 100 years. The first recorded use of a metal ring or band to mark a bird's leg was about 1595. For hundreds of years there was little or no coordination of marking efforts by those doing this work. In recent times there has been better cooperation and sharing of information. Today, the most common method of marking birds is a standardized system of uniquely numbered metal bands.

Since 1923, The North American Bird Banding Program has been jointly coordinated and administered by the United States Department of the Interior and the Canadian Wildlife Service. All North American banders use the same kind of bands, reporting forms and data formats. In Europe this activity is coordinated by the EURING database, where information is collected on all ringed (banded) and recovered birds in

The remarkably calm behavior of a recently banded healthy great gray owl about to released by my son Connor, age four.

Grounded! A northern hawk owl is caught in a Bal Chatri trap. This trap contains live prey in a protective case. The owl is stuck when a toe or foot gets tangled in one of the many nooses covering the trap.

ABOVE: **Trigonometry and timing. A hungry northern hawk owl eyes the lure while Patsy eyes the owl. The addictive nature of banding owls relates perhaps to the unpredictability of the outcome.**

RIGHT: **A color-banded burrowing owl about to eat a mouse. Color bands allow biologists and others to identify individual owls repeatedly in the wild without recapturing them.**

Europe. An estimated 100 million birds have been banded worldwide to date.

People band birds for a variety of reasons, but mainly it provides information about that individual bird if it is ever found again, dead or alive. How far did it travel? Where did it go? When did it travel? Did it return to the place where it hatched? How did it die? How long did it live?

When many birds are banded, generalizations about these questions for different kinds of birds can be made. This information is not only valuable to academics but also critical for conserving populations of endangered or threatened birds. It is important to have a thorough and accurate understanding of the dynamics and natural regulation of bird populations. Annual birth (productivity) and death (annual adult survival) rates, average life span and age structure (number of breeding-age adults and immature young in the population) are the most important types of information to monitor or to predict bird population changes.

Longevity Records

Although banding studies also give us information on the maximum life span, or longevity, of free living birds, biologists generally consider this statistic to be interesting but trivial and unrealistic. It is very hard to estimate how long wild birds live because birds frequently have fatal accidents in remote areas or are eaten by predators. Therefore, longevity records do not reflect the physiological maximum potential of species—birds in captivity live longer than birds in the wild. Longevity estimates are also sensitive to errors in record keeping, the number of birds banded and how long the study has been underway. The more birds banded and the longer the study, the greater chances of getting a more accurate longevity record.

Nevertheless, longevity records do have value. A detailed analysis of data from millions of banded bird species determined that the observed longevity of birds does significantly relate to annual survival rate (as calculated for individual species from mark-recapture/recovery data from tens of thousands of birds)—the longer a bird species lives, the higher its annual survival rate for both adults and young. This relationship is important for owl conservation, because banding relatively large number of owls is not often feasible. Therefore, owl longevity records may be the best approximation of annual survival rates.

The rotted-out top of a birch tree proves to be a tight squeeze for this nesting female northern hawk owl. The longevity record for this species was based on a nestling that Patsy and I banded in 1987; it was recaptured eight years and seven months later.

Calculating Longevity

Keeping a newly hatched owl in captivity until it dies is one way to determine how long it lives. But doing so may increase or decrease its natural life span and be less useful in estimating wild population dynamics for conservation planning. We need to know the longevity of owls living under natural conditions. Capturing, marking and recovering wild owls is the only way to get this information because banding many owls will increase the chance that someone will find a banded bird. As banding studies have progressed, the longevity records for owls have been increasing. At some point, in theory, longevity records will reach the maximum age possible for the species. Longevity is typically reported in years and months.The easiest way to calculate longevity is to band young birds and then note the time passed from when the owl was banded to when it was recovered. However, if the age of an owl banded as adult is known or can be reasonably estimated, that age can be added to the period between banding and recovery. The North American Bird Banding Office and others use a specific system to estimate longevity.

There is a positive correlation between

A fledged long-eared owl chick rests a wing after barely landing on a branch.

A Sample of Wild Owl Longevity Records

Species	Longevity
Long-eared owl	27 years, 9 months
Great horned owl	27 years, 7 months
Eurasian eagle owl	24 years, 9 months
Barred owl	18 years, 2 months
Barn owl	17 years, 10 months
Spotted owl	17 years
Boreal owl	15 years
Eastern screech owl	13 years, 6 months
Short-eared owl	12 years, 9 months
Great gray owl	12 years, 9 months
Western screech owl	12 years, 11 months
Snowy owl	10 years, 9 months
Northern saw-whet owl	10 years, 4 months
Burrowing owl	8 years, 8 months
Northern hawk owl	8 years, 7 months
Flammulated owl	8 years, 1 month
Eurasian pygmy owl	6 years
Elf owl	4 years, 11 months
Collared scops owl	4 years, 4 months

(Note that an albatross appears to hold the longevity record for wild-living birds—one Laysan albatross has lived to be at least 42 years and 5 months old. Another albatross that nested in New Zealand was reportedly at least 53 years old when it went missing.)

the body mass of an animal and its maximum recorded life span. This holds true for birds, even though the owl longevity table reported here seems to contradict this. The long-eared owl has the greatest reported longevity, but is a fraction of the body mass of several owl species with shorter longevity records. This anomaly may simply reflect the fickle nature of longevity records, especially when relatively few birds have been banded.

Compressed plumage, erect ear tufts and eyes closed to mere slits are features of the concealment display of smaller owls. This eastern screech owl keeps a peeping eye on the photographer at all times.

A Sample of Owl Studies

Many reference books on owls have done a comprehensive job of cataloging studies about owls. These works are valuable, but the volume of facts and information often omits detailed discussions about how the information was obtained. Reference books, by necessity, do not effectively convey the sense of excitement experienced by the investigator, nor do they have an in-depth, first-person perspective about the context in which the work occurred.

Who better to convey this spirit of adventure than the investigators themselves? I was therefore pleased when several friends and colleagues from around the world enthusiastically agreed to provide personal accounts of their work for inclusion in this book. The result is a sample of perspectives of contemporary owl studies. Keep in mind that there remains much to be learned about the biology and ecology of owls. More questions arise from each study undertaken. The unsolved mysteries of the natural world are a deep well from which students, scientists and curious people thirsting for knowledge can drink for their entire lives.

Female Desertion in Southern Boobooks

by Jerry Olsen and Susan Trost

Most ornithologists know about the separation of home duties that characterizes birds of prey during the

breeding season. The larger female does most or all of the incubating of eggs and brooding of small nestlings, while the smaller male does almost all of the hunting. While males travel long distances from the nest in search of food, females stay back and defend, often fiercely, the nest, eggs and young against predators, including humans. What is less known about some raptors, such as screech owls and Cooper's hawks in North America, and southern boobooks in Australia, is that they show an even more lopsided partitioning of duties after the young fledge. After the young have been out of the nest for some days, but are still dependent on parents for food and protection from predators, many females don't do their part—they refuse to defend or feed their dependent young, even when these fledglings sidle up to them on a

Southern boobooks are the smallest of the nine owl species that breed in mainland Australia.

A young southern boobook chick peers out from its nest cavity.

branch and beg. Females abandon all duties to their mates. Ornithologists call this "female desertion," and we can use what we know about southern boobooks to explain it.

Each year since 1993 we have studied three or four territories of southern boobooks in the Aranda Bushland and Black Mountain Reserve across the road from our houses in Canberra. We find the nests of three or four adjacent pairs, color-mark the adults and radio-tag those we can; we fit some with a backpack-style Sirtrack single-stage transmitter with a string harness and a weak link designed to break if entangled. The total weight of the transmitter and harness (.2 oz/6.4 g) is about 2% of the owl's body weight at capture. When we trap a breeding pair, we determine that the largest of the pair by weight, often with a brood patch (a bare patch of skin on the breast of the female used to heat the eggs during incubation), is the female. For our convenience, and because it is an active time for owls, we observe them during the 60- to 90-minute period beginning a half hour after sunset.

When first entering the forest we peer with binoculars up into the foliage where the owls hide by day. When we find them, the owls watch us, standing upright on the highest leafy branches, tilting their faces down at 90° and looking at us past their breast feathers and through their toes. Sometimes they roost in native cherries, at eye level, and walkers and joggers pass close by and don't notice them.

The fledged young sit next to their parents, sometimes along one limb touching shoulders, the brown-feathered adults next to the cream-colored juveniles. The just-fledged young are most endearing—creamy white and stubby-tailed with round bodies and dark bandit masks over their eyes, like raccoons. On partly feathered wings, they struggle to fly and they give persistent cricket-like trills as they move around through the forest after sundown, like children blowing little whistles in the dark. While the fledglings

After the young leave the nest, the male southern boobook becomes the primary caregiver.

The southern boobook sometimes utters a single hoot call when someone approaches its fledglings too closely.

follow their parents around from tree to tree, the female defends them; she sometimes gives a single "*hoot*" call as she swoops past our heads. The single hoot call is like the first half of the two-note call in a cuckoo clock, like the first half of the common boobook call, and adults give it with emphasis. Because southern boobooks in this study used the single hoot call most commonly after their young fledged, we heard it most often in December and January. Females used it first, after the young fledged, then males started to use it when they took over parenting responsibilities after the females "turned off" or left the breeding territory.

Female southern boobooks stop caring for their fledged young in different ways. In one variation, males move with their fledged young to different parts of the territory away from the family's nest and "camp out" during the day. As the male and fledglings move further and further from the nest, the female stays behind—she doesn't feed or defend her young because she is no longer with them—the family has left her. The adult male assumes total responsibility for feeding and defending the fledged young.

Breeding territories for southern boobooks are about half a square mile (1 square km), and these deserting females set up a home range of about the same size—on their own, and well before winter sets in. One female roosted in a cluster of three Mediterranean cypresses, all within 33 feet (10 m) of each other along a cul-de-sac—Sydney Avenue near Parliament House in Canberra. The three trees were about 23 feet (7 m) tall and had very dense foliage, so she was invisible to passersby. Her home range included a mix of native and exotic trees and shrubs arrayed across parks, gardens, houses, flats, offices, a school, shops, quiet streets and some main roads. On a typical evening she first moved from deep within the roost tree to the outer branches and peered out; she occasionally preened and sometimes disappeared back into the roost tree, to reappear a short time later. She might fly up to 100 feet (30 m) and perch before starting her foraging flights. On two occasions she did not leave the roost

A hungry, responsive owl flying in after an artificial mouse-like lure. Herb Copland kneeling in the snow, must time his net swing precisely in order to catch the owl. Robert Nero is a willing assistant controlling the lure with a fishing rod.

during the three hours that our colleague, Steve Taylor, watched the tree. Sadly, on May 12, 2001, seven years and four months after we banded her at her nest, she was killed on this winter home range by a domestic cat.

Many female raptors around the world migrate to more open places for the winter rather than the wintering grounds used by males of the same species. This could explain why female southern boobooks move away from their forest breeding territories and desert their families—to find more open ground that requires less agility for large females to fly in and hunt through, and lessens the risk of fatal accidents. However, this doesn't explain why some females remain on their breeding territories and simply stop feeding and defending their young. We look forward to spending many more hours watching southern boobooks and other birds of prey to learn more about female desertion.

To Catch an Owl

by Robert W. Nero

In response to a query as to whether we intend to use our tame shrew to lure great gray owls this winter, I'm undecided.

The shrew could probably stand the strain, and, of course, it's easy enough to obtain lab mice and other lure animals. But my long-time partner, Herb Copland, and I are now both 80 years old. It's mostly a matter of deciding whether it's useful to subject ourselves to the conditions under which we try to capture owls for banding in Manitoba, Canada.

Searching for owls is pleasant enough, even under severe winter conditions, though I can recall a few trying times when we had to be towed out of a snow-filled ditch. Fortunately, Herb Copland is a most agreeable companion, never scolding and seldom complaining. One late evening, heading home after a long

Herb Copland (left) and Robert Nero (right) introduce a captured and newly banded great gray owl to a curious but shy youngster. By attaching a yellow numbered tag to banded owls, they were able to identify wild individuals without recapturing them.

fruitless day in mid-winter, when out of frustration I tiredly suggested that we were wasting our time and money, Herb reminded me that some day we might not be able to do this.

This whole business of attempting to capture and band every owl we encounter (both great gray owls and northern hawk owls) involves some anxiety. Consider the scenario.

Upon sighting an owl, we get out of the car, and I cast out an artificial lure that, when reeled in, resembles a mouse or vole running on the snow. A hungry, responsive owl will fly in after the lure, coming close enough for Herb, kneeling in the snow, to capture it with a large fish-landing net.

An alternative technique involves using a live lure—gerbil, lab mouse or short-tailed shrew (the latter in a cage)—set down in front of Herb.

Usually, if an owl comes to the lure, Herb can net the bird, though occasionally we have watched in dismay as an owl slipped in and stole a gerbil or mouse. Netting an owl safely requires precise timing. I've netted owls a few times, but I lack the necessary patience and, especially, the nervous control that's necessary—it's simply too exciting for my blood.

So the ultimate responsibility for getting that net over the bird falls on Herb Copland. Poor Herb—he's caught a lot of owls, hundreds, but when he misses, he faces his own disappointment as well as my ill-advised scolding. Yet, Herb handles it all with seeming equanimity, and he's always keen to go after the next owl we come across.

I gained an even greater appreciation of his dedicated spirit while watching him kneeling in the snow on the side of the road early one morning near Lac du Bonnet, trying to entice a reluctant male owl with a hardy short-tailed shrew scrambling about in a small cage for 20 minutes at –43.6°F (–42°C). And yes, Herb netted the owl; that's how we learned that it was a male.

On another day, as Herb kneeled on a snow-covered road with a live mouse as a lure, the owl he was after glided in toward him, circled, hesitated and then, astonishingly, landed right on top of Herb's head! The owl stayed there, gazing down at the mouse, for four minutes.

On Wrangel Island, snowy owls usually lay eggs in old nest scrapes (shallow depressions in the ground) on wind-blown hills. A flamboyant ring of nitrogen-loving plants grows and flowers around such nests after years of owls defecating there.

Finally, it dropped down to get the mouse and was promptly netted. Herb, it turned out, had judged the owl's behavior by watching its shadow on the snow in front of him.

Still, the excitement of the game and the satisfaction of learning something more about the great gray owl is compelling.

The Snowy Owls of Wrangel Island

by Irina Menyushina

I first came to Wrangel Island in April 1982 as a research assistant for my husband Dr. Nikita Ovsyanikov, who was conducting a field study on the social organization of the arctic fox. We traveled for hours by a military track vehicle operated by two experienced border-guard drivers. Our destination was a biology field station situated at the most remote corner of the Northern Plain of the island—62 miles (100 km) from the nearest village. After several hours of bouncing over hard snow drifts, our "tundra tank" finally stopped at our destination. We emerged from the vehicle's dark narrow cabin and, after listening to the stunning rattling of the engine for hours, found ourselves in a "sea" of engrossing silence and glistening whiteness. I entered the small cabin, constructed from wooden boards and old plywood and covered on the outside with black roofing material, and found myself

An old female snowy owl returns to her nest with prey.

in a small dark room with a single tiny window and a very low ceiling. Along two walls there were plank-beds constructed from rough wooden boards, a wooden table by the window and a metal stove in the opposite corner. Everything was frozen; the plywood ceiling was covered with thick frost and decorated with icicles. Our drivers were appalled. They asked us how long we planned to stay, and our answer—"four to five months"—drew a whistle of surprise. They quickly assessed the situation, drank a cup of tea from a thermos, unloaded our baggage and then headed back to the village, where they could rest in warm comfortable apartments. Within a few minutes, their roaring vehicle disappeared across the snowbound tundra.

Meanwhile, Nikita got the gas stove running, whereupon ice from the ceiling turned into a cold dribble. Avoiding the cold drops, I cooked a welcoming meal. We opened a bottle of champagne and, feeling ecstatic, celebrated the beginning of our field season.

Tundra extended in all directions from the black cube of our cabin. It was covered by deep snow that hid not only the earth but also any relief. There was not a single black spot on this white, slightly hilly surface; nothing the eye could catch as a landmark. Only deeply frozen snow everywhere, on which our steps sounded loudly. Wind, frost and air hazed in bewitching ice-drifts beyond the edge of the northern coast. Seemingly, no life was possible in this frozen kingdom. Suddenly, a sonorous scratching sound could be heard from under the snow. The sound stopped abruptly, but in a few seconds it was repeated again. It was a lemming, invisible under the white carpet, but alive, making a many meters-long corridor in the densely compressed snow. Scanning the surroundings through binoculars, I discovered other inhabitants of this severe land. White arctic foxes were wandering everywhere, frequently stopping to locate lemmings under the snow by listening for their scratching. The foxes stalked the lemmings in deep tunnels dug down to lemming corridors, or by breaking through the sometimes-thin ceiling of snow above the corridor with a high pounce. Arctic foxes were not alone here; usually, not far from a foraging fox, I could see a silhouette of a watching snowy owl. After a few days, we learned how owls manage to survive the winter on Wrangel Island, where they don't have prey such as hares and ptarmigans, but only lemmings, which in winter live under the snow and are inaccessible to owls. Snowy owls constantly followed hunting arctic foxes, landing 500 to 650 feet (150 to 200 m) from a fox while it was digging or stalking. As soon as the fox took a lemming, the owl attacked, trying to strike with its sharp claws. To escape, the fox would jump and run, making sharp turns; if this was unsuccessful, the fox would drop its prey and lunge at the owl with open jaws. That was the moment the owl was waiting for; it would dive past the fox, snatch up the lemming and fly away. Sometimes foxes chased owls in such situations, but always without success. In May, as soon as lemmings started emerging onto the surface, owls began hunting on their own.

These observations were my first training in watching animal life in the High Arctic, and my first experience with snowy owls. For the following years, first arctic foxes and then snowy owls became the main focus of my research. From 1982 through 1986, Nikita and I studied the social life and ecology of the arctic fox. Severe Arctic conditions create enormous difficulties and hardships for field research, but there is one great advantage in working there: the open landscape provides excellent opportunities for observing wildlife at any distance, which is particularly important for behavioral study.

I started long-term observations on snowy owls in 1986, when large numbers of owls settled for nesting within my study area at a large snow goose nesting colony. Many of these owls nested in the vicinity of arctic fox maternity dens. That season was

The reproductive success of snowy owls depends completely on the availability of lemmings.

Male snowy owls provide most of the food to their rapidly growing young until mid-September.

the beginning of my focus on interrelations between snowy owls and arctic foxes. First discoveries were so impressive and intriguing that this aspect composed one of the main objectives of my study for all subsequent years. In 1988, I started my own project on behavioral ecology and intra-species relations of the snowy owl. During that first year this project was carried out in parallel with our research on arctic fox social organization, but from 1990 I was doing it completely by myself as part of the research plans of the Wrangel Island State Nature Reserve. By that time, Nikita had switched to studying polar bears, and thereafter we worked independently. I have been in the field in this part of the Arctic for a total of 55 months. In that time, I have studied 119 monogamous and six bigamous snowy owl families, traveled approximately 2500 miles (4000 km) on snowmobiles and ATVs (all terrain vehicles), and found and recorded a total of 713 snowy owl nests.

Each year my field season begins in May. To collect data on snowy owl numbers and distribution in different areas of the island, I conduct route surveys by snowmobile. To get representative data reflecting snowy owl population dynamics in different types of landscapes, I have to cover as large a part of the island 17.4 square miles and 1863 miles of trail (45 square km and 3000 km of trails) as possible each year. In practice, this task turns into a race against spring weather. Mass snowy owl arrival and the beginning of the breeding season coincides with the time of snowmelt. I need to complete these route surveys in only a few days—before rivers begin running and flood the land. And I can't start earlier, because the majority of the owls would not yet have settled in. Driving a snowmobile in wet "porridge" snow is difficult, sometimes even dangerous. One might get stuck far from the cabin or be cut off from the camp by running rivers. But this is not the only trouble; spring fogs often block work just when the schedule is particularly compressed. In mountain areas, streams running from slopes quickly form watertraps under the snow. Driving 31 to 50 miles (50 to 80 km) from the field station, I had to precisely estimate timing and my ability to get back home before any flood.

However, when rivers and lowland valleys are filled with water, driving or even walking far from the station becomes impossible. Fortunately, during this period, the beginning of the breeding season, I can observe many events from the field station. These include courtship demonstrations, competition for breeding territories, and formation of snow geese and eider duck colonies around snowy owl nests.

Snowy owl nests often become centers of activity in tundra communities. Arctic fox and snowy owl breeding habitats on Wrangel Island deeply overlap, and encounters between these two species often develop dramatically. I found their interactions to be a case of territorial competition between the two species, rather than simply competition for food. The tendency of geese and eiders to settle in colonies around snowy owl nests for better protection against arctic foxes resulted in conflicts among all parties and provided dramatic and exciting observations.

As spring advances, I begin surveying snowy owl nests at my model plot at the upper reaches of Neizvestnaya (Unknown) River, an area of 17.4 square miles (45 square km). Time is limited again. Some owls begin nesting after the flood and, if food is scarce, some may abandon their clutches very soon after egg-laying. To get a complete picture, I have to find all nests quickly.

In summer, I focus on work with owl nests; regular visits to the nests are necessary to estimate the age of owls, how active they are in defending the nests, family structure (monogamous or bigamous), timing of hatching and success of chick-raising. I follow the development of chicks by marking them with different colors and weighing them every five days.

Our daughter Katya experiences the wrath of a male snowy owl defending his chicks.

Katya with a young snowy owl. She regularly weighed young, described their development and learned how to age them.

Throughout the summer season I constantly combine observations from blinds with route surveys for counting animals and recording their life signs, thus obtaining quantitative parameters. I count birds, arctic foxes, clutches, litters, chicks at nesting sites, pups at dens, lemmings and their winter under-snow nests. Visiting owl nests and fox dens during seasons of lemming lows is most difficult. In such years, most young die from starvation. Moving from one nest or den to another, I find more and more losses. Sometimes owl chicks die right before my eyes, always a strong emotional blow.

Observations on owl chick growth allowed me to determine the minimum daily weight increase necessary for survival. I record chick personal life histories even when I know which ones would not survive. In some nests, starving 14- to 16-day-old chicks did not show any defensive reactions toward me. Instead of hiding, they gave a warble or heart-rending "*cheep*," solicited food from me, seized my fingers and notebook with their beaks, and tried to clamber onto my knees to get closer to my face. I observed how selflessly their parents struggled for the survival of every chick, trying to get food for them, even when the chicks were dying. The situation was the same at arctic fox dens. In such years, breeding was costly for rodent-eating predators; by the end of the breeding season the parents were looking much worse than non-breeders in their appearance and stage of molting.

Then the season of a lemming peak came and, despite miserable weather, it became a feast of life for owls and foxes. In that year, few chicks or pups died. Adults were very active; almost all owl males and some females heatedly attacked me on my approach to their nests. The season was filled with interesting observations on interactions between families and between different species. In August, while weighing female owl chicks, the 4.4-pound (2-kg) limit of my scale was not enough. Chicks were so well fed they began to fly later than on average. Arctic foxes were running about with a thick layer of fat visible and rippling under their skin.

At the beginning of my research, when I was leaving home to go to the Arctic, our small daughter Katya stayed with her grandmother at our cottage in the forest. Katya soon grew up and for the last two seasons she has worked with us on Wrangel Island. I am continuing my research on snowy owls, and Nikita is filming them. I am happy that Katya could see the world of the wild tundra, the pristine snow and crystal-clear water in the river at our field station, and could observe with us undisturbed arctic wildlife. It was a great experience for her to handle snowy owl chicks and follow their development. Now she understands why her mother and father are so eager to go back to Wrangel Island each spring. Our life on the island is filled with encounters with wildlife, and Katya could finally realize that.

We have successfully documented the influence of lemming populations, weather conditions and female age on nest densities, mean clutch size (5–8 eggs/nest), bigamy and survival of young. My life and work on Wrangel Island still continues after 20 years. I never imagined that Wrangel Island would become my home, and that two decades of field seasons would pass like a single happy moment.

Snowy Owl Migrants in New England

by Norman Smith

For the past 21 years, I have been studying the winter habits of the snowy owl at Logan International Airport in Boston, Massachusetts. One personally rewarding aspect of this work is that my daughter, Danielle, and son, Joshua, have spent countless days and nights with me since they were two years old. They have been with me in all sorts of weather while we gathered pellets, observed, captured, banded and marked snowy owls. As they grew up, they became increasingly involved in the research, with Danielle as a high school freshman starting her own project banding northern saw-whet owls in 1994. Their enthusiasm and resourcefulness over the years has been inspirational and has contributed significantly to this project.

Joshua (left) and Danielle Smith have been helping their father Norman with his study since they were 2 years old.

Snowy owls usually migrate from their breeding grounds in the Arctic to my study area in early November, but we have seen them as early as October 24. While we have seen some owls remain as late as July 7, they usually depart from our area in late April. Owls have been present in varying numbers each winter, from a low of one to a high of 49 with as many as 23 snowy owls seen on the airfield at one time.

This work stemmed from a desire to learn more about the nomadic snowy owls' habitat needs, diet, age, sex, physical condition, social interactions, nomadic existence and seasonal movements. Some have speculated, for example, that when large numbers of snowy owls come to New England, they are moving south due

A snowy owl newly equipped with a platform terminal transmitter. The transmitters send data to passing satellites, and, as information is retrieved and geographically mapped, the movement of each marked bird is tracked.

to lack of food. Indeed, some birds do starve to death, never making it back to the Arctic. However, by capturing and examining the condition of each owl as it arrived, we found out that in years when many owls were present, most of them were immature birds in good physical condition! This implied that there had been abundant food on their breeding grounds up north and that nesting pairs had produced many young owls that wandered south. Conversely, in years when few owls were seen, most were underweight adults.

Through many years of observations we have learned much about this predator's impressive hunting abilities. Perhaps surprising to many people is that, like many other owls in southern latitudes, snowy owls do most of their hunting after sunset. In fact, we needed to obtain a night-vision scope to keep track of them throughout the night. Roosting on the ground by day, they become increasingly active as the sun sets, stretching their wings and regurgitating a pellet from a previous meal. Then, after dark, they seek out a suitable hunting perch. While hunting, they often hover, much like a rough-legged hawk.

Snowy owls are skilled at taking a variety of prey items; over 35 prey species have been recorded to date. They are also strong fliers, hunting like large falcons, and usually pursing and capturing birds in flight. We saw snowy owls outmaneuver snow buntings and even overtake American black ducks! We also saw them capture meadow voles, muskrats, rock doves, brant and prey as heavy as the Canada goose and as large as the great blue heron. They have also been observed capturing and eating other raptors, including American kestrels, northern harriers, barn owls and short-eared owls. On one occasion I even photographed one snowy owl feeding on another. In addition to our observations, we have examined thousands of pellets, thus learning that the most common prey taken included the Norway rat, meadow vole and American black duck.

In 1986, we began color-marking owls on the back of the head in order to gather more information on owl movements. We wanted other people to be able to identify individual marked owls as they dispersed over wide areas during winter. We received reports of our color-marked snowy owls from all over New England, New York and Delaware. Many of these owls traveled over 93 miles (150 km) from Logan Airport, but many simply remained on our study site the entire season. Other recoveries included owls found dead in subsequent years in Canada (in Toronto and other parts of Ontario, in Quebec and in Nova Scotia). Not all multiyear recoveries are of dead birds—we have recaptured 13 snowy owls at Logan from one to 10 years after their original capture date. But even with many reports and recoveries of marked birds, their movement and summer destinations still remained a mystery—one that required modern technology to solve.

To identify longer-term and more continuous movements of individual birds, a multiyear research endeavor with the USGS Forest and Rangeland Ecosystem Center, Boise State University and the Owl Research Institute was launched in 1999. This collaborative project involved attaching platform terminal transmitters to snowy owls.

In January 2000, we placed satellite transmitters on three owls captured at Logan Airport. One owl was found dead on a sandy beach several weeks later. Curiously, the tail of this dead bird was first noticed sticking out of the sand by a bird biologist from Idaho while vacationing at Cape Cod, Massachusetts. He initially dismissed it as belonging to an immature gull, but later returned to uncover the carcass and transmitter of the owl. A subsequent autopsy determined that owl had died of a gunshot wound, something we did not expect here in Massachusetts. An investigation ensued, and the Massachusetts Animal Rescue League posted a $2000 reward for information leading to the prosecution of those involved.

In January 2001, four more transmitters were placed on snowy owls, and for the first time we were able to identify their local movements, migration routes and summering grounds (see map).

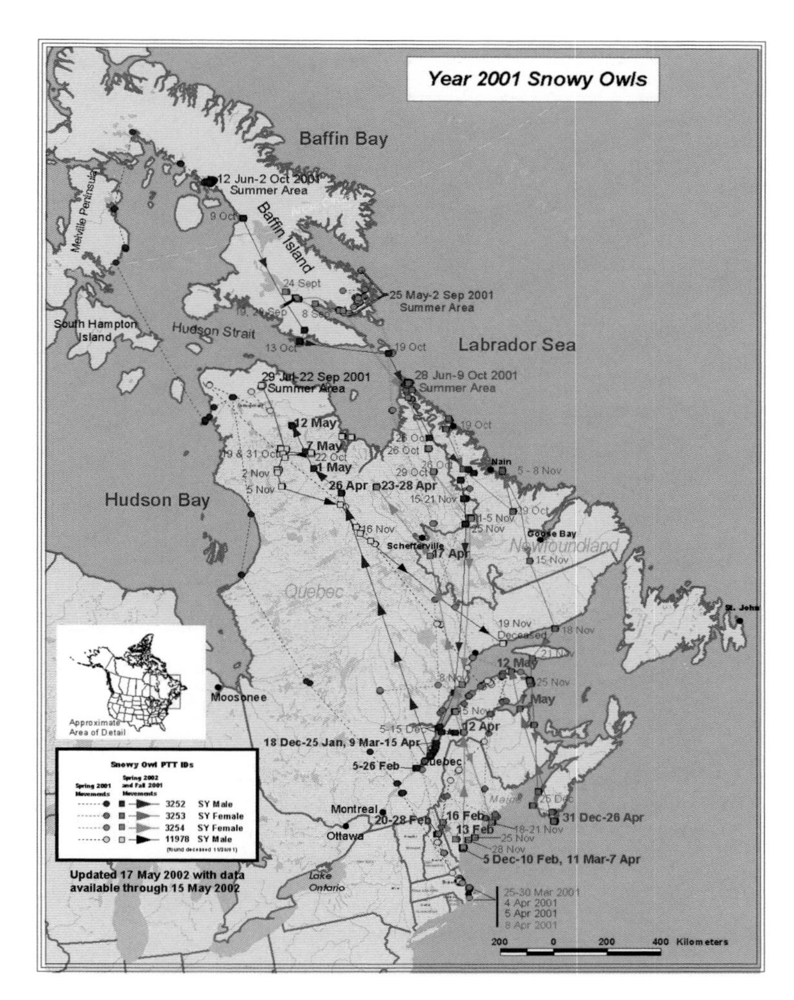

Movements of satellite-tracked snowy owls. This relatively new technology is shedding light on the mysterious migrations of this raptor, also called the "Prince of the Arctic."

We continue this project with anticipation to see what movements and unexpected data the satellite tracking devices will uncover as more birds are added to the sample group in years to come.

Even after 21 years, I still marvel at the fact that this airport, with all its intense activity, incredible noise and jet fumes, provides one of the best locations in New England to observe and study these magnificent raptors.

The Mystery of the Asio Owls on the Saskatchewan Prairies

by C. Stuart Houston

Two Saskatchewan owls of the genus *Asio*, the long-eared owl and the short-eared owl, are "ruled by a mouse," as the late biologist and author Fran Hamerstrom said of the northern harrier. All three of these mouse- or vole-eating species appear in large numbers whenever there is an irruption of voles.

Three major meadow vole irruptions occurred throughout southern Saskatchewan over 42 years—1960, 1969 and 1997. Each coincided with the unusual situation of an incomplete grain harvest, so that part of the crop lay under the snow all winter, cut but not harvested. In response, voles and deer mice proliferated amid the grain. In each following spring, northern harriers and long-eared owls and short-eared owls were plentiful.

A fourth year, 2000, was somewhat anomalous; voles were plentiful locally on dairy farms immediately north of Saskatoon, but this time, short-eared owls or northern harriers did not appear. My keenest nest-finder and banding subpermittee is Marten Stoffel, who as a gangly, 13-year-old tree climber, assisted bird ringers (associated with the Netherlands Youth Naturalists) in Holland in 1970. In his youth and young adulthood, he had seen only one successful long-eared owl nest in his native country. Imagine his excitement in 2000 when his thorough coverage of 40 square miles (182 square km) of dairy farmland north of Saskatoon resulted in his finding 34 breeding pairs, all but one of which were successful.

I took advantage of the four years of long-eared owl profusion by visiting every known nest. I have now banded 662 nestling long-eared owls—tops in North America! Banding peaks have correlated closely with increased sightings of these owls by nonbanders. In a 4319-square-mile (11,188-square-km) area around Saskatoon, long-eared owls were sighted during only 11 of the first 40 years of an all-day bird count conducted in late May. Some years, not one long-eared owl is seen within this area by any birder during the entire calendar year. Since 1964, there have been five years when I knew of no long-eared owl nest anywhere in the province, but normally there are one to three nests reported in a given year.

Nowhere else in North America, it seems, is an owl species common one year, absent the next, and relatively rare in intervening years. In Saskatchewan, these attractive but reclusive crow-sized owls must be nomadic. Direct proof of a

Mary Houston is ready to release a banded long-eared owl, October 20, 1965. In western Canada, we know why and when long-eared owls come to breed in relatively large numbers, but we do not know where they come from, nor where they go.

banded adult from a peak year nesting at a great distance the next year is as yet lacking, but no other explanation fits the known facts. Perhaps extreme nomadism is merely a function of a summer resident near the northern edge of its range. In striking contrast, Denver Holt in northwestern Montana and Mike Kochert in southwestern Idaho both have long-eared owls that are year-round residents; there, any tendency to cyclical number variations is at best considerably muted.

Color Variation in Owls: Albinism and Melanism

by Heimo Mikkola

The presence and structural arrangement of biochrome pigments account for feather colors of birds. Genetic and hormonal factors, however, influence the visible expression of these pigments. Three types of pigments are the basis for the biochrome colors, namely, melanins, porphyrins and carotenoids. Melanins are the most common pigments, while porphyrins are the rarest. Most owls show monochromatic coloration derived from melanins. Besides albinism (a condition in which there is a marked deficiency of pigmentation), there are other pigment abnormalities, such as owls that are blackish (melanistic), reddish (rutilistic) and yellowish (flavinistic). Melanism is caused by abnormal amounts of dark pigment in the plumage, making a bird appear almost black.

The correct identification of a particular genetic color mutation is difficult, so *leucism* or *isabelline* are terms that are often used to describe the abnormally pale color of birds. In the field, these birds are less prominent and apparently very rare, or perhaps overlooked more easily than the white mutations. Leucistic birds, for example, usually retain reds and yellows (carotenoids) while losing black and brown coloration. Some owl species have different color phases (polymorphic), but true mutations are rare. The screech owls often occur in three morphs or color phases: brown, gray and red or rufous. The tawny owl is typically brown, but also has a rarer gray phase. Color mutations occasionally appear in most animals, frequently in some species and rarely in others. For nearly all types of albinism, both parents must have an albinism gene in order to have a young with albinism. In birds, the most commonly noted color variation, albinism, has been separated into four main categories. These are:

1. **Total albinism—**a simultaneous complete absence of melanin from the eyes, skin and feathers. This is the rarest form of albinism, and is derived from a recessive gene, which inhibits the enzyme tyrosinase. Tyrosine, an amino acid, synthesises the melanin that is the basis of many owl colors.
2. **Incomplete albinism—**when melanin is not simultaneously absent from the eyes, skin and feathers.
3. **Imperfect albinism—**when melanin is reduced in the eyes, skin and feathers.
4. **Partial albinism—**when albinism or rather white color is localized to certain areas of the body. This type of albinism is most frequent. While some forms are known to be caused by a partial dominant mutation, white feathers may also result from an injury, shock, senility, diet, physiological disorder, circulatory problems or disease.

Of the 16 species of owls noted below, only in the great gray owl and the tawny owl are there 10 or more records of color variation. In contrast, color variants are much more common in other, perhaps more visible, diurnal bird groups such as waterfowl. Because so few albino owls have been studied, it is uncertain if albinism has a disproportionate negative effect on their survival or reproduction. The increased number of observers present over time, and better access to owl habitat, clouds our ability to determine if owl albinism has increased since the 1980s—an apparent trend in frogs and hedgehogs in the United Kingdom. Interestingly, the Chernobyl nuclear power plant accident in 1986 caused large-scale radioactive contamination, and was thought responsible for an increase (0% in 1986 to 15% in 1991) in the proportion of partial albino barn swallows compared to uncontaminated populations. To better understand color variation in owls, I would be most grateful if readers would contact me regarding any new records.

Barn owl—A private owl collector in Norfolk, U.K., bought a pair of barn owls from a breeder in Essex. The male is pure white, although its eyes are of normal coloration (incomplete albino).

Great horned owl—An incomplete albino great horned owl is known from U.S.A. in early 1950s.

Spectacled owl—The Antwerp Zoo in Belgium has one leucistic pair of spectacled owls, originally wild-caught in Central America. The pair has produced at least 14 leucistic young since 1976, although some young were normally colored. This pair's young have bred as well, and this bloodline is now widely represented in European zoos. Consequently, leucistic spectacled owls could now occur almost anywhere within the captive population.

Eurasian scops owl—A U.K. collection has one young leucistic female, whose brother has only one white feather in each wing.

Indian scops owl—A U.K. collection has a pair of Indian scops owls that produced one leucistic (isabelline) young in 1994, 1995 and 1996.

Western screech owl—Two incomplete albinos, one adult and one young, have been observed in Washington State, U.S.A. in the 1990s.

Eastern screech owl—A review on albinism in North American birds reported one total albino specimen. The Lincoln Children's Zoo, Lincoln, Nebraska, U.S.A. has an incomplete albino that is pure white except for a few tan feathers on the breast. One total albino eastern screech owl lived at least five years from 1982 to 1987 in Long Island, New York.

Barred owl—A total albino barred owl was reported from Texas in 1976.

Leucistic Eurasian scops owl from the United Kingdom.

Great gray owl—At least eight to 10 incomplete albino great gray owls have been recorded since 1980 in Canada, Finland, Sweden and the United States. In Manitoba, Canada, Copland and Nero reported five of 300 live and 80 dead adult great grays had some abnormal white feathers. Since then, two incomplete albinos have been observed: one in June 1990 near Norway House, Manitoba, and the second, distinctly different, north of Winnipeg, Manitoba, in December 1990. Mary Maj saw an imperfect albino great gray in 1980 in the Targhee National Forest, Idaho. Between 1990 and 1992, several observations of an incomplete albino adult were recorded in southeastern Idaho, some 70 miles (112 km) from the Targhee National Forest site. This white owl, later determined to be a male, occupied the

An incomplete albino female great gray owl nicknamed "Linda," Vesanto, Finland.

same breeding area over three seasons and, with a normal-color female, produced three normal owlets in two out of three breeding seasons. Two or three partial albino owls were also seen in Yellowstone National Park.

In Finland, an extremely light and large great gray owl was first seen on March 27, 1994 in Vesanto, where an invasion of great grays occurred. I concluded that it best fitted the definition of an incomplete albino, as its eyes and toes were not pink. The last observation of this bird was made in November 1994 about 1.9 miles (3 km) from where she had been seen the first time. In March 1995, a similar owl was seen near Kajaani, about 103 miles (165 km) north of Vesanto. By comparing good photos of both birds, they were thought to be the same individual. During the low vole population years of 1996 and 1997, this owl was not seen anywhere. In March 1998, however, she was found 124 miles (200 km) northwest of Kajaani at a coastal village called Liminka. When the owl was banded, she was found to be more than four years old and almost certainly the same owl as in Vesanto 1994; she was nicknamed "Linda."

In Liminka, Linda was paired with a normally colored male. Soon after ringing (banding), Linda disappeared, but in spring 2000, Linda was identified by her ring 124 miles (200 km) east in Puolanka. Again, Linda was paired with a normal male; they nested in an old goshawk nest and produced one normal-colored young. In Sweden, a blue-eyed and somewhat light-colored great gray owl was photographed on June 28, 1994 in Jämtland. Perhaps these owls originate from Siberia, where at least the eagle, Ural and short-eared owls are known to be unusually light-colored.

Brown wood owl—For several years a female albino brown wood owl paired with a normal-colored bird. Although she laid eggs, all were infertile.

Tawny owl—A wildlife rescue organization in Italy received a nestling albino tawny owl that was reared to independence and then released. That bird was a total albino. In spring 1996, the

R.S.P.C.A. Wild Animal Hospital at Somerset, England, received a young albino tawny owl. At least 11 incomplete albino tawny owls were reported in Germany.

Little owl—A local population of uniformly white little owls in Jerez, Spain, had normal eye coloring. Some of these were exhibited in a local zoo.

Burrowing owl—In 1912, one incomplete albino burrowing owl was reported from the United States.

Northern hawk owl—In December 1996, an adult owl caught 25 miles (40 km) north of Duluth, Minnesota, U.S.A., had some white feathers in each wing. In 1997, Ted Swem caught a partial albino northern hawk owl in Alaska.

Short-eared owl—A partial albino killed in Connecticut in the 1980s had white throughout its plumage. In Minnesota, U.S.A., one owl was reported with a perfectly white breast, and much white on its face and back. In England, two owls were reported in the 1930s as being gray and brown, apparently the first known. In 1967, a total albino owl was reported near Ketehaven, The Netherlands.

Northern saw-whet owl—At Hawk Ridge Nature Reserve, Duluth, Minnesota, U.S.A., a partial albino adult owl was captured on October 27, 1977. The number four and five primaries of both wings were pure white. The inner and middle front talons of both feet were also devoid of pigment. On August 10, 1983, the same research team captured an owl with a few white tail feathers.

A total albino tawny owl, Italy.

CHAPTER 4

Threats to Owls

Owls, like all forms of life, face many challenges in their struggle for survival. Starvation claims most young owls before they are a year old. But surviving youth is no guarantee of an easy life in the wild; some individual owls experience bizarre tragedies while entire owl populations or species are threatened by other factors.

Accidents

Perhaps the most intriguing causes of death in owls are those of an accidental nature. For example, one unfortunate eastern screech owl was found dead with a short-tailed shrew firmly grasped in its talons. Evidence suggested that the shrew bit the owl's toe before dying, paralyzing the owl, which then froze to death. How could a shrew bite kill an owl? The short-tailed shrew is a small insect-eating mammal that injects a neurotoxin into its victims! Great horned owls are known to attack a wide range of species, but one such owl under my care consistently refuses to eat dead short-tailed shrews when given a choice between them and other small mammal species. Yet other captive and wild owl species kill and eat this poisonous shrew without apparent harm.

Unfortunately for skunks, great horned owls ignore their potent chemical defense, an oily smelly spray from an anal gland, killing and eating them with no apparent aversion to the intense odor. Some owls are killed by snakes, poisonous ones and larger constrictors (constrictors can choke them if they loop their bodies around the owl's neck or body), while trying to subdue them.

Eating has other perils; one eastern screech owl chick, seemingly in a rush to eat a dead red-breasted nuthatch (owl siblings often try to steal food from each other), swallowed its meal such that the nuthatch's short but pointed bill punctured the roof of the owlet's mouth. The owlet's painful struggle only drove the nuthatch's bill deeper, ultimately killing the owlet when it penetrated its brain.

OPPOSITE: **For every owl found caught on a wire we can only guess at how many more die unseen.**

Deadly decision; careful examination of a dead eastern screech owl chick revealed that the bill of its intended meal, a red-breasted nuthatch, had punctured its upper mouth and its brain.

The Perils of Flight

Adult birds make the art of taking off, flying and landing seem easy. However, in-flight accidents occur, especially among young owls. Hunting agile prey in complex three-dimensional environments takes considerable practice and skill. One young great horned owl was found dead, floating in the water of a large metal-rimmed fish-farm tank containing trout. The owl had a dead trout locked in its talons. It apparently had swooped into the open container, grabbed a large fish by its back and subsequently could not gain enough height to avoid fatally colliding with the metal wall of the tank.

A tangled wing caused the death by starvation of two great gray owls during my study of this species. One was a young bird found dead dangling 15 feet (4.6 m) above the ground in a large dead tree; the other, an adult male that appeared to have gotten stuck in a willow clump as it dove down to catch a meadow vole.

The number of owls found dead or injured implies that the hunting life is perilous. Pouncing on and extracting prey from tangled, overgrown and concealed places such as thick vegetation or snow cover is hazardous! Hidden twigs and branches can puncture eyes, and tree limbs and rocks can break bones. Human-altered environments are no safer—owls have been observed hitting high-tension power wires (sometimes getting electrocuted), windows and barbed wire fences.

Road Mortality

Relatively open and grassy roadsides offer owls and other predators good hunting opportunities. Owl prey are attracted to sun-warmed paved surfaces at night but also to salt, garbage and easy food such as mowed grass seeds found beside roads. These are also attractive habitats for owls because prey are relatively vulnerable on or near open roads—prey have few hiding places and owls have clear flight paths to catch prey. This is a fatal mix because owls are so focused when pursuing prey that they do not notice or react quickly enough to cars, trucks or even trains.

In a patch of scrub habitat in Far North Queensland, Australia, vehicles killed at least 13 southern boobook owls in one evening. The owls were struck while feeding on grasshoppers at night on a single-lane highway during a full moon. A year later, in the same habitat, over 30 southern boobooks were noted dead on the highway. One study in France likewise noted that specific habitats associated with certain highways were death traps for

A snow-plunging great gray owl risks injury, especially to its eyes, when plunging after prey into snow-covered weedy fields.

hunting barn owls. In Manitoba, Canada, one birder, Reto Zach, was moved to action when he repeatedly found dead great gray owls along a highway near his workplace. This area appears to be a forest migration corridor for great gray owls. Reto contacted me and other provincial government officials and urged us to design and erect an "owl alert" sign at either end of the "death zone." Anthony Wlock, a Manitoba Traffic Sign Technician, responded swiftly and enthusiastically and designed the sign, and we are all hoping that it makes a difference. Similar signs have been posted in specific areas in other countries to alert drivers, slow vehicles and reduce owl mortality.

A sign alerting motorists of high concentrations of great gray owls in southeastern Manitoba. Sign by Anthony Wlock.

Pesticides, Shooting and Trapping

The development and use of chemical agents to control pests has been heralded as a hallmark of achievement in man's unrelenting struggle against the forces of nature. Indeed, human suffering from disease and relief from starvation has been alleviated by our ability to capitalize on the natural abilities of other life forms to produce chemical agents that act as pesticides. This allows human populations to continue to grow, which in turn requires us to grow more food and control more disease. We manipulate and apply these chemical tools or pesticides to our advantage, but sometimes they have unintended effects that harm nontarget species such as owls. The catastrophic impact of organochlorine pesticide use from the 1930s to the 1960s is well documented for the peregrine falcon and fish-eating raptors. We do not know to what extent these pesticides impacted owl populations. Anticoagulant rodent poisons have had significant impact on owls around the world. One such compound, Klerat, has reportedly caused population declines of over 70% in the common barn, masked and eastern grass owls in northeast Australia.

Pesticides can also have far-reaching and long-lasting effects on wildlife. Scientists have firmly documented that industrial compounds (e.g., dioxins) used or produced in tropical and temperate latitudes have been transported by air and ocean currents and are accumulating in the Arctic. More recently, these have also included brominated fire retardants, chlorinated paraffins and naphthalenes. Organohalogen compounds are persistent, especially at lower arctic temperatures, and build up or concentrate over time. A process known as "bioaccumulation" describes how diluted fat-loving (hydrophobic) compounds in air and water become increasingly concentrated in the fat of predatory species at the top of food chains, including owls and humans.

The sublethal and lethal effects of DDT and dioxins are well documented, but the effects of newer compounds are not fully understood. In recognition of this problem, over 120 countries have adopted the Stockholm Convention on Persistent Organic Pollutants (aka "POPs"), which commits them to reduce and eliminate 12 pesticides and commercial chemicals. Many cities and urban areas are recognizing the hormone-disrupting effects of pesticides by banning the noncommercial use of such compounds.

Over the years, the perception of owls as "vermin," bad omens or predatory pests to be exterminated has been largely replaced with a better understanding and appreciation for the role they play in the environment. Strong and effective laws exist to protect them and are enforced in many countries. However, in some areas they are still persecuted. In the 1980s, owls in one part of northern Canada were referred to as the "practice bird" because their bold nature made them good objects for target practice. In other parts of the world, rare and endangered owls are still trapped in nets strung between poles or trees and killed for food.

Some farmers still shoot owls or trap them in deadly leg-hold traps on raised poles, reportedly to protect chickens or even fish in stocked ponds. Owls use tall perches to hunt from, and many can be captured on pole-trap sets. Ironically, while individuals of some owl species may learn to take chickens on a regular basis, most are attracted to such sites because of rodent prey that often thrive in barns and buildings where livestock and feed are stored. Livestock can be protected from owls by simply putting them inside buildings or secure pens at night. Another creative solution to this problem is stringing up a tight web of fishing line over livestock pens—owls and other predatory birds apparently learn to avoid such areas after flying into the hard-to-see web.

Traps set on the ground for furbearers such as fisher or lynx occasionally capture owls. Small mammals and shrews, attracted to baits or scents on traps, are not heavy enough to set off traps. When a hungry owl pounces on such prey it is caught by its legs and is either eaten by predators or dies a painful death. Many trappers take precautions and ensure that traps are well covered or hidden. But even special traps set in boxes attached to branches or trunks with the opening angled downward catch hungry owls that crawl up into the boxes in pursuit of small mammals attracted there. Both the boreal and barred owl have been caught in such traps in Canada. The extent to which this occurs is unknown.

Predation and Competition

While owls are formidable predators, they in turn are captured and eaten by many species of mammals and birds, such as bears, eagles and weasels. When an owl kills and eats another owl, it is hard to distinguish if this act is simply predation or the exclusion of competing predators. In northern Minnesota, a pair of nesting great horned owls killed and ate at least three adult great gray owls and over 10 of their recently fledged young. Although larger owls usually kill and eat smaller owls, smaller owls are not necessarily intimidated by larger species. I heard a northern saw-whet owl calling from some large trees in a cemetery one night, so I stopped and listened for a while. I wondered how it would react to my imitated call of a great horned owl. It immediately became quiet, and I thought it might be afraid. A few seconds later I was hit on the head by the small irate owl with such force that my wool hat fell off.

Cannibalism is perhaps not rare among owls— This snowy owl killed and ate another on its winter hunting territory near Boston.

Habitat Loss

Researchers in France found breeding pairs of short-eared owls in only one year of their five-year study. The ground-nesting owls were discovered clustered within cereal crop and rye grass fields that contained high vole densities. Although the owls produced over five fledglings per pair, chicks had to be temporarily removed from the fields to avoid being killed by agricultural harvesting or

Ground-nesting owls face many risks. Habitat loss results in greater predation because it is easier for predators to find nests in smaller patches of suitable nesting areas. Farm activities destroy many ground nests each year and heavy pesticide use contaminates and reduces prey species. Even though this short-eared owl nest is in a protected area, fire used to manage grasslands may destroy it before the eggs hatch or the young are able to fly.

mowing activities. The cumulative impact of this mortality on an irregular vagrant breeder is unknown across its European and North American range. Such mortality may help explain why some populations of the short-eared owl have declined over the last few decades. This, together with a loss of its relatively safe nesting grassland habitat, prompted its designation in Canada as a species of Special Concern.

Habitat loss is globally recognized as the leading cause of owl species endangerment and extirpation (see Chapter 6). Over one-half of the world's owls (e.g., the spotted owl) depend on large mature forests for their survival while others (e.g., the burrowing owl) require prairie grasslands with functional ecosystem processes and healthy, pesticide-free prey populations. An ever-growing human population that requires new places and more resources to live and extensive areas to grow food has devastated both these habitats. Stopping habitat loss was extremely important for the conservation of the critically Endangered Christmas Island hawk owl, a permanent resident of a small island.

Different owl species react to habitat change in different ways. Some species, such as the great sooty owl, are sensitive to habitat changes and are used as indicators of mature forest health in Australia. Conversely, the common barn owl has expanded its range over the last 50 to 100 years in North America and Europe as humans converted forest habitat to open agricultural landscapes. More recently, however, this northern range expansion has contracted somewhat with the changes in farming to more efficient agricultural practices. The result is fewer nest structures (old trees and buildings) and a reduced prey base (larger fields result in fewer hedgerows, and increased pesticide and fertilizer use results in fewer summer fallow fields).

While nest boxes are successful in mitigating some of these changes, they are labour intensive because large numbers are needed and relatively few are used for breeding by the target owl species.

Captive breeding efforts are controversial, and early uncontrolled programs for the common barn owl in Britain documented that most of the thousands of birds released each year died before breeding. Releasing birds not trained to hunt on their own or in unsuitable habitat can waste precious conservation resources and cause undue suffering of the birds themselves.

The fate of the burrowing owl rests in our hands. This species is threatened with extirpation from the Canadian prairies.

Invasive Alien Species

A second leading cause of species loss in the world is the impact of invasive alien species, such as the yellow crazy ant introduced to Christmas Island. This rapidly spreading insect eats nestling birds, including young of the Christmas Island hawk owl. Globalization and free trade have increased opportunities for species to expand to new geographic areas. Some of these prove fatal to both owls and people.

The West Nile Virus was accidentally introduced to the eastern United States as early as 1999. This virus is transmitted from animal to animal when a mosquito feeds first on an infected individual and then bites a new animal. Birds of over 120 bird species in North America have died from the virus. Infected humans sometimes die from it too. By 2002, the virus had spread outward from New York state north to Canada and south to Texas and Mexico and as far west as Manitoba by millions of migrating birds. Because of the number of owls and daytime raptors that have died in wildlife rehabilitation facilities, these species are thought to be susceptible to the virus. Hippoboscid flies, also known as louse or flat flies, feed on the blood of growing feathers and are also suspected of

The critically endangered Comoro scops owl is found only in a 62-mile (100 km) square area of high elevation forest on the volcano of Mount Karthala. Comoro is the largest of four islands between Madagascar and mainland Africa. Threats to its survival include ongoing habitat loss and an introduced invasive bird, the common myna, which competes with the owl for limited nest sites.

transmitting the virus from bird to bird. Time will tell if this new invasive virus will have significant impacts on owl populations in North America.

Owls in Captivity

It is illegal to capture and keep owls in captivity in many countries unless the proper permit is obtained by a governing authority. These measures are essential because young owls are vulnerable to capture due to their bold nature, appealing looks and early fledging (before they can fly), and because popular literature and stories continue to promote them as interesting pets. Even with the best intentions, young owls raised by humans often suffer in two ways. An inadequate diet during an owlet's rapid growing period results in overall stunted growth, deformed limbs and fragile, breakable feathers. Owls thus raised, whether released or kept, usually die a miserable death. But even when a human-raised owl does survive to its first birthday, then another, more subtle psychological illness affects it.

Young birds and mammals imprint on their parents (their own species) during a "critical period" in their development. If a person is raising a young owl during this imprinting period, the owl grows up considering itself to be human. This is cute until the owl is sexually mature and wants to select a mate. Then the owl's normal territorial behavior kicks in, and it tries to exclude competitors from its selected or desired territory. Instead of acting aggressively toward members of its own species, it attacks humans (who are usually avoided by normal, healthy owls). Sometimes one human may be selected by the owl as its object of affection, and it will fly to this person to seek personal grooming, prey exchanges or copulations as part of its seasonal courtship cycle.

When an aggressive or an amorous courting owl approaches unsuspecting people, the owl is usually killed in self-defense. Even if the owl is "rescued" by local wildlife rehabilitation or animal control center staff, it is typically euthanized because animal shelters or zoological gardens are either at capacity or lack the appropriate facilities to properly house the owl. A rare exception to this normally tragic outcome is the story of a human-imprinted, aggressive great horned owl currently under my care for research purposes under the authority of the Manitoba government.

"Ed" was captured in spring 2002 after he had terrorized a woman living in rural Manitoba, trapping her in her car after she got home from grocery shopping. The woman was attacked by the fiercely territorial owl, which was defending its newly formed territory from its "competition." Ed attacked the woman every time she tried to sneak to her house from the car. Using her cell phone, she called her neighbor, who arrived in his pickup truck, but he too was attacked by the owl and could not exit the truck. They then called the police. The officers likewise were attacked and trapped in their vehicle by the owl. Finally, to the relief of all involved, an experienced wildlife rehabilitator arrived to net and remove the aggressive owl.

Ed now lives in a suitable outdoor enclosure where he can live out his days in safety, but owing to the misdirected good intentions of someone who raised and inadvertently imprinted him to humans, he can never be released. Ed is not a pet, and I have a healthy respect for his privacy, navigating carefully with a watchful eye when I have to enter his territory to maintain his pen. The key message here is that should you encounter one or more flightless young owls in the wild, the best course of action is to ensure that they are in no immediate danger and then leave them alone. The parents are likely nearby, and may attack you if you attempt to "rescue" their young.

Apparently, few owls have been captured for falconry, an ancient practice that uses live birds of prey to hunt and capture wild game. Falconry techniques are used by those working in amusement parks and other venues that host "Wild Birds of Prey Shows" in which trained hawks, eagles, falcons, vultures and owls are trained to sit on a gloved hand. Sometimes these trained birds are flown from one perch to another, near and even over the heads of awestruck audiences. The northern hawk owl has apparently been used for falconry in Fennoscandia with considerable success, perhaps owing to its fast, hawk-like flight. (Wild, free-living northern hawk owls have been observed to capture ducks and grouse in flight.) Conversely, owl species such as great horned and Eurasian eagle owls that "sit and pounce" on prey have been described as "hard to train." Owls are not suited for falconry in the traditional sense of the sport, and therefore the impact of it has not likely ever been significant.

The Threat of "Postmodern" Conservation Strategies

In some countries there is a trend away from science-based or modern protectionist conservation to "postmodern" community-based conservation approaches. The modern approach to conservation uses science-based decisions to select and protect habitat that supports species of concern, such as the endangered spotted owl in western North America or the Christmas Island hawk owl. However, one consequence of this protectionist approach is often the exclusion of human activities or people that benefit from the consumptive use of a natural resource critical to the survival of the species at risk.

The alienation of the affected or displaced humans can fuel political pressure to instead allow local communities to decide what is the best relationship between them and the resource upon which the rare species depends. In this manner, science-based conservation has been marginalized—the relative objectivity and truth of scientific knowledge is challenged in light of traditional or common knowledge ("postmodern" community-based conservation). Ecology, the foundation of modern conservation science, is publicly discredited because of its inability to forecast consequences of events with precision owing to its complexity.

The greatest danger in the adoption of "postmodern" community-based conservation practices is downplaying the central issue of human population pressure on habitats and ecosystem degradation. Simply put, many wild species must be protected from people and human-induced habitat modifications. In some cases, the only way to do this is to set aside areas excluding human activity. The transfer of authority over natural resource use to local communities is rationalized in the hope that such use will be sustainable. However, few community-based programs have accurate ecological indicators, quantifiable objectives, or the financial or human resources by which to monitor species of concern or overall ecosystem health.

Supporters of community-based conservation assume that human-caused ecological collapse is an invalid forecast of modern conservation science. This support is spreading, as the media and public appear to ignore the repeated plea of thousands of scientists who have publicly stated that further human population increase will accelerate the ecosystem collapse of major global biotic communities. In the last 100 years or so, agricultural development has devastated North America's Tall-grass Prairie Ecoregion—less than 0.001% of it remains. As in musical chairs, there are fewer places left for the myriad displaced species that call this habitat home. Owls, and other species near the top of the food chain, are often recognized as excellent indicators of ecosystem health.

Many forest-dwelling owls around the world depend on cavities found in large, mature and decaying trees for nesting. This white-browed hawk owl nest was associated with an old growth forest in a privately protected reserve on Madagascar.

The late Raymond Tuokko watches over banded young northern hawk owls near a nest he discovered in 1996. The young owls later perished during a prolonged period of intensive rain storms.

Ignorance as a Threat

One sadly repeated feature of the owl species accounts in Chapter 6 is the phrase "not known" with respect to the basic biology of many owl species. Most research on owls has occurred in temperate regions; however, increasing efforts are being focused on lower latitude and tropical species for which there is little or no information on owls' conservation status. Of greatest concern are distinctive forms confined to small islands; to what extent the lesser-known races of common species constitute actual species is also unknown. Examples of recent discoveries of unique forms include the West Indian ashy-faced owl.

Basic knowledge of where owls occur and the habitat they need to survive are required before their conservation status can be assessed, threats identified, and conservation stewardship and planning actions implemented. BirdLife International has spearheaded efforts to acquire this information around the world for birds with restricted ranges. Ironically,

A disproportionate number of owl species are at risk in the southern hemispere.

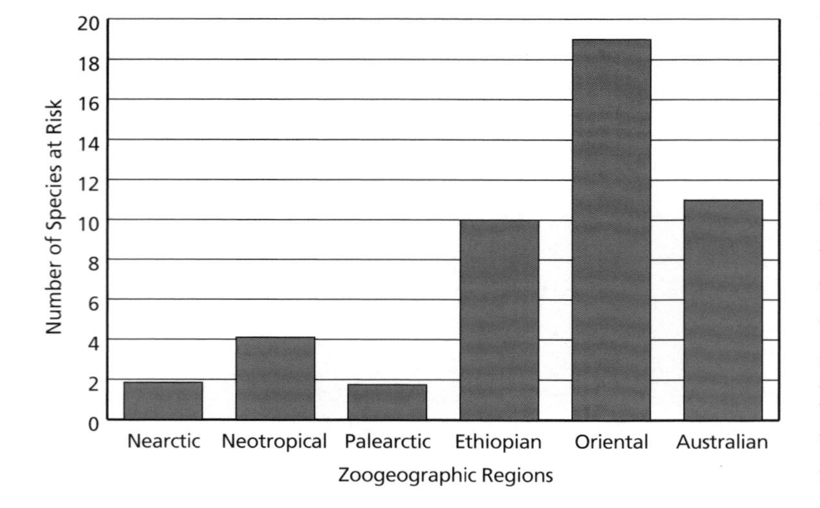

some species we know little about are protected, at least in the short term, because they occur in remote habitats with restricted access.

The following invited contributions provide additional insight into the plight of owls in Canada, Mexico and the world and provide an informed basis for seeking conservation solutions.

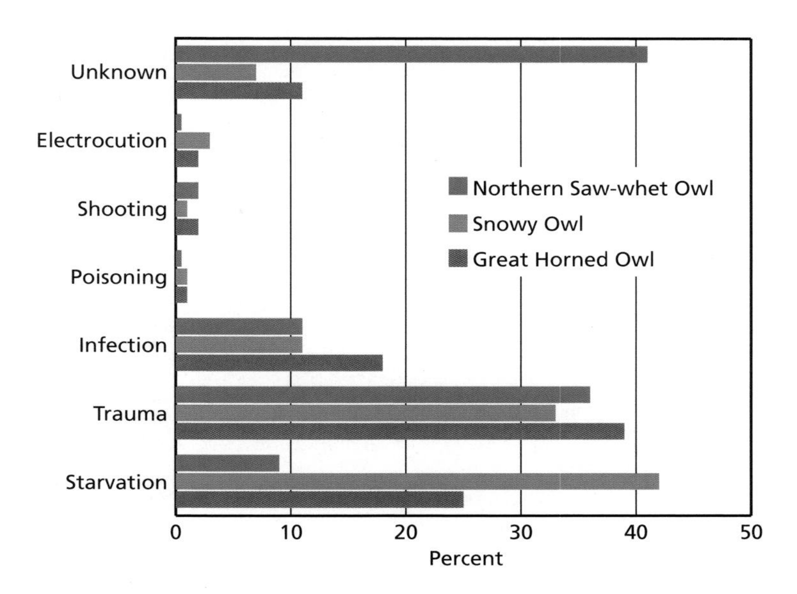

Causes of owl mortality in Canada.

Causes of Owl Mortality in Canada

by Ted Leighton & Tyler Stitt

Disease, generally defined as any departure from health, is a component of the natural history and ecology of all species, and owls are no exception. Owls are host to all manner of infectious parasitic organisms: viruses, bacteria, fungi, protozoans, mites, lice, maggots and worms of many kinds. Noninfectious diseases—starvation, various forms of trauma, poisoning, cancer, drowning, electrocution and metabolic disorders—also affect them. The survival of any species requires successful strategies to avoid disease, whether through the evolution of effective immune and physiological systems (e.g., detoxification processes) or through various behaviors that reduce disease risks. Thus, disease has been and remains an important aspect of natural selection in evolution, and owls are what they are today, in part, because of disease.

What are the important diseases of owls? Too little is known to answer this question definitively. However, some data available from Canada provides an overview and some insights. The Canadian Cooperative Wildlife Health Centre maintains a national archive of disease surveillance data for wild animals in Canada. This archive contains records of the causes of mortality in a sample of 584 owls of 13 species found dead in the wild from 1993 to 2001. These data represent owls found by chance, and hence are not a statistical survey. Nonetheless, they provide a window through which to glimpse the diseases of wild owls across a large segment of the northern hemisphere.

Two species predominate in these 584 mortality records: the great horned owl (215 or 37%) and the snowy owl (148 or 25%); third place goes to the northern saw-whet owl (61 or 10%). A summary of mortality factors is presented for these three species only.

Starvation and trauma were by far the major causes of mortality within this sample of 424 owls. Starvation is the natural outcome of reduced food supplies or failure of young birds to hunt

A young great gray owl found dead in early September. The owl's wing had become entangled in the branches of a dead tree 15 feet (4.6 m) above the ground. Unable to free itself, the owl likely died a slow death by starvation. How many owls die from similar "natural" accidents is unknown as most go undetected.

successfully on their own. Snowy owls migrate to and overwinter in southern Canada and the northern United States every year from their Arctic breeding grounds. High production of young and a subsequent shortage of prey is probably the reason large numbers of snowy owls move south in some winters, and it is not surprising that in those years substantial numbers of starved owls are found and submitted for examination. Trauma, mostly from collision with motor vehicles, is the other big killer. This may be artificially overrepresented in this sample of owls because victims are readily seen and retrieved from roadsides. Nonetheless, it is clear that trauma accounts directly for a large proportion of total owl mortality. Some of the owls killed by trauma also were in poor nutritional condition.

Infections and other minor causes of mortality reported here are likely underrepresented in this sample as many owls that die from these causes are less likely to be found by people. Infectious disease accounts for less than 20% of total mortality. Some important infectious conditions that killed owls included pneumonia caused by the fungus *Aspergillus*, a bacterial infection of the feet commonly called "bumble-foot," which limits owls' ability to hunt, and a small number of acute lethal infections caused by viruses (e.g., herpesvirus) and bacteria (e.g., *Pasteurella multocida*). Poisonings were rare in this sample of owls: one case each of death by consumption of strychnine and of the insecticide carbofuran.

The Plight of Burrowing Owls in Winter

by Geoff Holroyd, Helen Trefry, Enrique Valdez and Jerry Batey

The email "Sad Sunday" from Enrique summed up the events of the day. He had just returned from checking on the location of five radio-marked burrowing owls that he was following near Guadalajara, Mexico, and all was not well: numbers .099 and .663 were dead.

Earlier in the winter of 2001–2002, Enrique had assisted in trapping seven owls in Mexico in order to attach backpack radio transmitters to them to track their movements through the winter. Our objective was to determine the overwinter survival of these owls, and now two were dead.

The impetus for the project is the decline of the migratory burrowing owl in Canada. With annual population declines of 22% per year, we do not have long to determine the cause of the decline. Studies in Canada indicate that over 50% of the owls do not return from one breeding season to the next and, as with many species of migrating birds, nothing was known of the winter lives of this species. In the winter of 2000–2001, by flying aerial surveys, we located six Canadian radio-marked owls in Texas and Mexico. We wanted to compare the winter lives of these owls in different regions.

The hunting owls were trapped at night on the Zapopan military training air base of the Mexican air force, an island of grassland in a heavily populated region. Enrique had first joined our efforts to locate wintering burrowing owls by reporting owls he had seen near his home outside Guadalajara in 1994. Now, years

The burrowing owl is a migrant from southern Canada to central Mexico, where it lives in burrows and rock holes. Educational material, a key component to protecting this migratory Endangered species, therefore must be multicultural.

later, working from his home near the secure air base, we were set to learn about the lives and habitat use of these wintering owls. During the daytime, several owls roosted in an abandoned quarry where rock squirrels had burrowed holes into the fine soil and erosion gullies in an adjacent arroyo; the owls were waiting for the arrival of night under the sound of daytime training flights between the main runway and the fuel supply depot. Not all owls used roost holes; one chose to spend the daytime under tall grass, and one even roosted in a tree at times. The transmitters allowed us to relocate roosting radio-marked owls and over time we discovered that wintering burrowing owls were fairly faithful to their roosts. Finding owls also allowed us to collect their pellets to determine their diet.

Locating owl .099, however, had been Enrique's biggest challenge. We had trapped .099 on a dirt road used by soldiers to travel to and from their barracks. We owe apologies to the soldier who, while cycling along the road, ran over one of our bow-net traps, causing it to spring into his pedals, catapulting him off his bike! After that incident, we put the traps on the edge of the road. We released .099 with its transmitter, but this owl was not detected on the air base over several days. Enrique was excited when .099 returned to forage on the air base one night, but the owl was not detected in the daytime. After about a week of evenings backtracking the owl's approaching signal, he found the owl roosting by day in a rock squirrel hole. This was in the middle of a construction site consisting of an old quarry, over a half a square mile (1 square km) in size, which was being converted into a golf course. This owl was roosting in a leftover pile of earth with a survey

Burrowing owls must be on the lookout for other grassland-dwelling predators like this short-eared owl.

marker on the top. Rather than flying away, the owl would disappear into the hole whenever we visited the site to get pellets. The owl appeared to be habituated to disturbance from trucks and workmen passing nearby. Enrique found this unsettling—and it was to be the owl's undoing. Each night this owl flew 2.8 miles (4.5 km) from the emerging golf course over the suburbs of Zapopan and Nuevo Mexico on the outskirts of Guadalajara to the air base.

Every evening at dusk, as the sun was setting, black-shouldered kites would converge on the air base to roost just when the roosting owls would start their search for food. Upon arriving at the air base, owl .099 foraged in a mowed grassland until another, unmowed area was burned. Afterwards it foraged in the new burn. Two owls, including .663, foraged outside the air base in a cornfield, but only after it had been harvested in late November. Another owl foraged in a mowed lawn for many weeks before leaving the base at night to forage in abandoned suburban weed-filled lots. Nightly, our work settled into a routine of locating the "beep, beep" of each signal as we followed the hunting owls throughout the dark hours. When a signal remained stationary for too long, we knew something was wrong.

Enrique's son, Yael, discovered the carcass of .663. The owl was on the ground, seemingly partly eaten by some raptor. There was no head, chest or abdomen; only the legs and wings were still intact, attached to the remainder of the body and the transmitter. The body was fresh, and Enrique estimated that the owl had died only a couple of hours earlier. The previous night, .663 had foraged on the same cornfield.

On the aforementioned Sunday, Enrique decided to take his son with him to check on the owls. Upon arriving at the golf course construction site, Enrique discovered the work had quickly advanced to the hill where .099 roosted. On the Thursday and Friday before, the dump trucks arrived with soil to fill the area around the hill. Owl .099 retreated into the hole as it had so many times before when approached by trucks. The truck driver did not know he was burying .099 alive. That Sunday, Enrique found that the radio-signal was coming from beneath the pile of soil at his feet. Several efforts with a bulldozer and shovels were required to dig out the dead owl, as the radio-signal was hard to locate precisely in the underground burrow.

One more owl would disappear, its fate unknown, before mid-March, when the three surviving owls in Zapopan left the region for their breeding grounds. Time will tell if they return to the air base next winter. Three of the four owls tracked near Corpus Christi also survived until they migrated at about the same time; the fate of the fourth remains unknown.

Over the next two winters we plan on radio-marking more burrowing owls to better understand the winter ecology of the migratory population of this Endangered owl.

Our thanks to General Roberto Huicochea Alonso and General Sergio Parra Estrada for permission to work on the Zapopan Air Base and to Chris Fisher for comments on an earlier draft of the manuscript.

SALVAGED BOUNTY—OWL ECTOPARASITES

by Terry Galloway

The great gray owl perched motionless on the branch, intent on the almost imperceptible sounds of a small mammal beneath the snow. Its patience paid off, and it silently took flight, down low across the snow, and over the road. The pickup truck came around the corner quickly, and the owl didn't have time to respond. It glanced off the windshield, spinning crazily into the willows. The driver stopped the vehicle and ran to where the bird sat glaring at him with yellow eyes. Distressed at this unexpected turn of events, the driver returned to his truck, pulled out the blanket from behind the seat and quietly covered the owl, wrapping it carefully, without a struggle. It was a two-hour drive back to the city and the rehabilitation center where he hoped they could help the stricken owl.

Unfortunately, this story, or ones similar to it, are common occurrences. Humans have so modified the environment, invaded wild spaces with fast-moving vehicles, introduced predators and constructed obstacles from buildings to fences to communication towers, that we have created significant risk of injury and death for wildlife species. Although we may try to minimize this risk, the number of casualties can be astronomical. Wildlife hospitals have sprung up in many regions to provide skilled services for rehabilitation and release of these animals. However, many animals succumb to their injuries, or, in some cases, their injuries are so severe that the only humane solution is euthanasia.

For the animals that are successfully returned to the wild, gratification is immense. Just to see these animals regain their place in the ecosystem is a tremendous reward. But what about the cases where animals are lost? Can we learn anything from them? From my experiences since 1994, I can say that the answer is categorically "yes."

In 1990, I was involved with a project derived from the Biological Survey of Canada. It was our task to review the status of ectoparasites of vertebrates in Canada.

Most people see birds and mammals as individuals, walking, running or flying about. I have come to accept them as a vehicle for a complex community, a home really, for unimagined numbers of parasitic animals. To my great surprise, as a result of our investigations, we reported that the ectoparasite fauna of birds in Canada was in a deplorable state. Though our general knowledge about fleas on birds was not too bad, clearly, there was much to be learned about the lice and mites. Less than 50% of the species of lice expected to occur on birds in Canada had been found, and less than 1% of the feather mites.

When the opportunity arose to collaborate with the staff at the Manitoba Wildlife Rehabilitation Centre at Glenlea, I couldn't wait to see what we might discover about the ectoparasite fauna of birds in Manitoba. In conjunction with the support of the Canadian Wildlife Service and Manitoba Conservation, I began to process the animals, most of which were birds, which either died or were euthanised at the Centre. Each animal is individually bagged, frozen to kill all ectoparasites and then washed in warm, soapy water. The water is strained through a fine sieve, and the ectoparasites are sorted under a microscope, and where possible, counted, sexed and identified.

Since the beginning of this study, I have examined over 4000 animals for ectoparasites, of which over 400 have been owls. I have examined 10 species of owls, with only the burrowing owl not being represented in the sample of the Manitoba owl fauna. Salvaging and examining birds is a slow process, with owls coming in from various locations of the province at all times of the year.

We have seven families of Mallophaga (chewing lice) known in the province, all but one of which are ectoparasites of birds. Species in the remaining family are specifically restricted to mammals, and include some of the most common parasites of domestic animals. All species are small (except for the giant species of *Laemobothrion* found on eagles, hawks, vultures and coots), wingless and dorso-ventrally flattened, with chewing mouthparts. These insects use their mandibles to hold their position.

Most species seem to feed on skin, feathers and debris on the body surface of their host (*mallos* = wool; *phaga* = eater), but there is clear evidence that many feed on blood. It is uncertain how they actually obtain the blood. Chewing lice look quite different from sucking lice and use completely different mechanisms for avoiding host grooming activity. First, they rely on their powerful mandibles, not only for obtaining food but also for maintaining their position. Typically, they use the mandibles to grip the skin or feathers, and they hold on tenaciously, in some species even after the lice have died. Chewing lice also avoid host grooming by moving rapidly through the feathers. Most of these species are found on the body and wings of their hosts. Other species of lice are rotund and move much more slowly. However, they are often confined to the upper neck and head regions of the body; places that are more difficult for the host to reach effectively.

Chewing lice cement their eggs onto the feathers. The tiny immatures hatch from the eggs and take up their position on the body, shedding their skins as they grow. Although they are active in all

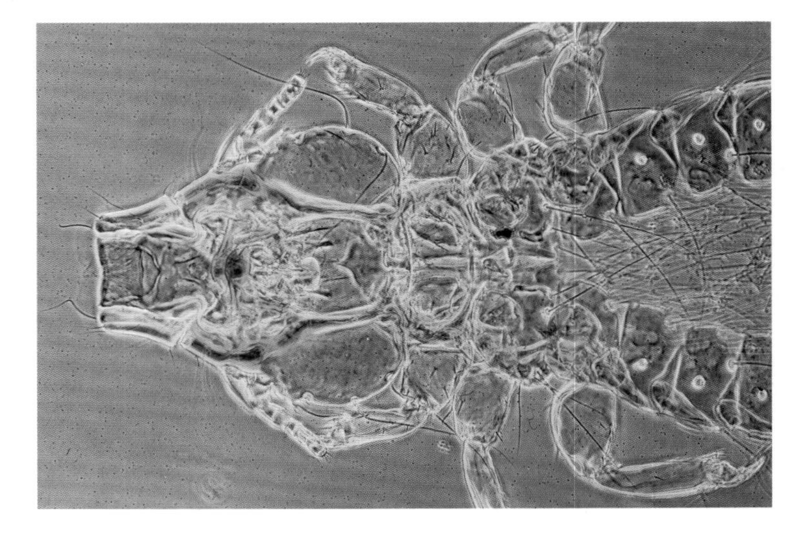

Head of the parasitic feather louse *Strigiphilus acadicus* (Emerson and Price). The genus name, *Strigiphilus*, means "owl loving," and the species name *acadicus* mirrors that of its specific owl host, the northern saw-whet owl.

A hippoboscid fly collected from a live great gray owl captured and banded in winter. Commonly called "flat flies," they can survive below freezing temperatures by living in the warm air trapped under an owl's thick insulative plumage.

immature stages, and pass directly from the juvenile stage to the adult, no species ever possesses wings. The only opportunity they have to gain access to a new host appears to be by direct contact. Therefore, mating, roosting, reciprocal grooming and brood care are critical activities for transfer of lice throughout the host population. This is also one reason for the high degree of host specificity that we see among lice. In fact, there may be other dispersal mechanisms available to lice on birds. There are numerous records of chewing lice hitchhiking aboard the bodies of hippoboscid flies. It is also possible that if lice become dislodged from their host, they may be able to crawl off and locate a new potential host that is nearby.

There are essentially two types of data that are collected from salvaged owls. In the first case, I have been able to build up a good database on the biodiversity of the ectoparasite fauna of owls. In the second, there are now data on the infestation parameters of each of the species of ectoparasites on their respective hosts. For example, I have found at least 12 species of chewing lice on the 10 species of owls. One of the two species of chewing lice on Manitoba great gray owls—*Strigiphilus remotus* (Kellogg and Chapman) and *Kurodaia magna* (Emerson)—is of particular note.

The latter species has been previously collected on other species of owls, but has never been reported to infest great grays. The former was originally described in 1896, from collections in the Pacific Northwest. The first or type specimen for *S. remotus* appears to have been lost or destroyed.

Subsequently, another species of *Strigiphilus* was reported from a great gray owl in the eastern United States, *Strigiphilus syrnii* (Packard). The name, *S. remotus*, has existed somewhat in a state of limbo without the type material, and with few additional specimens being collected from this host. About two-thirds of the great gray owls I have examined were infested with an average of more than 70 *S. remotus* per bird. There is now available an abundance of specimens on which to consolidate the status of this species of louse as part of the North American fauna, and, in addition, we know a great deal about its association with its host.

The quantitative data for lice on the salvaged owls has also proven to be extremely interesting. Not all species of owl seem to suffer from louse infestations to the same degree. For example, having examined 46 eastern screech owls, I have found only one juvenile louse, essentially no lice at all. At the other end of the spectrum, more than 90% of the 39 snowy owls I have examined were infested with *Strigiphilus ceblebrachys* (Denny), a louse specific to this host. There has been an average of 430 lice per infested snowy owl, the greatest number being 5353 lice!

The feather mites collected from owls also have proven to be a very interesting problem. Very few people study the taxonomy of these mites, and it has only been in collaboration with two specialists (Sergei Mironov and Andrei Bochkov) at the Zoological Institute in St. Petersburg, Russia, that our feather mite fauna is

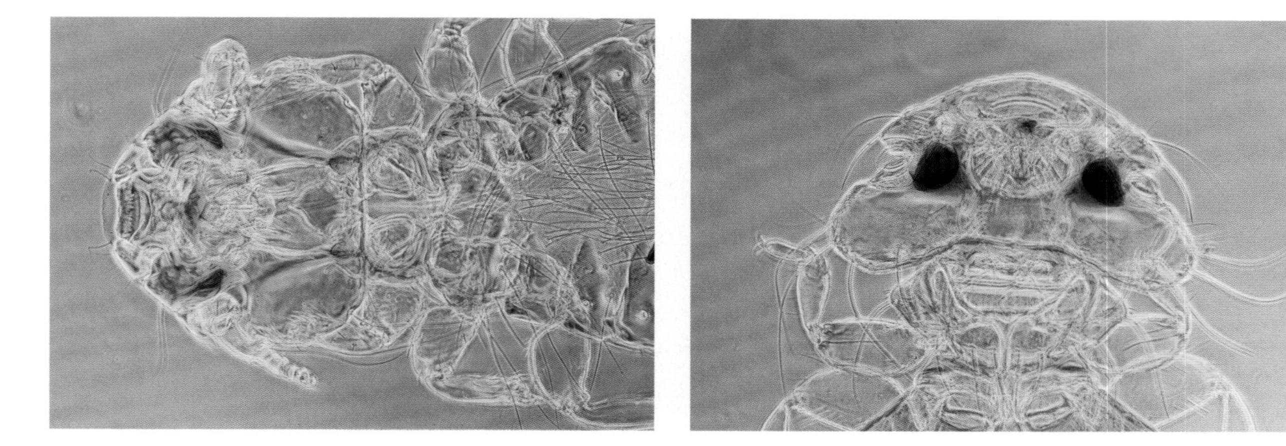

Portraits of two feather lice:

ABOVE LEFT: ***Strigiphilus remotus*** **(Kellogg and Chapman). Note the swollen antennae segment, characteristic of the male of this species. This species typically parasites the great gray owl.**

ABOVE RIGHT: ***Kurodaia magna*** **(Emerson), a parasite of numerous species of owls, including great gray owls, from which this specimen was removed.**

being expanded. These tiny mites live on the skin surface and on the feathers of birds, where there has been incredible niche separation and specialization by different species of mites. In general, feather mites seem to cause little direct damage to their hosts. However, I have seen specimens of great gray owls that were infested with hundreds of thousands of feather mites. Birds carrying such heavy feather mite loads surely must be affected by the irritation of these mites crawling through the feathers and over their skin. However, there is no information about whether or not these mites affect the fitness of their hosts.

In the family Kramerellidae, there are two genera of feather mites that have been found specifically on owls, *Kramerella* and *Petitota*. In our preliminary assessment of these mites, there are at least nine species (six *Kramerella* spp. and three *Petitota* spp.) on owls in Manitoba, two of which are probably new species. Only one of these species of feather mites has been previously recorded in Canada.

The one truly essential value of examining these salvaged owls is that we gain knowledge about the ectoparasite fauna of these magnificent birds without having to resort to active collecting of host specimens. With this knowledge of the ectoparasite fauna, perhaps consideration can be given to conservation of these interesting species as well as to the hosts. We are seeing many bird populations in decline, often as the result of habitat modifications and other pressures from human activity. As their numbers decline, persistence of those host-specific ectoparasites that are typically present at low prevalence, or that have less efficient host-to-host dispersal mechanisms may be at risk. We always assume that parasites are detrimental to their hosts, or at least that they are of no benefit. We have only hints about their beneficial roles in maintaining the fitness in host populations, or affecting mate selection.

The next time you see an owl perched on a branch, or hear one calling in the night, stop and admire the beauty of the bird, but pause, too, to reflect on the complex community it represents.

CHAPTER 5

A World for Owls

As the host country, Canada adopted this logo, by Bernard Pelletier, depicting the snowy owl for the October 1996 meeting of the 20th General Assembly of the World Conservation Union (also known as International Union for Conservation of Nature and Natural Resources—IUCN) (http://www.iucn.org/wcc/index.html), and the first IUCN World Congress on Conservation. Part of the stated rationale for selecting this species for the logo was the hope that its hypnotic gaze would continue to remind us that our very survival depends on the conservation of natural resources. The snowy owl is an important link in the food chain of harsh northern ecosystems. It demonstrates behavioral and morphological adaptations to extreme environmental conditions that are an inspirational testimony to the will and strength with which life exists on earth.

Around the world there are many people working hard for conservation against the formidable tide of a burgeoning resource-consuming mass of humans. The actions of individuals working in trades are often stereotyped as anticonservation without compassion for wildlife. However, I read in a newspaper about some loggers, who after discovering a barred owl nest in a massive tree they had just felled, rushed to check its contents. On discovering that the young inside had survived, they took great care in tying the portion of the tree containing the nest high up in an adjacent live tree. They then continued cutting, but left a sizeable buffer of trees around the nest site. One fellow even stayed and watched from a respectable distance until the adult female returned and located her brood.

Despite all the trauma owls experience at the hands of men and their machines, they can be amazingly resilient. One trucker told me about a remarkable encounter with a great horned owl near Kingston, Ontario. It was the middle of a cold winter's night when, much to his dismay, his truck windscreen was struck by an owl flying across the highway. Stopping, he got out and searched the area

OPPOSITE: **The long tail and pointed wings of the northern hawk owl make it look falconlike in flight.**

A nesting great gray owl is well hidden in the hollow of this fire-killed aspen. Fire plays a significant role in the creation of habitat and nesting structures for owls around the world.

for the dead owl, but failed to locate it. Three hours later he pulled into his final destination, a large heated commercial garage in Montreal, Quebec. Many hours later, he returned to his truck and heard a strange scratching noise coming from behind the truck's cab. Imagine his surprise when a pair of intense yellow eyes stared back at him! This remarkable bird had survived the collision, flipped up and over the cab, became wedged between the cab and the truck's cargo box, and endured a frigid three-hour trip. After being examined by wildlife rehabilitators, it was eventually released back to the wild.

Many more stories can be told about the remarkable ability of owls to endure the travails of life as a wild creature, and of the equally heroic efforts of humans dedicated to their conservation. But clearly owls and other wildlife need our help to survive. I fondly recall the thoughtful comments of Richard J. Clark when asked to provide the summary and concluding remarks at two major owl conferences in Winnipeg, Manitoba, in 1987 and again in 1997. He urged researchers to become more involved in conservation in general, but also pointed out the great need for research and conservation efforts in the southern hemisphere, especially in tropical Africa and the countries of the southwestern Pacific Rim. Richard's advice did not go unheeded, as a subsequent major owl

Providing opportunities for young people to have hands-on experiences with wildlife may be as important as research in conserving owls and their habitats.

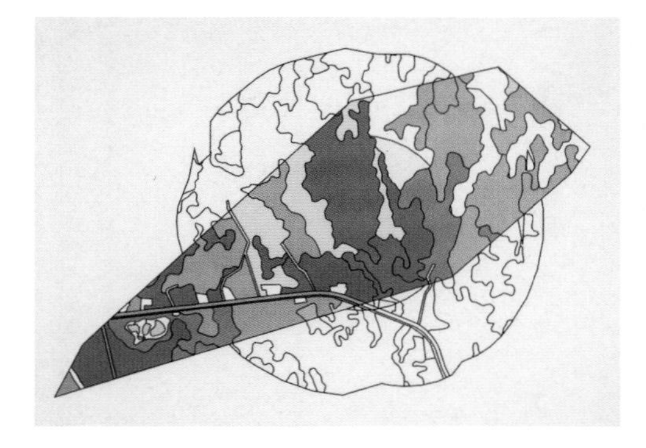

conference was held in Canberra, Australia, in 2000.

Hope for the future of owls lies in the continued renewal of people involved in three basic areas: research, education and conservation. Wildlife managers regularly refer to the greater need to manage people, not wildlife, to deal with conservation issues. However, we need to greatly increase our selective investment in viable conservation solutions and in people. To borrow again from Richard Clark's aforementioned presentations, the fate of owls and indeed people can be depicted in his figure entitled the "Circle of Survival."

By examining habitat types used by owls (colored areas, left) wildlife biologists can develop habitat management guidelines, like this forest cut block (above), to maintain owl populations.

The Roseau Bog Owl Management Unit in northern Minnesota, U.S.A.—experimental forestry allows biologists and habitat managers to determine what level of resource use is sustainable while maintaining viable owl populations.

ABOVE: **It pays to advertise, especially on money—a recent coin depicting the great horned owl. The use of owl imagery on everyday items increases people's awareness and appreciation for nature.**

RIGHT: **An owl mascot welcomes scientists and others to an international symposium on owl conservation in Winnipeg, Manitoba. Such gatherings help focus new research on owls around the world where it is most needed.**

RIGHT: **Robert W. Nero (right) and a great gray owl. Education is an essential component of conservation. Owls are among the most popular ambassadors of wildlife school programs.**

Once again, I am delighted to introduce relevant contributions from friends who have dedicated good portions of their lives to better understanding owls. What is remarkable about these people is the extent to which they freely share their knowledge and wisdom about owls with both colleagues and the general public. They are fulfilling their dreams while furthering conservation in remarkably divergent but complementary ways. To these people, owls are more than just another animal to study; rather, they are part of a chosen lifestyle.

Adventures with Saskatchewan Great Horned Owls

by C. Stuart Houston

I am lucky. I grew up in Yorkton, Saskatchewan, population 5000, in beautiful aspen-parkland habitat, in the center of "The Great Duck Factory." I have banded over 120,000 birds of 204 species, resulting in over 3100 recaptures and recoveries (dead birds found and reported), not counting repeats and returns. The largest number of recoveries from one species, 555, have been from over 7000 nestling great horned owls that I have banded, tops in North America by a factor of four.

In 1943, I was approaching my 16th birthday. Biologists were unavailable in wartime, and Ducks Unlimited (DU) wanted someone to band ducks. My cousin and I rode our bicycles 6 miles (10 km) to Rousay Lake, where we had a 4-mile (6-km) "trap-line" and then rode home. We banded 556 ducks that year, and were paid 10 cents a duck for the hard work. That winter, our blue-winged teal were shot in Cuba, Jamaica and Dominican Republic. What a thrill for a Grade 10 student! The second summer we banded 1248 ducks, now at the lucrative rate of 20 cents a duck—and the winter brought reports from Colombia and Venezuela. I was hooked! Banding remained my summer employment through high school, pre-med and into medical college.

Stuart Houston holding a young great horned owl at Adamson Lake, Saskatchewan, June 17, 1961.

Great horned owl chicks in a nest, May 2001. In the Prairies owls can see banders coming from miles away!

I married in 1951 and my bride, Mary, joined me in the hobby. Seth Low, head of the banding office in Maryland, visited me in Yorkton, where I was a family practitioner, taught me how to use mist nets, and left me two of them. We kept mist nets open from dawn to dark throughout spring and fall migration, banded over a thousand "dickybirds"—but not one recovery resulted from all that effort. It dawned on us that there was more satisfaction in banding species that had a higher recovery rate—colonial birds and raptors.

From opportunistic banding of 12 nestling and fledgling great horned owls through 1956, I had two recoveries, one of them from an unexpected distance for a nonmigratory species. This owl was found with an injured eye in Bluffton, Minnesota. It was a harbinger of things to come, and stimulated my interest in banding more owls. In 1958, 15-year-old Bill Horseman found 10 active nests of the great horned owl, and together we banded 22 nestlings. This received accolades that year in the continent-wide summary in *Audubon Field Notes* (12:394) as "a fine example of field work." This spurred us on, and in 1959, Bill and I banded 70 nestling owls.

In 1960, before coaxial cable and relay transmitters were available, our local television station received its footage by mail, broadcasting it a week late. The Brooke Bond Tea Company, which placed a Roger Tory Peterson bird card in each package of Red Rose tea, wanted someone local to do a live TV program. It was titled "Watching Parkland Birds with Stu." I was paid $10 for each 15-minute program. The station offered a *Peterson Field Guide* as a prize to the farm child who found the most owl nests, but they gave out three.

George Chopping, age 15, walked 30 miles (49 km) near Dubuc to find five nests and the next week walked 25 miles (40 km) to find three more. Eleven-year-old Myles Ferrie found 16 nests near Invermay, and Ron Hilderman found 10 nests at MacNutt.

Again, I was lucky. That was "the year of the owls"—the peak of the 10-year snowshoe hare cycle; because of an early snowfall the previous fall, the grain crop lay unharvested in swaths beneath the snow, causing an irruption in numbers of meadow voles and deer mice. The result: I banded 150 nestling great horned owls, together with 74 long-eared and 69 short-eared owls. That's a lot of owls to band in spare time during evenings and weekends. Clearly, one person could not locate that many nests; one must enlist nest finders if one is to band large numbers of any raptor.

I decided on an academic career in teaching and research, so left small-town practice to train in radiology in Saskatoon and Boston. In 1963, I joined the staff of the medical college in Saskatoon. The result was more control over my time and more weekends free to band owls. A friend, Doug Gilroy, had a weekly wildlife column in the *Western Producer*, which reached 200,000 farmers across the West. Doug asked farmers with an active owl nest to telephone me.

From 1964 through 1992, my spare time every May was spent banding owls. Some mornings Mary and I would go to three nests near the city before morning fluoroscopy at 8 a.m. Some evenings after work, five nests could be visited. Our weekends began at 4 a.m. Saturday and ended after midnight Sunday.

We were hell-bent on these expeditions. A nest every 50 miles (80 km) was sufficient for us to lay out a route, though on two unrepresentative occasions there was a gap of 100 miles (160 km) between far-flung nests. The first weekend we went south, where owls fledge a week or two earlier, and the final weekend we went north. We traveled up to 1118 miles (1800 km) during the three-day Victoria Day weekend near the end of May; sometimes our most distant point was 298 miles (480 km) from Saskatoon, at either the Manitoba boundary or, rarely, the Alberta boundary. We had no time for tenting or cooking. We stayed in small-town hotels, and watched the price for a room rise from $4 to $40. The four-wheel-drive vehicle consumed prodigious quantities of gasoline. I covered all expenses. Mary packed our lunch kits; sandwiches, apples and oatmeal cookies were eaten "on the run," between nests, but a five-minute stop was required to down the iced tea. We stopped only for "lifers and whooping cranes."

I called out songs of small birds near the nests, and, when in grassland habitat, introduced most of my helpers to the songs of the Sprague's pipit and Baird's sparrow. Some climbers soon surpassed their teacher in ability to identify bird songs; some went on to distinguished careers in biology.

My motive was to learn about owls, and to publish my results. We have the best data anywhere on the relation of great horned owl productivity to the snowshoe hare cycle. At the height of the cycle, there was a pair in every 2 square miles (5 square km) of parkland, we knew of no non-nesting pairs, and the owls raised up to 2.5 and even 2.6 young per successful nest. At the bottom of the hare cycle, many pairs made no attempt to nest. In 1965, near Yellow Creek, the Nemeth family had no successful pairs; one female was on her nest on May 9, but had deserted by our visit on May 19. Other owls moved out of aspen parkland. Productivity dropped to 1.6 to 1.8 young per successful nest, almost exactly equal to the long-term steady mean in the Cincinnati area of Ohio where Jack Holt has banded about 1600 nestlings.

Survival rates were higher for all age classes in years when hares were abundant than in years when hares were scarce (58% vs. 37% for immatures, 74% vs. 59% for yearlings, and 88% vs. 81% for adults). Dispersal distances are unexpectedly high from Alberta and Saskatchewan, and the southeast direction is remarkably consistent, to the Dakotas, Minnesota, Nebraska and Iowa. This is in the direction of prevailing winds and follows the configuration of aspen parkland into Minnesota. It is not a measure of human density because the eastern Dakotas have low human density equivalent to that where the owls originate. The greatest distance traveled was 1278 miles (2058 km), from Alberta to Illinois. These banding results confirm the prediction by Myron Swenk in Nebraska (1937) that the nearly white great horned owls must be originating in the Canadian Prairie Provinces. For owls recovered as adults, the mean dispersal distance was 93 miles (149 km). Not only

Almost 50 years later and still banding birds, Stuart Houston examines an owl pellet at an active great horned owl nest, May 2001.

young of the year but also adults move southeastward.

The greatest pleasure of this endeavor came from the people we met. I guaranteed to each newcomer that he would meet the nicest farmers in the province. But the special farmers, who were observant, interested in nature, and kind enough to phone a strange doctor in the city, were a highly selected group, and they found nests year after year. They kept track of red-tailed hawk nests from previous years, and watched to see whether a great horned owl took over occupancy. Four farmers each took us to well over 100 successful owl nests, and another four had over 50. At Yellow Creek, when we began our visits, in the minds of some farmers the "only good owl was a dead owl." After 25 years, we merited a page on owl banding in their local history book; owls had become a respectable interest.

Precious Moments—Why Filming Owls is Worth the Effort!

by Jeff Hogan

Breaking trail through light powder snow over 3 feet (1 m) thick in the Yellowstone backcountry, we pause often to catch our breath and take stock of our whereabouts. Our goal was to position ourselves on a rock outcropping, located directly across a narrow canyon from a den where Dan Hartman had observed a black bear preparing and entering the previous fall. With luck, the bear would emerge from its winter shelter and expose itself to our cameras. This was a typical day in late March with the warm sunshine interrupted far too often by sudden wind laced with stinging snows.

Stuart (left) and Mary Houston (right) banding young great horned owl near Yellow Creek, Saskatchewan, with five Simon children, May 25, 1967. The owl nest, found by the Simon children, was on father Mike Simon's land.

Then from the north a single note broke through the weather and thick forest about every other second. Dan turned to Leine Stikkel and me and said, "Pygmy owl . . . it's up ahead." We continued on, the sounds of the forest lost to the sounds of traveling humans complete with chafing winter clothing and heavy breathing. We stopped to rest and listen. Owls have an incredible ability to evoke a sense of wildness to a forest or place with their presence, detected almost always through a call and nothing more. Every species of owl has its own unique "hoot," "whistle" or "scream." It's not necessary to see the owl to feel its wonder, or to identify its character or rarity. We studied the very tops of the lodgepole pines, firs and spruce where pygmy owls often perch to exclaim their presence. Suddenly the owl whistled above our heads, not in the treetops but on a dead branch draped in old-man's beard, just a dozen feet away.

A pygmy owl poses for Dan Hartman's camera.

Our excitement was shared as we moved slowly to set up the Arri 16mm film camera, hoping to capture a few seconds of this whistling pygmy owl on film. As it turned out, I shot all the film needed and was content observing through the viewfinder. As difficult as it may be to catch a glimpse of a pygmy owl, once you have located one they can be very cooperative, and this one posed beautifully for the camera. Although the owl's head and shoulders filled my frame, it was only 6 inches (15 cm) tall. I was amazed that this little owl can withstand the extremes of the Yellowstone winter; very few birds can. In an instant, the owl turned its head and was off, disappearing into the forest across the narrow canyon. Its whistle drifted away with the wind. The bear never did show itself that day.

A serious search for owls requires traveling for miles through habitat that appears favorable. This is when I leave the cameras home and move through the forest slowly and peacefully, studying the trees, branches and forest floor for any evidence of an owl's existence. A scolding nuthatch or Clark's nutcracker may direct me to the edge of an aspen grove surrounded by spruce and fir. Whitewash splashed under a likely perch demands closer inspection for a dropped pellet or maybe a feather. These items will help identify the owl species before ever seeing the bird. The perch could have been used just once by a transient bird or daily, depending on the amount of whitewash and pellets below. A water hole for bathing is a huge discovery and may provide for great footage down the line. A plunge mark in deep snow is a telltale sign of the presence of an owl diving for prey in mid-winter.

After an extensive search and a lot of luck, I'll find an ideal situation for filming at the nest. In northwestern Wyoming

Jeff Hogan on the job filming a great gray owl fledgling, one of his favorite subjects.

An attentive female great gray owl on a nest after a rain—a study in devotion to her unhatched young.

where I live, the great gray owl represents a true wilderness bird. The habitats in which I've located nests to film have been visually spectacular, allowing for stunning cinematography. Like many owls, the trusting great grays are very cooperative in front of the camera and perfect for filming behavioral sequences and extreme close-ups necessary for captivating an audience.

Opportunities to film exciting owl behavior run year-round; the excitement starts with courtship behavior in spring and continues until the young become independent and disperse by late summer. Highlights are viewed while perched precariously in a blind as high as 45 feet (14 m). Often it rains, soaking an incubating female great gray owl and making for incredible images.

The struggle for life as the newly hatched chicks cry out in competition for the next feeding sends shivers down my spine. The anticipation is torturous as the flightless chicks peer over the edge of the nest, ready to make that leap of faith. After leaving the nest, the chicks climb up leaning trees. Many times they make great progress in gaining altitude, only to slip and tumble back down to the ground.

Humans have always known there is something mystical about owls. Their calls can be haunting, their stare penetrating. I'll never tire of watching them. My hope is that by sharing my enthusiasm for owls through my films and photographs I will invigorate a sense of wonder in viewers, who then may in turn be active voices for the conservation of wilderness upon which owls and all wildlife depend for their survival.

The Ural Owl and Forest Management in Japan: A Conservation Solution *by Aki Higuchi*

The relationship between wildlife management and forestry in Japan is a relatively new research topic that is increasingly important, given the growing demands of society upon our natural resources. There is increased awareness of the need to conserve natural resources, but increasing economic development is occurring with little concern for wildlife. Raptors, such as the golden eagle, have become endangered as a result of habitat loss or degradation, e.g., road building, dam construction, etc. The lack of information about the basic ecology of raptors or even more common wildlife means that few conservation solutions are available for increasingly complex environmental problems. Effective conservation planning for wildlife species must be based on sound information. This is partly my reason for studying the Ural owl, a forest-dependent raptor, in a human-altered landscape.

Although 70% of the land surface of Japan is forested, almost half (41%) of these forests are managed coniferous plantations that were established as a result of government policy following World War II. The availability of cheaper imported wood and a decreasing demand for fuel wood for heating and energy have enabled these plantations to mature. Now 30 to 50 years old, these forests are known to support fewer native wildlife species than more natural broadleaf and mixed forests.

Ural owl study areas in Japan.

Aerial view of the Niigata study area, Japan.

My main study area is adjacent to the Sea of Japan along the northern coast in the Niigata Prefecture (another study area is farther south in the Yamanashi Prefecture). A long time ago, this lowland area was developed for cropland using irrigation. The area may once have been forested, and hence would have supported Ural owl populations prior to cultivation at various times between the 17th and 20th centuries. Later, black pine forests were planted in sandy soil along the seashore to prevent wind erosion of cropland soils and sandy beaches, and to protect adjacent crops from strong winds and wind-blown sand and saltwater spray. The resulting mixture of habitats, from sand beaches, pine forests and cropland to villages and rice paddies, comprises a unique human-altered landscape. The pine plantations form a NE-SW forest belt-like zone about 12 miles (20 km) long and 656 feet (200 m) wide.

Elsewhere in its range, the Ural owl depends on mature forests for its survival. For example, southwest of this study area

Female Ural owl incubating eggs.

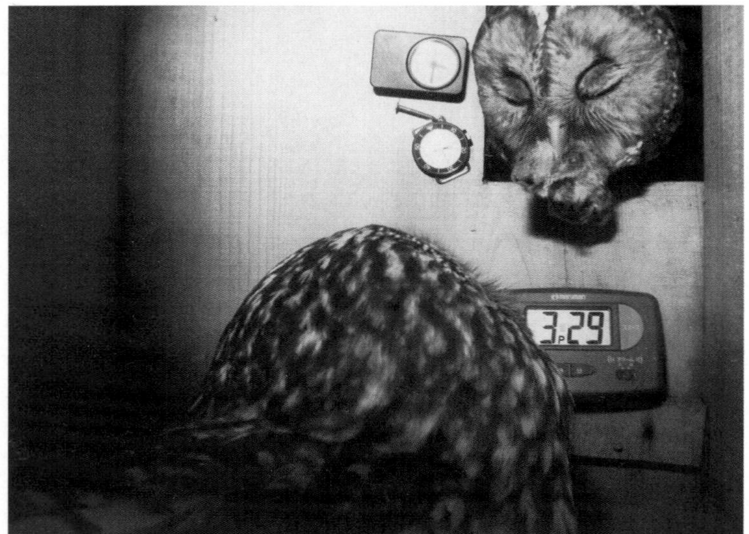

Male Ural owl delivering a vole at 3:29 p.m. to the female brooding a chick.

there are forested mountains in which breeding populations of Ural owls can be found. It had also been reported that Ural owls sometimes nest in smaller forests on the grounds of temples and shrines. Every village, town and city in Japan has at least one Shinto shrine and one Buddhist temple with a number of big trees. Some of these trees are huge and have many cavities that serve as good nesting sites for Ural owls. Therefore, at least some Ural owls were formerly found in villages and towns in Japan. However, they were not present in the Niigata study area prior to my study because trees in the pine plantations—which are now 30 to 50 years old—are too young and slender to have large natural cavities suitable for nesting Ural owls. Starting in 1994, 100 nest boxes were erected in woods near the main campus of Niigata University to find out if and how Ural owls could survive in the mosaic mixture of human-altered habitats. This was part of my Ph.D. thesis research supervised by Dr. Manabu T. Abe, who kindly guided me throughout the work. Fortunately, many pairs of Ural owls were attracted to this managed habitat and our nest boxes. About eight

Small mammal and bird prey of breeding Ural owls.

to nine pairs nest every year in each of the Niigata and Yamanashi study areas.

Using nest boxes equipped with cameras, it was possible to record the male delivering prey to the nesting female. The Ural owl preys on a great variety of species, including forest mice, voles, other rodents and birds. However, although the pine plantations in which they nest host a rich diversity of migratory birds in spring and fall, they do not support abundant small mammal populations. Therefore, in order to find out where and when the Ural owls were hunting, I captured and radio-marked one male and three young. What I discovered surprised me, as I had expected them to hunt mostly in cropland habitat. By day during the breeding season they roosted within the shelter of the pine plantation, near the nest boxes. At night they moved inland, away from the shoreline. The owls spent some time in the early evening and early morning hunting in wooded areas around houses and farms, and, to a limited extent, in cropland. But for most of the evening they hunted mainly in and around the roads, houses and other buildings of a town.

I followed and observed the radio-marked male Ural owl for 10 nights as he hunted around houses and along roads in the town. Eurasian tree sparrows, which roost at night in garden trees and trees lining the street, were hunted from power poles, wires or trees. When hunting pigeons, which roost under the roofs of various buildings, the owl perched on a building's terrace. The owl would sometimes wait on the corner of a roof to catch bats (the Japanese pipistrelle) as they emerged at dusk from their roosts under roof tiles. Even insects, attracted to village

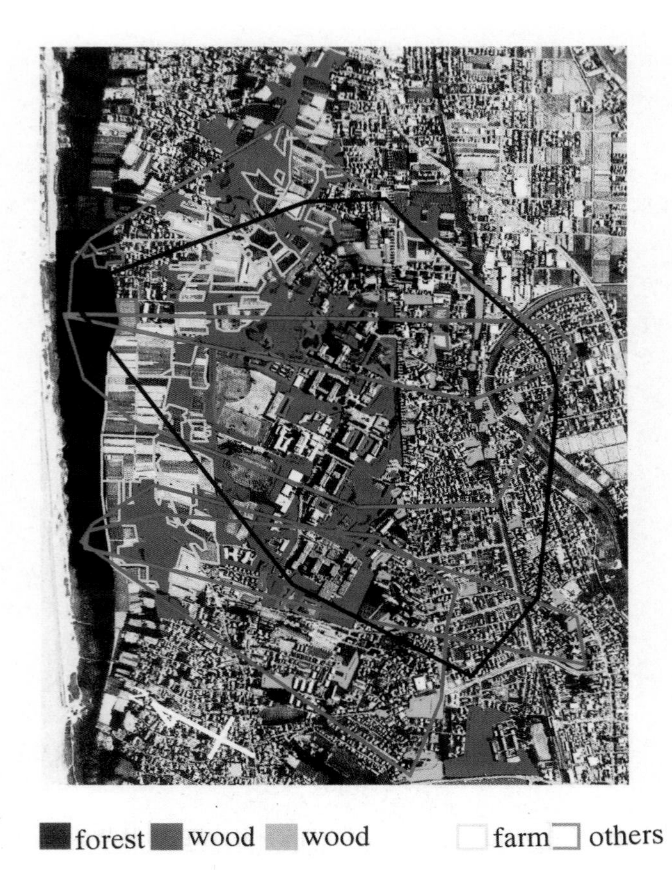

Five daily home ranges shown in red or black for a radio-marked adult male Ural owl. The left black strip is the pine plantation and the bottom third is the town area. Between them are the rarely used cropland, depicted in green.

lights, were fair game, especially when they lay exhausted or injured on the ground beneath street lamps. The owl sometimes perched on trees or buildings near restaurants, waiting for Norway rats that were attracted to restaurant garbage. It would try to capture a rat as it crossed a road or other open area.

One consistent feature used by the owl, regardless of habitat, was a tall perch from which to wait for prey. Its avoidance of cropland and rice paddies seemed to be due to a lack of perches. Farmers aware of this need sometimes erect wooden poles in their fields to attract owls in order to reduce vole populations that damage their crops.

While many pairs of Ural owls were able to breed successfully in this severely altered habitat, their long-term survival depends on the availability of natural nest cavities; nest boxes are only a temporary solution and used primarily as a research tool. Only recently have we begun to recognize the importance of broadleaf forest habitat to the survival of a great variety of life, including Ural owls, relative to the conifer forest plantations. If managed properly, broadleaf forests can also provide a sustainable supply of firewood, thus enabling people to return to a more traditional Japanese lifestyle and a better relationship with nature.

The Japanese government has changed its policy and is attempting to replace coniferous forests with a mix of tree species including broadleaf trees, which will help restore the Ural owl population in forested areas of Japan. My research has shown that Ural owls breed successfully in a mixture of natural and artificial habitats, using trees, electric power poles and even houses as hunting perches. The Ural owl can thus persist in such areas until pure coniferous forests are replaced by more diverse forests, and natural tree cavities once more become available.

A Wilderness Experience—Studying the Backcountry Boreal Owl

by Gregory and Patricia Hayward

Wilderness—to some it's the distant call of a wolf, to others the mournful cry of a raven, the whisper of a breeze through the pines, or a gale crashing on a cliff and

rumbling across the treetops. For me, the staccato call of a boreal owl ringing through a cold, snowbound spruce-fir forest will be forever tied to a vision of wilderness.

A wilderness owl—unnoticed, quiet, remote and of little recognized economic value—the boreal owl inhabits harsh spruce, fir, aspen and birch forests throughout the north from Finland to Canada, and sings its song in the high mountain forests of the United States Rockies. As late as 1979 the boreal owl remained an unknown breeding resident of the United States. The best information available suggested that the southern limit of its distribution was over 400 miles (644 km) north of our study area! Certainly trappers, miners and a few hardy souls who ventured into the high mountains during the dead of winter might have cocked an ear at the sound of a distant winnowing. But just as the value of wilderness went unrecognized for centuries, the boreal owls roosting among the fir boughs remained unknown by ornithologist and birdwatcher alike.

As a graduate student, and then as a professor working for the University of Wyoming, I have dedicated over two decades to exploring the natural history of the boreal owl. On February 26, 1980, my first night in a remote mountain basin

Nesting in tree cavities high above the forest floor, boreal owls are safe from many predators. This presents a formidable challenge to curious researchers determined to study them.

of Idaho's Frank Church–River of No Return Wilderness, I skied out from my cabin under a star-filled sky. The nearest household was 50 miles (81 km) away, the nearest plowed road, 70 miles (113 km). When I stopped skiing, silence filled my ears. Total silence. I felt a frigid breeze touch my face and heard the pines tremble under its touch. Still, I felt alone. Then I heard it for the first time. High, distinct notes ringing down the hill and across the snow-covered meadow. I was here to study the lives of owls in this area. This call obviously belonged to an owl, but which one? Back in my candle-lit cabin, I played my tapes over and over. Although difficult to believe, it became clear that a boreal owl had been calling, 400 miles (644 km) south of its expected range. That winter I heard seven boreals. The next year I found a nest 83 feet (25 m) up in an old dying ponderosa pine.

In the winter of 1984, my wife, Pat, returned with me to the wilderness to study the boreal owl in more detail. Trepidation and excitement alternated as we stuffed ourselves and our gear into a single-engine ski plane and raced the lengthening shadows into the wilderness. Landing after sunset on the powdery, snow-covered strip at Chamberlain Basin, Idaho, we left society to begin a simpler life with one goal—to understand the boreal owl. Learning from the wilderness and the owls went hand in hand. To locate, trap and follow the owl, we traveled the wilderness, searching its stream bottoms and ridge tops, experiencing every season. In four years, we and our assistants hiked and skied 16,000 miles (2575 km) of its remote trails and ridges.

Our study centered in Chamberlain Basin, a 100 square mile (259 square kilometers) plateau in central Idaho. Here we could study the owl's biology in an undisturbed environment.

Winter: Here in the high mountain forests winter and silence are synonymous. A mid-February evening finds us skiing through miles of unbroken powder.

Slipping sweaters over our woolen shirts and adjusting headlamps as the first few stars appear, our survey begins abruptly. Pat stiffens and strains to make out a soft call that I fail to hear. We charge up the hillside, top a ridge and stop to listen. The melodic repeated call of a male boreal owl drifts down the ridgeline. Feeling like monsters crashing through the forest, skis and poles crunching with every step, we invade the silence from which the owl sings. After 20 minutes of madly rushing toward the owl, stopping every few moments to determine his direction, we stand beneath the owl atop Lucky Ridge. Somewhere in the darkness overhead, the boreal sits among fir boughs and crusted snow continuing his song, relentlessly calling for a mate.

From February through April this male boreal owl, and dozens of other boreals, sing throughout the night. Each male sings from the treetops near a potential nesting snag. When a female approaches, sounding a shattering "bark" or series of soft "peeps," the male flies to the cavity he has selected and continues

A boreal owl is discovered roosting during a snowstorm. Perching close to the trunk gives the small owl some protection from the wind and makes it less conspicuous to arboreal predators such as lynx, marten and fisher.

calling in a softer, continuous staccato call. Rendezvous between the male and female may continue every couple of nights for two or three months prior to nesting; each encounter centers on the invaluable commodity—a nest cavity in a snag or dead-top tree.

One secret the wilderness still holds from us is the communication system between male and female. As we tracked mated pairs of boreal owls wearing radio transmitters, we realized that they rarely roosted together. Each wandered its own way, roosting 1 to 5 miles (1.6 to 8 km) from the future nest. Somehow, though, every few nights they joined one another at the nest snag to reinforce their pair bond. The encounter, usually only a few minutes in duration, involved courtship feeding and sometimes an examination of the potential nest cavity.

The nest cavity ties the boreal owl to the forest and other wilderness wildlife. Although some people view an old forest as a wasted resource, the numerous dead trees harbor a plethora of fungi, lichen and insects. Forest insects feed pileated woodpeckers, three-toed woodpeckers and northern flickers that, in turn, excavate cavities in the standing snags and injured live trees. Rotting fungus softens the heartwood, facilitating the woodpeckers' excavations. After a year's use by woodpeckers, the abandoned cavities are available as nests to northern saw-whet, pygmy, western screech, flammulated and boreal owls.

Mature forests hold other keys to the boreal owls' survival. The owl's primary prey, the red-backed vole, is most abundant in older spruce-fir forest. Rotting logs on the moist forest floor support fungi, an odd but important food for these small forest rodents. Arboreal

Thick snow cover does not stop wilderness mailman and pilot Ray Arnold from checking on owl biologists who express their gratitude with home-baked cookies.

lichen, or "old-man's-beard," which is abundant here, clings to tree branches and gains its sustenance from the minerals that accumulate in water dripping through the canopy. Like the fungi that decompose old logs, the arboreal lichen, some of which falls to the forest floor, is important food for the red-backed vole, and therefore indirectly important to the boreal owl.

Even in the undisturbed forests of the wilderness, dramatic fluctuations in prey populations affected the owls. In 1986, following two years when many owls nested and produced young, small mammals—most importantly red-backed voles—became scarce. Despite mild winter conditions, several of our radio-marked owls died or left the area. In January, one male moved 50 miles (80.5 km) from his former home range. That winter the forest was exceptionally quiet. The boreal owls that remained did not sing to attract mates. In fact, we heard few great gray, great horned, pygmy, or northern saw-whet owls, all of which had called during previous winters. In spring, few owls could capture sufficient prey to raise even a single nestling. In the following year, however, we located more nesting boreals than during any of the three previous years, thus demonstrating the resilience of the wilderness.

One night in April, with the aid of night-vision goggles, Pat and I watched a radio-marked male owl as it moved silently through the forest. With rapid, jerking head movements, he surveyed the forest floor from perches 12 to 15 feet

(3.7 to 4.6 m) high for three to four minutes before flying 25 to 30 yards (23 to 27.4 m) to his next perch—allowing us to relocate him before the next flight. During an extended vigil in a dark spruce stand, the owl intently watched the ground, turned and cocked his head, moved nervously on his perch, and bending forward, comically, almost fell off the perch. Then he dropped like a parachute, talons spread, into the beargrass and onto his prey.

The color of bark and lichens, a recently fledged boreal owl chick almost disappears against the trunk of its tree roost. Young boreal owls depend on the male for food for at least a month before they hunt on their own.

Spring: For four months we have been alone with the wilderness, giving Pat and me the opportunity to come to know one another like we never could in the hustling world on the "outside." Forcing us to work together to overcome its challenges, this wild country has taught us a degree of cooperation and companionship we may never have otherwise developed.

Weeks slip by. As the streams swell with runoff, and nighttime temperatures less frequently dip into the low 20s Fahrenheit (less than −10°C), the female boreal owl begins spending most of her time near the nest tree. Each evening the male's call rings from the nest stand, and the female boreal fattens with the courtship food brought nightly by her mate. Before the ground becomes completely bare of snow, the female begins occupying the cavity day and night. She may remain in the cavity, receiving all her food from the male, for as long as two weeks prior to laying eggs. Ultimately, she lays a clutch of two to four eggs, each egg added two days after the previous one. For 30 days she will incubate the eggs day and night, remaining in the nest cavity constantly except for a 10- to 15-minute break each morning and evening. She leaves on these short excursions to defecate, regurgitate a pellet and to stretch her wings and preen. Her mate delivers one to four mice or voles through the night, which she neatly arranges along the edge of the cavity or in one corner. As the hatching date approaches, excess prey accumulates. Soon, however, the growing owlets' demands will nearly outpace the male's ability to deliver.

Summer: Female boreal owls have spent the past 45 days in the nest cavity. Quarters are tight, however, with growing

Greg and Pat Hayward continue their studies on boreal owl ecology with the help of their children, Phil and Isaac, and a small group of enthusiastic graduate students.

young and the adult maneuvering for space in an 8-inch (20-cm)-diameter cavity. At 24 days old, the young can swallow whole mice and voles delivered by the male. He may bring eight to 12 prey a night, including shrews, red-backed voles, pocket gophers and even birds. The female is no longer needed at the nest and she leaves to hunt for herself after two months of being completely dependent upon her mate. Within two weeks of leaving the nest she may have moved 10 miles (16 km), to the other side of the basin, leaving her young while she molts and grows new plumage. The male brings food to the nest for another week before the young leave the confines of the cavity.

Autumn: By late November, when the snowpack begins to thicken, many young boreals have been predated by Cooper's hawks, goshawks and great horned owls, and some boreals will disperse from the basin. Given adequate prey populations, however, a majority of the adult boreal owls will remain in their wilderness basin. Winter snow 4 feet (1.2 m) thick and sub-zero temperatures are not enough to drive the owls from the forest with which they are familiar.

After four years of hiking its trails, skiing its ridges and climbing its trees, we too have become familiar with the basin. The wilderness has been our laboratory, where predator and prey meet, struggle, thrive and die; a laboratory whose drama has remained largely unrecognized. Herein lies another value of wilderness, a landscape from which to discover the relationships between soil and sun, plants and animals. A place that offers hope for mankind if we conserve it. The boreal owls' song, which called Pat and me to search for truth in this untouched land, continues to be sung, inviting others to explore this unique natural laboratory.

Bringing Back the Barn Owl in Ontario

by Bernt Solymár

The common barn owl—a flagship species for healthy grassland habitats—is considered an important ally of farmers due to its voracious appetite and almost exclusive diet of rodents. It nests in tree cavities, as well as in sheltered corners of barns, silos and abandoned buildings. In Canada, it breeds regularly only in lower mainland valleys in British Columbia and along the north shore of Lake Erie in Ontario. Since 1999, the Ontario population has been designated as Endangered by the Committee on the Status of Endangered Wildlife in Canada.

Southern Ontario is the northernmost range in eastern North America of the cosmopolitan barn owl, which is found on every continent except Antarctica. The species requires large, open tracts of contiguous grassy fields in which to hunt for voles and mice. Although there were scattered grasslands in southwestern Ontario prior to European settlement (tall-grass prairie represented about 9% of the landscape), it was the clearing of forests and subsequent planting of pastures and hayfields that probably first attracted barn owls to Ontario. Barn owls are not well adapted for cold weather: their legs and feet are not feathered and they do not store fat. However, livestock, corn silos and haylofts in most barns provided supplementary heat and food sources during cold, snowy Ontario winters. Old-timers speak about barn owls as common residents on farmsteads along the north shore of Lake Erie prior to the 1970s.

These barn owl chicks, photographed in Kent County in 1988, are among the last young to be produced in Ontario in the wild.

Ontario's barn owl population declined so dramatically in the last 30 to 35 years due mainly to the reduction of foraging habitat. The conversion of pastures and hayfields to more intensive monocultures such as soybeans, corn and horticultural crops; the disappearance of corn silos from farms in favor of central storage depots; and the elimination of hedgerows and fencerows have all had negative impacts on the barn owl. Less grassland habitat meant lower densities of voles and mice, and subsequently fewer prey. Considering that barn owls, because of their diet preferences, have always been considered an important ally of farmers, it seems paradoxical that intensive agriculture may have contributed to their decline.

In fall 1997, the Ontario Barn Owl Recovery Project was established to take on the daunting task of recovering this endangered species. The development of a formal recovery plan lent structure, targets and credibility to the project. Today, the recovery team consists of a diverse volunteer committee membership including conservation organizations, a fish and game club, farmers and government biologists. The goal of the project is to restore levels of barn owls in southern Ontario to historic levels by increasing grassland habitat along the north shore of Lake Erie.

Project priorities include the following:

- To foster community volunteerism and partnerships by involving individuals and groups in barn owl nest-box building, installation and monitoring programs;
- To identify, enhance and protect grassland and wetland fringe habitat along the north shore of Lake Erie through conservation agreements and creation of grassland reserves. With

Ever hopeful, a dedicated volunteer with the Ontario Barn Owl Recovery Project inspects a nest box for signs of nesting.

the support of farmers and rural landowners, the project is exploring innovative approaches to revert marginal and inactive farmland, public lands and corporate-owned lands along the north shore of Lake Erie into productive grasslands. New and expanded grasslands are expected to benefit a wide variety of wildlife, including other species at risk such as the Karner blue butterfly, Henslow's sparrow, bobwhite, short-eared owl and American badger;
- To develop public awareness, appreciation and grassroots support for barn owls, other grassland species and grassland habitat through public seminars and workshops, and development and distribution of educational materials to schools, parks, conservation organizations and interested members of the public.

By accomplishing the above, the recovery team hopes to encourage the return of a sustainable population of barn owls to Ontario. Restoring and conserving grassland habitat will benefit other flora and fauna, conserve biodiversity, link many partners in rural community in a common goal, and increase awareness and appreciation of the importance of grasslands as an ecosystem in southwestern Ontario. As of March 2002, 300 nest boxes have been installed on farm structures along the north shore of Lake Erie. Although barn owls have not yet nested in them, since December 1999, there have been five confirmed and many reported sightings of barn owls in Ontario, thus providing hope for those of us who dare to care.

Remaining Choices—Understanding the Needs of Injured Owls

by Katherine McKeever

The Owl Foundation is a place where one can watch the development of intimate relationships between individuals of most of Canada's owl species. The by-product is that recycled genes go back to wild populations in released progeny.

The physical property, about 4 acres (1.6 ha), is equally divided between a forested slope (remnant Carolinian vegetation zone) and new grassland recovered from former orchards. Geographically, we are on an ancient lake bed between the Niagara Escarpment and the south shore of Lake Ontario. Our forest slopes down from 100 feet (30.5 m) to a wide river estuary. Many of our old trees are 120 feet (37 m) tall with a girth of 12 feet (3.7 m). Our latitude is the most southern in Canada (in line with Roseburg, Oregon) and our climate the most moderate in Eastern Canada. This permits all Canadian owl species, except the insectivorous flammulated owl, to be maintained outdoors year-round.

In 1965, we embarked on a modest plan to attempt rehabilitation of injured owls to the point of responsible release. Now, 3600 owls later, owls arrive from across the continent already assessed as permanently damaged, in the hope that

Katherine McKeever, founder of The Owl Foundation, introduces an unreleasable blind northern saw-whet owl to a young apprentice. However, Kay wisely insists that human contact with injured owls destined for release must be minimal.

some use can be made of their lives.

Our challenge is to attempt the best recovery of lost faculties and broken spirits by providing the opportunity for making their own choices in every aspect of their lives. These choices include the ability to move from one defined space to another, and to yet another, through overhead corridors; to meet others of their species; to choose a territory; to select every size and type of roost, of all heights for exposure or seclusion; to select from two to four available nest sites in each territory; and to be alone or in company.

All of this involves withdrawal from public visitation and opens the need for private funding, since these are inherently wild but traumatized owls, and spaces for recovery must be very large. This need is not only to promote the likelihood of ultimate breeding but also to provide a suitable habitat for young owls' early

Owl species	Number of successful pairs	Span of breeding years per species	Number of young raised
Snowy owl	7	1978–2001	77
Great horned owl	6	1984–2002	53
Great gray owl	8	1985–2002	41
Barred owl	6	1990–2002	45
Barn owl	5	1974–2002	236
Long-eared owl	2	1995–2001	27
Northern hawk owl	7	1987–2002	76
Eastern screech owl	12	1976–2002	141
Burrowing owl	23	1982–2002	169
Boreal owl	3	1983–1993	9
Northern saw-whet owl	8	1979–2001	73
Flammulated owl	2	1992–1993	1
Northern pygmy owl	3	1985–1990	9

Note:

- Short-eared owls have produced eggs but not living progeny.
- Three spotted owls have been residents but never two at the same time.
- One male snowy owl, wing-crippled as an adult in Alberta in 1965, sired his first chick in 1996!

experiences. A breeding pair of snowy owls, even when the female is flightless, needs a minimum of 1200 square feet (115 square m). A pair of great gray owls requires 1000 square feet (96 square m), an open forest setting and cage heights to 18 feet (5.5 m). Even a wing-crippled northern hawk owl will promptly climb 16 feet (4.9 m) to a semi-cavity nest site. In forest owls, height is security, and we must provide "furniture" for their use in reaching acceptable levels. Finally, since these enclosures will be the first exposure to surroundings for the offspring, they must also be able to contain live prey species, in suitable cover, that the male can catch. This is critical to the development of a memory that food moves, makes noises, is brown and fights back!

The following are two examples of very different cage designs for encouraging pair bond formation and/or breeding of the two main groups of native owl species—those that build lasting pair bonds, and those that have seasonal affiliations only. (Snowy owls form bonds, but require size and uncomplicated open space only.)

Owl Enclosure Design 1: Breeding Complex for Owl Species That Form Lasting Pair Bonds

This design continues to be successful in establishing two breeding pairs of great gray owls, a species that, at least in captivity, develops long-lasting bonds between potential partners as a requisite to annual breeding. Adult great grays, admitted with permanent damages, are assumed to have formed a bond with a wild mate, a relationship now put asunder by trauma to the one. Our challenge is to create an environment conducive to the formation of a new bond with an existing captive potential mate.

In this design, pictured above, accommodation is offered for six individuals to ultimately form two pairs through mutual selection. Initially, all corridors are left open, allowing free flow

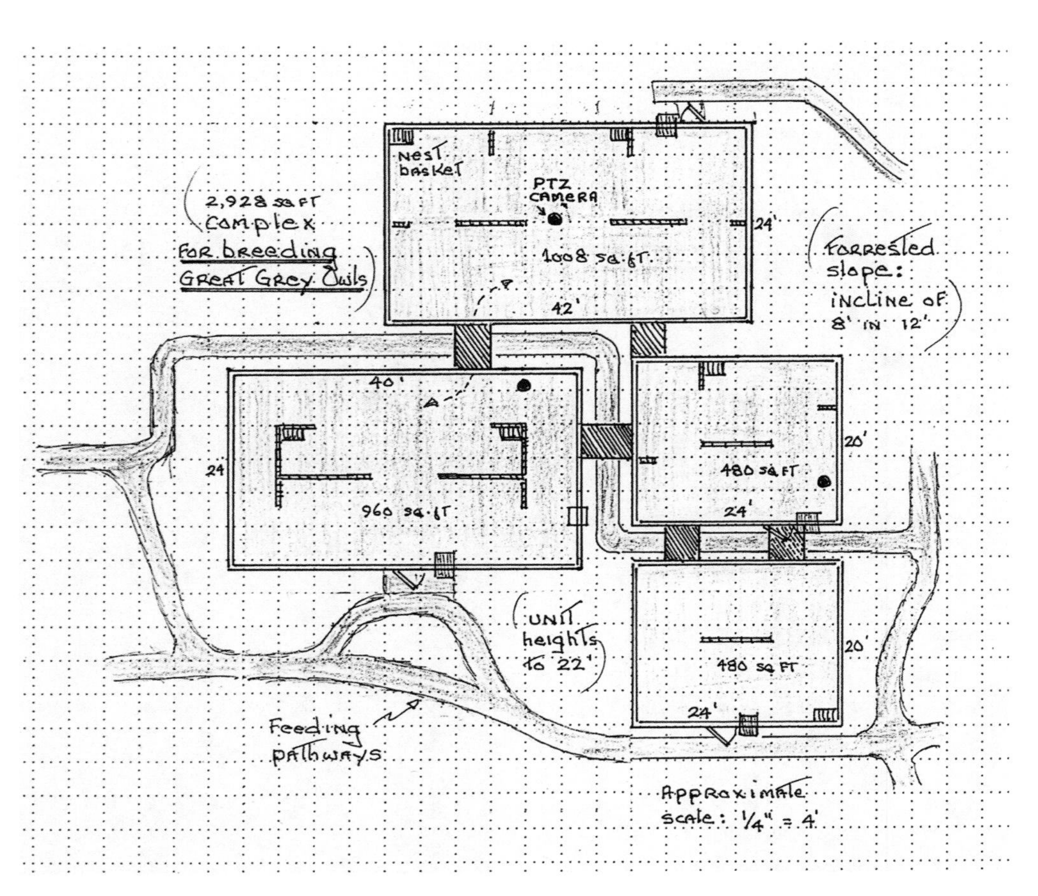

Well shaded by mature trees, this 2928-square-foot (281-square-m) great gray owl breeding complex is the cumulative result of decades of experience and observation.

through the units with slatted visual baffles offering further choices for concealment or socializing. Usually by the third year the observer will notice a growing relationship in one of the units. When territorial defense is exhibited by one of the occupants against a third party, the two connecting corridors are closed, thus isolating the incumbent pair from the other four owls.

As this pair progresses to grooming and copulation in the ensuing season, a second pair may be developing a similar relationship in another unit, and may soon bring this growing bond to fruition as the remaining two are excluded by closure of the final gates. This third "pair" is unlikely to form a partnership as they and the unit to which they are relegated are the residuals of the first two choices. They should be removed from the complex and put into a new rotation elsewhere. The now empty third unit is then reserved for the progeny of the first two pairings.

Mutual separation of parents and young is usually in August of the hatch year as juveniles from both pairs venture from home territories via the reopened corridors and find the third unit. Fathers may accompany their young to the new unit, but soon return home, after which the corridor is closed again. (Mothers seldom or never accompany their progeny away from the home unit.) Once both sets of juveniles are isolated in the third unit, they are fed increasingly on live rodents only. In early winter they are netted and removed to the exercise and overwintering complex for their species for training on live prey in preparation for release. During the winter and spring, the empty third unit is available for temporary housing of great grays undergoing rehabilitation.

Owl Enclosure Design 2: A Breeding Complex for Nomadic or Migratory Open Forest Owls

Northern hawk owls develop brief, transitory pair bonds in early spring with any convenient and/or soliciting partner. But even a successful partnership will dissipate before September, when males usually depart. Females have a much stronger attachment to a nest site and probably the surrounding area, but not to a specific male. In the design illustrated, we close most corridors by late November, regardless of which, or how many, owls may be in the various units at the time. The gates are opened in late February, and those residents that can fly will cruise the corridors again, apparently testing social opportunities. Flighted females are the most interesting to watch since their expeditions appear to be "shopping" trips for males! A male solicited by a female usually follows her back to a territory familiar to her, and copulation may occur soon after his arrival. These are always the earliest nestings. The non-flighted females are obligated to "stay home," but they solicit vocally, loudly and often, and are visited by what appears to be nervous males! An interesting comparison of strategy.

Of note with this species and this design is that while gates are open in connecting corridors between units, they are also open to adjacent complexes for compatible but larger owl species (in this case great grays) that are unlikely to enter hawk owl corridors themselves. In this design, the two top hawk owl units are

Besides simply providing opportunities for exercise, flight corridors between enclosures permits female northern hawk owls to go "shopping" for new mates each spring.

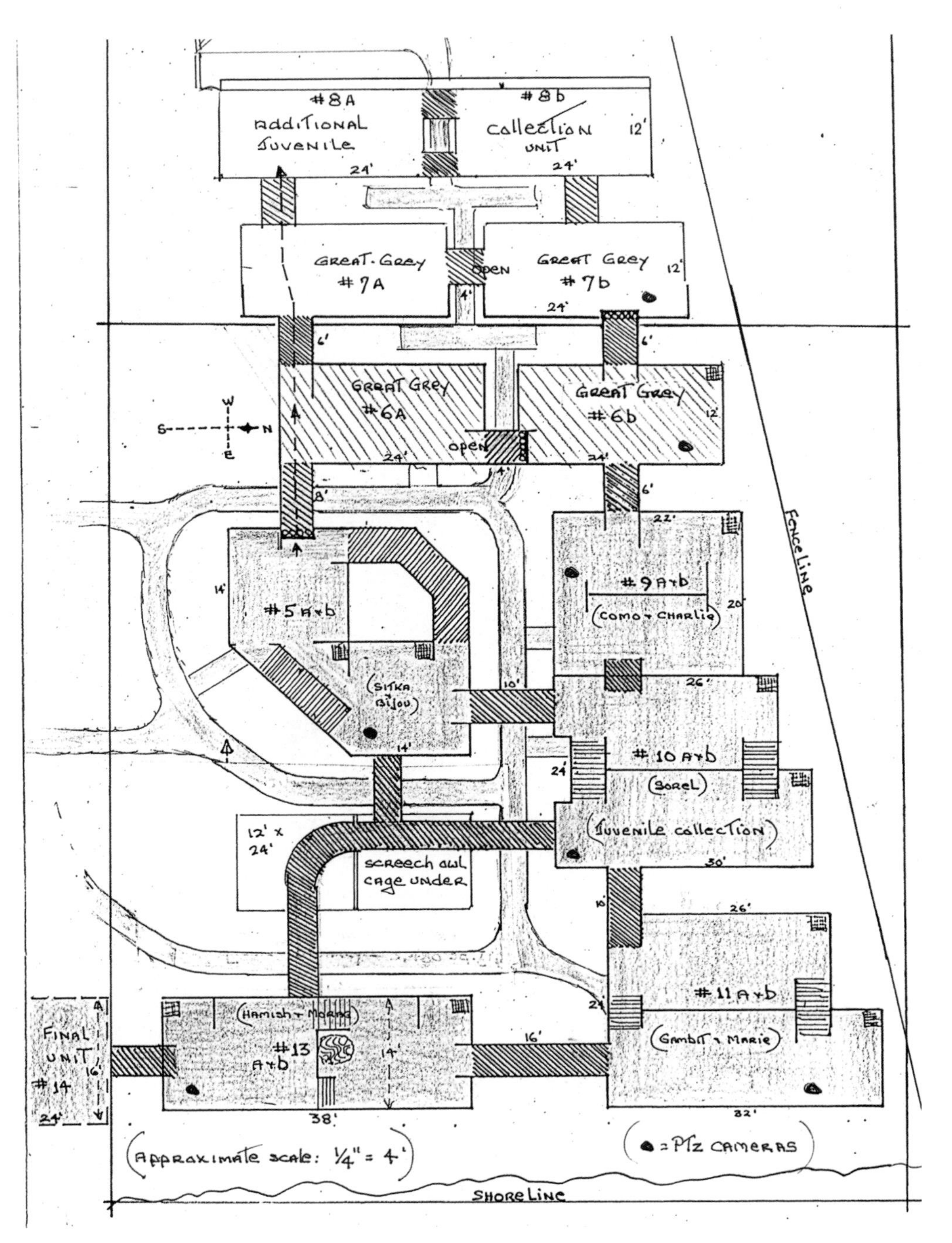

accessed by flying through the great gray units. They are seen in flight from a camera in B unit, the great gray in residence showing only mild interest!

As in other owl species, separation time between parents and young always occurs in August. Because of the length and complication of their corridor system—for instance, one corridor is 48 feet (15 m) long, climbs a steep hill and turns a corner—we streamline dispersal routes for young hawk owls and direct them to the units most suitable for the collection of juveniles. Here again, the male parent often accompanies his progeny initially but very briefly. The mother never does; but even she may have wandered elsewhere by late September. By October, we will have netted the juveniles in the collection cages and moved them to the overwinter exercise and hunting units. We have evolved the policy here of never (except in a medical emergency) molesting or interfering with young owls in their parents' compound to avoid creating a memory of danger in subsequent use of that place for nesting.

Monitoring the activity of up to 30 owl enclosures simultaneously by remote cameras would be a challenge for this young girl were it not for the video recording equipment

Conclusion

Finally, the surprising, even astonishing, demonstrations of bonding between self-selected individuals can be seen, year-round and through their remaining lifetimes, only on a remotely controlled monitoring system. Thus up to 30 video cameras, rotated by season amongst our 52 enclosures, routinely relay images of private lives. Unlike the confrontation inevitable with human presence, the owls are unwitting of this surveillance.

Of course, springtime is the season when one hovers hopefully around the monitors as the breeding behavior unfolds, watching for the first glimpse of irresistible fuzzies and admiring the total devotion of both parents. Anticipation of these exciting moments is part of all our winters, even as we review the tapes of previous seasons.

But for the most memorable moments, there is nothing so touching, relayèd through the camera's eye, as a pair of middle-aged great gray owls, sitting close together in December sunlight (when egg follicles and gonads are at their lowest ebb) grooming each other quietly and solicitously, keeping the pair bond in good order. For the watcher who has kept the dream alive for so many years, trying and failing and trying again, these are the golden moments. Against all predictions, such permanently damaged wild owls have overcome their physical deficits, left fear and stress behind, and are utterly absorbed in each other. Over the years it has been a privileged look at ancient relationships, still enduring, still strong, even after such calamitous lives.

CHAPTER 6

Owls of the World, Their Global Conservation Status and General Distribution

A brief summary of 205 owl species is presented below. Each summary presents the owl's common and scientific name. The measurement given is for body length. Range maps and some photographs are also included.

Conservation Status—In 1994, the International Union for Conservation of Nature and Natural Resources (IUCN) released a set of criteria for use in evaluating the likelihood of a species (or a similar group of related organisms called a taxon) becoming extinct. IUCN's categories of risk of extinction stated in their 1994 report are generally defined as:

Critically Endangered—facing an extremely high risk of extinction in the wild in the immediate future.
Endangered—facing a very high risk of extinction in the wild in the near future.
Vulnerable—facing a very high risk of extinction in the wild in the medium-term future.
Lower Risk—Considered conservation dependent, near threatened.
Data Deficient—inadequate information to assess conservation status.
Not Evaluated—not yet assessed.

These categories are based on rigorous and numerical criteria (population size and trend, range size, threats). However, as with the species lists, problems with the quality of information and its application suggests that the list is intended to serve as a warning signal. Species' extinctions are high stakes to be gambling with, and it would be a catastrophe to designate a species as safe when it is in trouble. As with many other conservation status assessment systems, the species categorized as data deficient are typically ignored due to limited funds and resources. It is these species that perhaps warrant immediate and greater attention.

Conversion Chart

To convert metric unit	multiply by	to obtain imperial unit
centimeter	0.3937007	inch
square centimeter	0.1550003	square inch
meter	3.2808398	foot
square meter	10.764262	square foot
meter	1.0936132	yard
kilometer	0.6213881	mile
square kilometer	0.3861302	square mile
hectare	2.4709661	acre
gram	0.0352733	ounce
kilogram	2.2045855	pound

OPPOSITE: **The Verraux's eagle owl is one of the heaviest owls in the world. Its pink eyelids highlight its large dark eyes.**

1 Greater Sooty Owl, *Tyto tenebricosa*

Description: This owl (35 to 50 cm) is undoubtedly the most distinctive member of the barn owls. It has an awkward appearance, with a large head, thin body, gangly legs with large feet and a very short tail. Its eyes are huge and black. Its plumage is completely sooty brownish-black, with the underparts varying from entirely brownish-black to somewhat lighter. Both upper- and underparts are flecked with small whitish spots.

Habitat: Inhabits pockets of rainforest and wet eucalypt forests with old, hollow trees. Remaining suitable old-growth

habitat is found mainly in gullies and valleys. It is sometimes found in younger, regrowth forests providing there are suitable nesting trees present. In New Guinea, sooty owls are found from sea level to about 2000 m.

Natural History: The greater sooty owl is strictly nocturnal, roosting in crevices or hollow tree trunks during the day. It also hides in the thick foliage of tall trees and occasionally in caves. Hollow trees and caves are also used, sometimes repeatedly, as nest sites. Nest sites have been found as high as 30 m above the ground. Only 1 to 2 eggs are laid on a layer of debris, often with only 1 egg hatching. It hunts within forests and seems to avoid clearings. Preferred prey includes rats and bandicoots, captured by the owl with a quick pounce from a perch. Other prey items include possum, gliders and small birds. These owls require a larger territory (200 to 800 ha) than the lesser sooty owl (50 to 60 ha). Their calls have been described as having an unearthly quality; one in particular is an eerie descending whistle.

Conservation Status: Not Globally Threatened (IUCN). Due to its secretive nature, its status is perhaps best described as unknown, although some authors suggest that it is likely rare and endangered. Habitat loss continues to threaten local populations, but surveys have shown that some logging operations have not reduced population densities. An estimated 50% of its reported critical tall old-growth habitat remains intact.

2 Lesser Sooty Owl, *Tyto multipunctata*

Description: True to its name, this owl has sooty-gray plumage, with paler coloring below and many whitish spots above and below. It is similar in appearance to the greater sooty owl, with similarly dark plumage, but is smaller (32 to 38 cm) and more heavily spotted. Its eyes are larger compared to the other barn owls. The facial disk is grayish-white with sooty coloration around the eyes.

Habitat: Found in rainforests and wet eucalypt forests. Tall trees and trees with hollow trunks characterize these forests.

This species seems to prefer areas with extensive tracts of rainforest, but it can also be found hunting along roadsides and from the edges of clearings created by logging.

Natural History: There is no information on the dispersal movements of this species. It is nocturnal, roosting by day in dense foliage, in any available crevice, and among aerial roots. It hunts mainly small mammals from the ground or from perches in trees. It is able to catch prey in the rain and in almost complete darkness. Breeding biology of this species is relatively unknown. Hollows in old trees up to 30 m above the ground are used for nesting, the female laying up to 2 eggs; usually only 1 egg hatches. Territory size is reported to be 50 to 60 ha. Its vocalizations are similar to those of the greater sooty owl, most notably a strident, descending whistle.

Conservation Status: Not Globally Threatened (IUCN). Some authors report that it may be threatened or possibly even endangered due to the cumulative impact of logging.

3 Australian Masked Owl, *Tyto novaehollandiae*

DESCRIPTION: The Australian masked owl, the largest owl of the Tyto group (37 to 47 cm), most closely resembles the common barn owl. It is named for its dark facial disk. Its strongly marked upperparts are grayish-brown with white and black spots; yellowish or orange-buff patches may occur. Underparts are white to pale orange-buff, usually with arrow-shaped spots. Both pale and dark morphs are found. The powerful legs are feathered to the base of the toes.

HABITAT: Found in forested areas, including eucalypt forests, as well as open woodlands and clearings. Uses open country to hunt.

NATURAL HISTORY: The masked owl is apparently sedentary, and is never common. Because it is shy, secretive and nocturnal, it is rarely observed. During daylight hours it roosts in the thick foliage of tall trees, in hollow trunks, or occasionally in holes between rocks or in caves. The female lays 2 to 4 eggs in the hollow trunks of tall eucalyptus trees, often as high as 20 m. Eggs may also be laid on bare rock or sand in caves. Prey items consist mainly of rabbits and smaller mammals; small birds and lizards are also taken. In Tasmania, it feeds extensively on possums, rabbits, bandicoots and sugar gliders. Its vocalizations are comparable to the common barn owl, but are much louder and more rasping; calls have been described as "strange and wild cackling"!

CONSERVATION STATUS: Not Globally Threatened (IUCN). This species is believed to have declined and disappeared from certain areas in Australia. Early records, however, are questionable owing to its resemblance to the barn owl. The decline may be due to the decrease in native mammal populations following introduction of alien species by European settlers. The status is uncertain and is difficult to determine due to this species' secretive nature. Some experts treat the Tasmania population as a separate species, the Tasmania masked owl (*Tyto castanops*), based on enhanced size and color differences between males and females; this population is considered Endangered. Furthermore, the Tiwi Islands subspecies is recognized as Endangered, and both the northern and southern Australian subspecies are listed as Near Threatened.

4 Golden Masked Owl, *Tyto aurantia*

DESCRIPTION: One of the smaller owls (27 to 33 cm) in the Tyto group, the golden masked owl is a particularly handsome owl. Overall coloring is pale golden or orange-rufous with dark, delicate markings. The upperparts have dark V-shaped markings that become smaller around the head. Brown spots cover the underparts. No other *Tyto* owls have golden-rufous coloring with V-shaped markings on the back and wing coverts.

HABITAT: The golden masked owl is endemic to the island of New Britain. It is reported to use tropical rainforests with clearings, as well as ravines with shrubs and trees. It occupies both lowland and mountainous areas up to 1830 m.

NATURAL HISTORY: Very little is known of this owl's biology. Like other *Tyto* owls, it is nocturnal. It is apparently sedentary, but there is no information regarding movements. It is reported to prey on small rodents, but may also take other small vertebrates and insects. There is no information on its breeding behavior. This owl is known locally as "a kakaula" from its call. Screeching and hissing calls have also been reported.

CONSERVATION STATUS: Data Deficient (IUCN). This species is uncommon to rare. It may be negatively impacted by changes to its habitat, especially because it is confined to the island of New Britain. Species recognized as Data Deficient should be recognized as a high priority for conservation research. It is listed as Vulnerable by BirdLife International.

5 Manus Masked Owl, *Tyto manusi*

DESCRIPTION: The Manus masked owl is quite large (49 cm). Its upperparts are grayish-brown with white and black spots and ochre patches. Underparts are whitish, turning to a pale ochre-buff toward the belly and feet, and are heavily spotted with brown from the breast to the thighs. It is most similar to the Australian masked owl, but lighter above.

HABITAT: Forests with open areas.

NATURAL HISTORY: Unfortunately, extremely little is known about the basic biology of this species. More information is needed regarding its vocalizations, movements and breeding biology. It feeds on small rodents and other vertebrates; possibly larger insects are included in its diet. Its biology is probably comparable to that of the Australian masked owl.

CONSERVATION STATUS: Data Deficient (IUCN). Possibly endangered, with no records available since 1934. Species identified as Data Deficient should be recognized as a high priority for conservation research.

6 Lesser Masked Owl, *Tyto sororcula*

DESCRIPTION: Similar in appearance to the Australian masked owl, but smaller in size (31 cm).

HABITAT: Thought to inhabit lowland forest.

NATURAL HISTORY: Almost all knowledge of this species' biology is lacking, as there are no observations of the lesser masked owl in the wild. In fact, only 3 specimens of this species exist. Its call is described as 3 to 4 high-pitched slow screeches followed by 3 fast whistle-like screeches.

CONSERVATION STATUS: Data Deficient (IUCN). Probably rare and

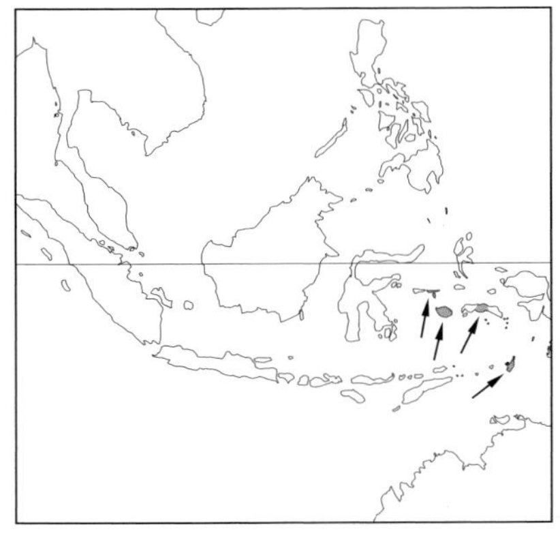

endangered. This virtually unknown species requires study of its taxonomy, distribution, ecology and biology. The habitat of the lesser masked owl on the islands of the lesser Sundas must be protected in order to ensure the existence of this species.

7 Taliabu Masked owl, *Tyto nigrobrunnea*

DESCRIPTION: Typically similar in appearance (31 cm) to other barn owls, but differs in that it is brown with uniformly dark brown wings. The secondaries have whitish tips. The facial disk is a pale reddish-brown. The upperparts are flecked with white, and underparts are a deep golden brown with many dark spots. Unlike other *Tyto* owls it has uniform brown wings with no markings on the primaries.

HABITAT: No information available.

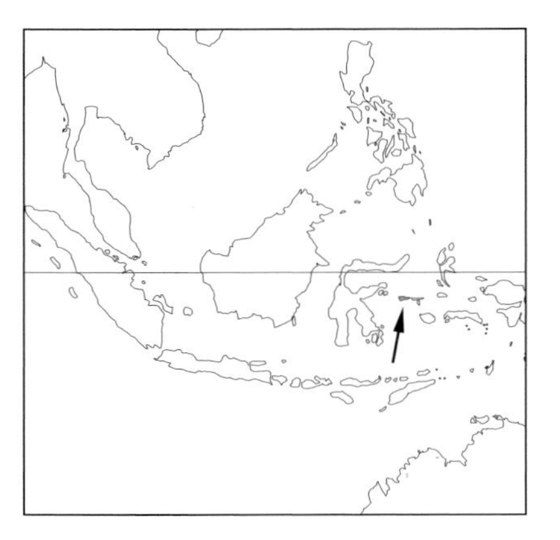

NATURAL HISTORY: The Taliabu masked owl is virtually unknown. A specimen was collected on the island of Taliabu in 1969, and there is a recent sighting from 1991.

CONSERVATION STATUS: Endangered (IUCN). Listed as Vulnerable by BirdLife International.

8 Minahassa Masked Owl, *Tyto inexspectata*

DESCRIPTION: The Minahassa masked owl is a relatively small barn owl (27 to 31 cm). Its rounded wings are relatively short; the carpal area is dark grayish-brown with white and orange-brown spots. The upperparts are an attractive pattern of grayish-brown with orange-yellow to rusty-red patches with large white spots. The underparts are fulvous (reddish-yellow or tawny)-white to pale ochre with small blackish spots. Notably, the facial disk is small compared to its body size.

HABITAT: Found in both tropical rainforest (a demanding habitat of high heat and humidity) and drier, degraded forests, from 250 to 1500 m.

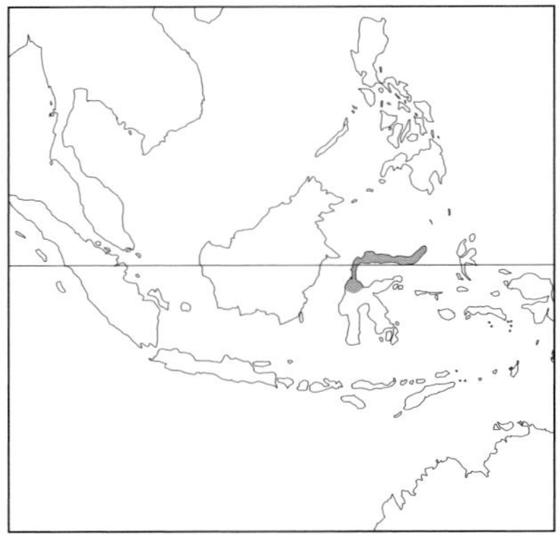

NATURAL HISTORY: The natural history of this species is almost totally unknown. It most likely preys predominantly on small mammals. Nesting is thought to occur in early April in hollow trees.

CONSERVATION STATUS: Vulnerable (IUCN). Population size is unknown; sightings of live birds are very rare.

9 Sulawesi Owl, *Tyto rosenbergii*

DESCRIPTION: A large (41 to 51 cm), heavily marked owl. Its whitish facial disk has a prominent reddish-brown rim. Its upperparts are brownish-gray with scattered black spots, the lower half having white spots. Its underparts are pale fulvous with brown spots, and the breast feathers have brownish edges.

HABITAT: Inhabits both rainforest and lowland areas (sea level up to 1100 m). This species adapts easily to areas in the rainforest that have been deforested. It uses clearings, cultivated areas and wooded and semi-open landscape, including forest edges, for hunting.

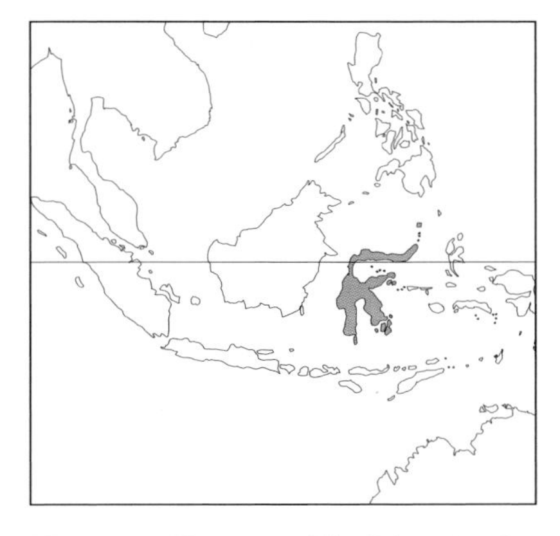

NATURAL HISTORY: The Sulawesi owl is little studied. It is known to be nocturnal, found near villages, and probably preys upon small vertebrates—rat and shrew remains have been found in pellets. Its call is a raspy, quavering 1-second screech.

CONSERVATION STATUS: Not Globally Threatened (IUCN). Population size unknown, but it can apparently easily adapt to changes in habitat caused by deforestation.

10 Common Barn Owl, *Tyto alba*

DESCRIPTION: An owl with a heart-shaped facial disk, blackish small eyes and long legs. It is medium-sized (34 cm) with a 90 to 100 cm wingspan. Its yellowish-brown

to ashy-gray plumage varies considerably across its worldwide range—over 30 subspecies have been described. Lacking ear tufts, it is similar in appearance only to other species in the genus *Tyto*.

HABITAT: Predominantly open countryside with scattered trees, often near villages or towns. Appears to avoid areas with more than a month or so of permanent snow cover, which restricts it to temperate or warm climates.

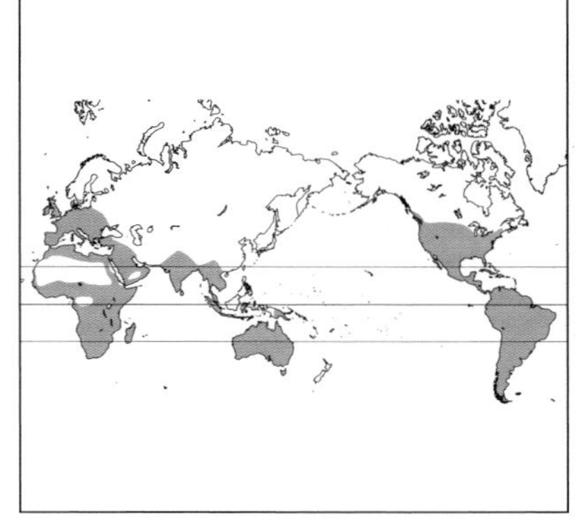

NATURAL HISTORY: The common barn owl is primarily a permanent resident, with young known to disperse as far as 1600 km. It lives singly or in pairs, and is active at night from dawn to dusk. To see one flying about in the daytime is rare. It typically roosts during the day and nests in nooks and crannies of old abandoned buildings, but it will also use occupied buildings. It eats mainly small mammals, even catching bats in flight! Not a fussy eater, it will also consume smaller birds, reptiles, amphibians and insects. It finds its prey using its remarkable sense of hearing, and then seizes it in its powerful talons. Common barn owls appear to mate for life, and can live over 20 years in the wild. After an elaborate aerial courtship, a female will normally lay 4 to 7 eggs. In years of extreme prey abundance, up to 15 eggs may be laid in a nest and a pair may nest up to 3 times in a year. It is best known for its frightening long, harsh and screeching call, uttered mostly by the male, which is repeated at irregular intervals.

CONSERVATION STATUS: Not Globally Threatened (IUCN). In temperate climates, entire populations may starve and perish during severe and snowy winters (the barn owl seems to lack the ability to store fat). The removal of nest sites, typically old buildings, can be mitigated by installing nest boxes. Owls hunting along busy roads are sometimes killed by vehicles, especially in winter. Rodenticides and other pesticides have been implicated in the death of barn owls and can reduce local populations.

11 Ashy-faced Owl, *Tyto glaucops*

DESCRIPTION: A medium-sized owl (33 cm) similar in appearance to the barn owl, except that its facial disk is ashy-gray with an orange-brown rim. Its upper body is yellowish-brown with fine black streaks and dark arrow-shaped spots below.

HABITAT: Found in open areas interspersed with bushes and trees; often occurs near human settlements, as well as in open forests.

NATURAL HISTORY: Apparently many of the habits of this species are similar to that of the barn owl, including its call. It is another little-known *Tyto* species requiring study.

CONSERVATION STATUS: Not Globally Threatened (IUCN). Because it is confined to Caribbean islands, this species is possibly endangered by human settlement.

12 Madagascar Red Owl, *Tyto soumagnei*

DESCRIPTION: The Madagascar red owl is relatively small (28 cm) with a heart-shaped facial disk. Its plumage is ochre-yellow to reddish-ochre above, speckled with fine, blackish spots. Its underparts are ochre with fine blackish spots. Most of the information on this owl has been obtained recently from a single radio-marked bird. Otherwise, very little is known about this species.

HABITAT: Inhabits rainforest between 900 and 1200 m. Also inhabits primary forest with openings, secondary shrub and forest edges. This owl can be found in human-altered habitats such as semi-open forests due to deforestation, plantations, agricultural plots and rice fields.

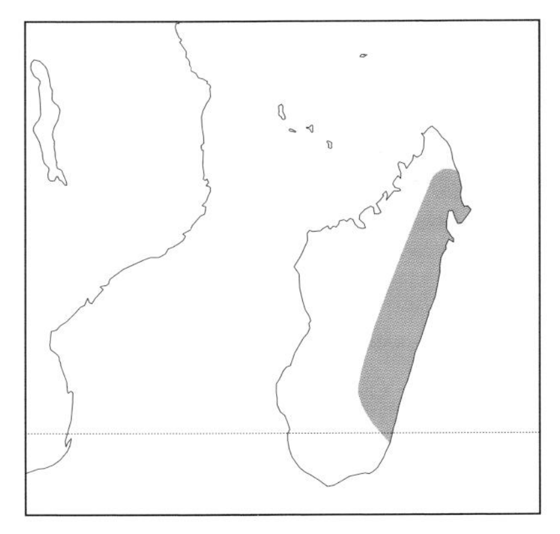

NATURAL HISTORY: Strictly nocturnal, roosts by day in ravines with secondary growth and bananas. It also roosts in dense canopies about 3.7 m above the ground. Prey includes tenrecs (*Microgale*) and black rats (*Rattus rattus*), but other small invertebrates and insects are probably part of its diet. Almost no information is available on its breeding biology, but it apparently uses tree cavities for nesting. Its 1- to 2-second call is barn owl-like, but louder and falling somewhat in pitch.

CONSERVATION STATUS: Endangered (IUCN). Very rare and endangered; it is listed as Endangered by BirdLife International. It is possibly nearing extinction. It was reportedly found in captivity (town of Andapa) and was said to have been captured 482 km north of its previously known distribution. There are only 6 records of this bird since 1934, but perhaps because it may be confused with the barn owl.

13 African Grass Owl, *Tyto capensis*

DESCRIPTION: Similar in appearance to the barn owl but differs biologically. It is a medium-sized owl (38 to 42 cm), weighing between 355 and 520 g. Its upperparts are

sooty brown with white flecks or spots, while its underparts are pale with small dark spots. Its long bare legs are dull gray, its wings are very long, and its tail is relatively short.

HABITAT: Prefers moist grassland and open savannah up to 3200 m. Although moist patches of rank grass, bracken and heath are favored, it is sometimes found on dry grassland. Generally the habitat is characterized by long, dense grass in both types of areas, hence its name.

NATURAL HISTORY: The African grass owl is primarily a permanent resident; a pair may be found in the same patch of grass year-round. They will disperse to areas of food abundance when necessary. African grass owls hunt at night and rarely fly during the day. They often roost in pairs or small groups on the ground in tall grass. They form domed platforms by trampling the surrounding grass, and this area may also be used as a nest site. The African grass owl even makes tunnels in the long grass! Small rodents and mammals, especially vlei

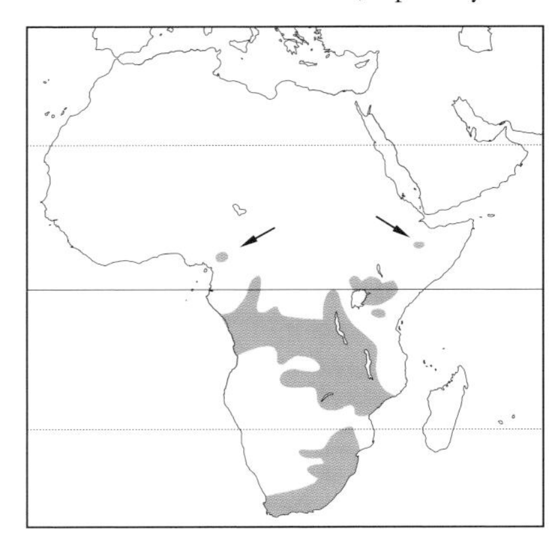

rats, are its preferred prey. These owls usually hunt with a wavering flight low over the ground. They will also hunt from a perch. Bats, small birds, frogs and larger insects are taken from the air or ground. Nests are on the ground in a shallow hollow lined with grass, typically at the end of a tunnel. The female lays 2 to 6 eggs, most commonly 3, in late summer and autumn. She is the sole incubator, and there is usually only 1 brood. Initiation of incubation reports are conflicting. The African grass owl makes clicking sounds in flight while hunting and, rarely, a single harsh screech; otherwise it is normally a quiet bird.

CONSERVATION STATUS: Not Globally Threatened (IUCN). Locally it is common. Details of post-fledging period require study.

14 Eastern Grass Owl, *Tyto longimembris*

DESCRIPTION: The eastern grass owl is a medium-sized (38 to 42 cm), long-legged owl with very long wings. The lower half of the tarsi are unfeathered. Although plumage is variable, upperparts are generally dark brown with yellowish-ochre to white, with black spots. The rim of the pale facial disk is not very pronounced. This species is similar in appearance to the African grass owl.

HABITAT: Found in open tussock grasslands with tall, rank grass, on dry ground or wetland areas.

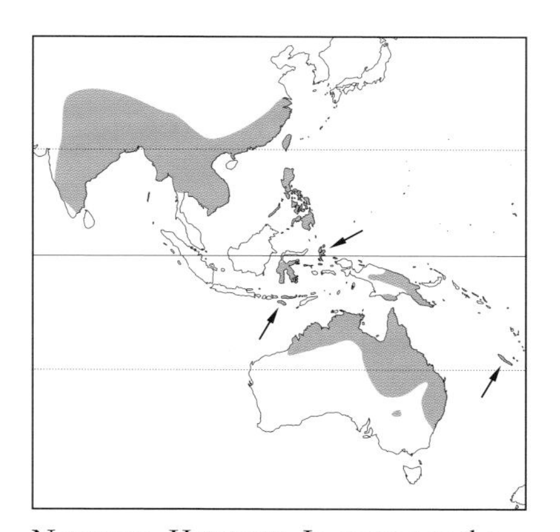

NATURAL HISTORY: In contrast to the African grass owl, the eastern grass owl is nomadic, although coastal populations appear to be sedentary. Where they occupy inland habitats, they are reported to follow "plagues" of long-haired rats and dusky field rats. In years of prey abundance, these owls are known to breed semi-colonially; one area may support several dozen hunting owls. Typically a quiet owl outside the breeding season, but more study may show otherwise. Adapted for life on the ground, this owl nests in a shallow depression under a dense tussock of grass; tunnels in the grass result from walking back and forth to the nest. It lays 3 to 8 eggs, with the female alone incubating, starting from the first egg. Although primarily a nocturnal species, it occasionally flies during the day. Its preferred prey are mice and rats, but it also takes insects, birds and reptiles; it hunts on the wing. The vocalizations of this normally silent owl are poorly studied, but it appears to utter a quiet screeching call.

CONSERVATION STATUS: Not Globally Threatened (IUCN). Common locally; agricultural pesticides (e.g., owls eating rats poisoned in sugar cane) may threaten local populations.

15 Oriental Bay Owl, *Phodilus badius*

DESCRIPTION: A somewhat small owl (20 to 30 cm) with an elongated pale pinkish facial disk that rises on each side of the face and above the head, giving the impression of rounded ear tufts. The bay owl has dark eyes set in a dark brown-colored area within its facial disk. It has short and rounded wings and is dark brown above with yellow and black specks; it is lighter brown to rosy below with narrow buff streaks and black flecks. Its legs are feathered to the toe joints of its strong feet. Barn owls are larger and have a more rounded heart-shaped, white facial disk.

HABITAT: Hunts near water in dense evergreen primary and secondary forests in lowlands, and especially in foothills up 2300 m above sea level in montane forest. It can also be found in patchy forested areas between agricultural lands, including rice farms and fruit-tree orchards.

NATURAL HISTORY: The Oriental bay owl is very vocal during the breeding season, and pairs may frequently sing together. It nests in tree cavities, decayed stumps or between layers of palm leaves. Three to 5 eggs are laid at 2-day intervals. Incubation period and fledgling behavior are not known. These owls are commonly seen roosting by day about 2 m above the ground on branches sheltered by leaves or trunks. Unlike many other owls, it is described as sleeping so soundly during the day that it can be captured by hand from its day roost. An elusive and strictly nocturnal owl, it appears to hunt mainly by sound rather than by sight. The prominent folds of its facial disk can be moved to help focus faint sounds. Its large feet enable it to perch sideways on vegetation while hunting. It eats small rodents, bats, birds, reptiles, amphibians, fish and insects. Highly maneuverable short and rounded wings permit it to fly through thick and tangled vegetation. Its vocalization is a repetitive series of 4 whining whistles: "woo-woo-wee-wee." It may call suddenly during the night when disturbed or alarmed by humans. It has a dramatic bluff-and-escape behavior consisting of a side-to-side rocking motion, followed by a deep bowing so low as to face backward between its own feet! It then slowly shakes its head from side to side before suddenly flipping it up with dark eyes and bill open wide.

CONSERVATION STATUS: Not Globally Threatened (IUCN). Locally threatened by forest harvesting and other habitat loss. Appears to be rare and spread out, resulting in low population estimates. Where necessary, it responds well to captive conditions.

16 Congo Bay Owl, *Phodilus prigoginei*

DESCRIPTION: A small rufous-colored owl with a heart-shaped facial disk. It appears similar in color and size (one specimen—length 24 cm) to the Oriental bay owl, but has relatively small feet and a more compressed bill. Its underparts are russet-cream colored and more heavily spotted. A dark gray wash is visible on the scapulars and wing coverts. The taxonomy of this owl varies in the literature, and some consider it closely related to the barn owls of the genus *Tyto*.

HABITAT: Seemingly restricted to montane forests and grass and bush-covered upper slopes (1830 to 2430 m elevation) of the Itombwe Mountains of eastern Zaire. This and other aspects of its biology require more thorough study.

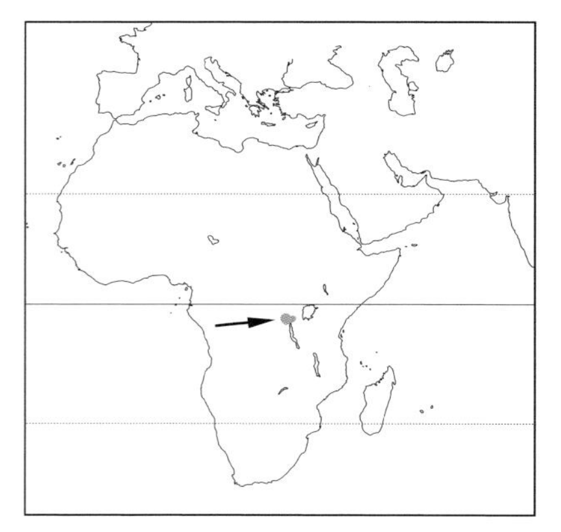

NATURAL HISTORY: All that is known about this species comes from a collected specimen (1951), 2 sight records (1970 and 1989), and 1 suspected female that was trapped, banded and released in 1996. There is no information on its diet, vocalizations, breeding behavior or movements. It is likely strictly nocturnal. It may nest in tree cavities.

CONSERVATION STATUS: Endangered (IUCN). BirdLife International considers it Vulnerable. Its status is perhaps more accurately described as undetermined, hence more research is urgently needed.

17 White-fronted Scops Owl, *Otus sagittatus*

DESCRIPTION: The white-fronted scops owl, a medium-sized owl (25 to 28 cm), is the largest of all the southeastern Asian *Otus* owls. It has an unusually long tail, rounded wings, a prominent white forehead and large ear tufts. Its upper plumage is chestnut-rufous, with small spots, and it is pale rufous below. The throat and breast are gray-brown with small, round black spots, the belly having the largest of these spots.

HABITAT: Occupies primary rainforest or tall secondary forest in hills and lowlands of the Malay Peninsula (at 600 to 700 m). Also occupies degraded swampy forest. It is reported to be a lowland habitat specialist.

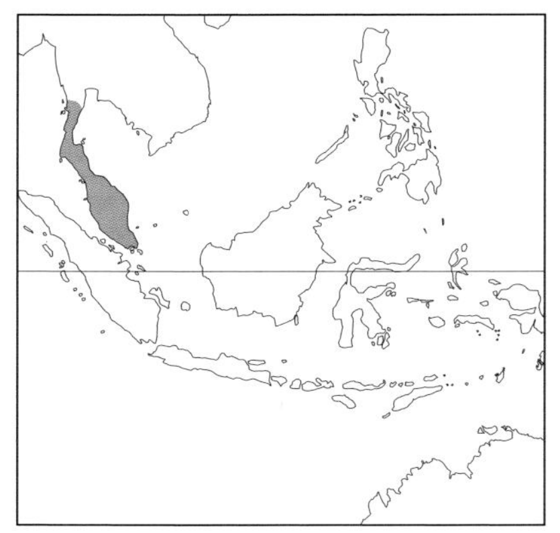

NATURAL HISTORY: An elusive bird, for which little information is available. It is probably a permanent resident. Females nest in tree holes and lay 3 to 4 eggs. From stomach contents, it is known to eat insects, primarily moths. Its call is a terse and hollow "hooo" whistled at well-spaced intervals.

CONSERVATION STATUS: Vulnerable (IUCN). Also designated Vulnerable by BirdLife International. Although rare, it is said to be common in suitable habitat. Widespread deforestation of lowlands within its small range is a threat to this species. Needs study in all areas of its natural history.

18 Reddish Scops Owl, *Otus rufescens*

DESCRIPTION: A small owl (15 to 18 cm) with ear tufts, its rufous-brown plumage is fairly uniform with some black spots. The upperparts are marked with triangular pale fulvous spots. These spots are larger on the mantle and wing coverts; toward the back and rump are shaft-stripes. The underparts are orange-brown with a few small black dots. The eyes are bright chestnut-brown or orange-brown.

HABITAT: Preferred habitat is lowland primary forest. Also inhabits foothill and submontane rainforest (up to 1000 m), and logged primary and secondary forest (most often found in the thicker lower and middle areas of these forests). Has been recorded at 1300 m.

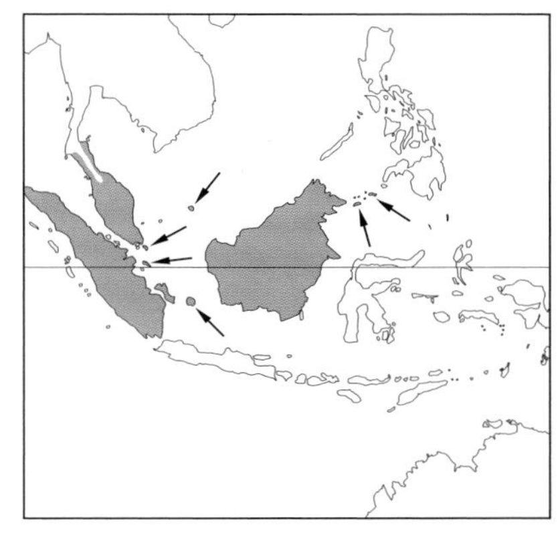

Natural History: Little known. Reportedly calls more often on moonlit nights—its call a drawn-out empty-sounding "heeooh" uttered at well-spaced intervals.

Conservation Status: Lower Risk—Near Threatened (IUCN). Locally, populations are threatened by deforestation and forest fires.

19 Sandy Scops Owl, *Otus icterorhynchus*

Description: A small owl (18 to 20 cm) with pale yellow eyes. It has buff to sandy-colored plumage with blackish-edged whitish spots. There is much individual variation in this species, and lighter and darker morphs occur.

Habitat: Inhabits lowland forest.

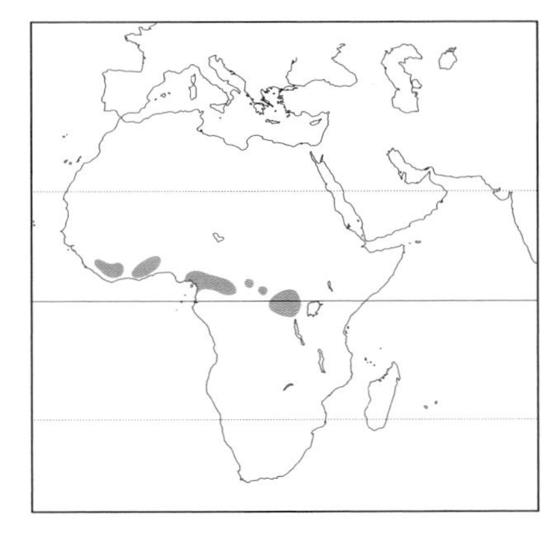

Natural History: There is little information on the natural history of the sandy scops owl. It is reported as a resident, seemingly nocturnal bird that is hard to observe. It utters a drawn out whistle-like "kweeah" spaced about 7 seconds apart. It roosts and nests in holes in trees and eats insects.

Conservation Status: Not Globally Threatened (IUCN). Deforestation may result in this species' becoming endangered.

20 Sokoke Scops Owl, *Otus ireneae*

Description: The species is small (16 to 18 cm). Three morphs are present: gray, rufous and dark. Its underparts are spotted and its facial disk frequently shows concentric rings.

Habitat: Inhabits forests with trees taller than 3 to 4 m. It is found in Tanzania at elevations between 50 and 170 m, and between 200 and 400 m.

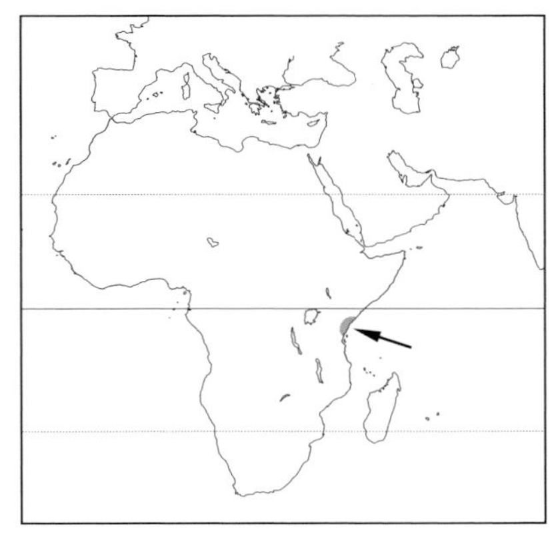

Natural History: A permanent resident. The species is nocturnal and roosts in the lower canopy of trees. Probably eats small insects, and nests in holes in trees. The natural history of this species is not well known. Its call is 5 to 9 high-pitched "goohk" notes.

Conservation Status: Endangered (IUCN). Rare and only recently discovered (1966); its existence was considered threatened as it was thought to be confined to the remnant Sokoke-Arabuko Forest on the coast of Kenya. However, it was subsequently discovered in northeast Tanzania, indicating that its range along the east African coast may be wider than previously thought. Deforestation and bird collecting are known threats to this owl.

21 Andaman Scops Owl, *Otus balli*

Description: Two morphs occur in this small (18 cm) scops owl: brown and rufous. The rufous form is not as boldly patterned as the brown. The crown has blackish buff-white flecks, and the ear tufts are short. The underparts are a blend of light and dark feathers bearing whitish spots with lower black tips that are often arrow-shaped. This species is an island form of the mountain scops owl, which, through isolation, has evolved differences sufficient to earn specific status.

Habitat: The Andaman scops owl appears to be versatile in its habitat use. It is found in semi-open areas clumped with trees, in cultivated areas with trees, close to settlements, and even in gardens with trees and shrubs. It frequently perches on

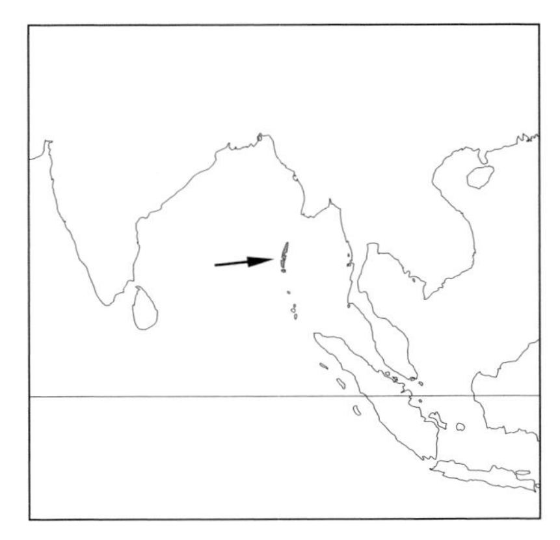

rooftops and will occasionally roost inside buildings.

NATURAL HISTORY: This species is strictly nocturnal, uttering a rolling series of "curro" notes. It eats mainly insects and their larvae, preferring caterpillars. It has been observed capturing caterpillars by sidling along tree branches in a parrot-like fashion. It nests in old tree cavities made by woodpeckers or barbets, where the female lays 2 to 3 pure white eggs on the floor of the hole. The species is not uncommon in its range, but it needs more study.

CONSERVATION STATUS: Lower Risk—Near Threatened (IUCN). Because this species is tolerant of human activity, it is less threatened than other tropical owls.

22 Flores Scops Owl, *Otus alfredi*

DESCRIPTION: This is a relatively small owl (20 cm) with dark rufous-brown plumage and small ear tufts. The outer webs of the scapulars form a white band across the shoulders. The underparts are finely barred with cinnamon coloring.

HABITAT: It is found in humid mountains above 1000 m. Recorded at 1400 m in 1994 in humid montane forest.

NATURAL HISTORY: The Flores scops owl is one of the least studied tropical owls. Information on all aspects of its ecology and biology are simply unknown. Since its vocalizations are completely unknown, many attempts to locate this owl by voice have been unsuccessful.

CONSERVATION STATUS: Endangered (IUCN). Three specimens were collected in 1896. Not until 1994 were 2 more specimens collected; it was recognized as Flores scops owl in 1998. One bird was observed in 1997. Due to its restricted range, it is most likely threatened by habitat loss.

23 Mountain Scops Owl, *Otus spilocephalus*

DESCRIPTION: This small owl (18 to 20 cm) has short, rounded wings and short ear tufts. Its overall plumage is tawny-rufous, ochraceous-buff or dull gray-brown. The plumage is covered with blackish or dark brown spots and is densely vermiculated. There are many races with wide color variation. It resembles the Oriental scops owl but is more spotted, with a rufous face and bare lower legs.

HABITAT: The Mountain scops owl is found predominantly in humid forest. In the temperate northern range of its hilly habitat it prefers dense evergreen forests comprising oak, pine and chestnut. In the south, it inhabits montane tropical rainforest. It is found in the foothills of the Himalayas at 600 to 2700 m, but more often above 1500 m. It nests in dense jungles and gullies between 1000 and 3000 m. After nesting, it descends to warmer lowland valleys. Tall primary forest is preferred, where it forages in the lower areas of dense vegetation.

NATURAL HISTORY: Primarily a permanent resident, but in winter the northern subspecies will descend to lower elevations. The species is nocturnal, with activity starting during and after dusk. It will respond to playback. Eats mainly Coleoptera, small moths and other insects (mantids and cicadas), which it probably hawks in the air. Is also thought to take small rodents, birds and lizards. Nests in natural cavities in dead trees (1.5 to 7.5 m above the ground), but will also use old woodpecker or barbet holes. Nests are unlined, and 3 to 4 eggs (sometimes 2 to 5) are laid. Vocalizations are metallic-sounding and have been compared to a hammer hitting an anvil, or mule bells, giving this bird the name "Himalayan bell bird."

CONSERVATION STATUS: Not Globally Threatened (IUCN). Locally common, but threatened in some areas due to habitat loss.

24 Rajah Scops Owl, *Otus brookii*

DESCRIPTION: The Rajah scops owl is a rare and attractive owl with striking orange-yellow eyes. It is medium-sized (22 to 25 cm) with long ear tufts. Its rufous to brownish plumage is heavily speckled with black and brown. The hindneck is broadly collared, with a smaller collar on the nape. Its lighter underparts are also heavily marked with rufous vermiculations and dark brown shaft markings. It gets its name from the Rajah Brooke, one of the White Rajahs in Sarawak.

HABITAT: Inhabits tropical montane rainforest to cloud forest. It seems to prefer the center story of dense forests, between 900 and 2500 m, but usually 1200 to 2400 m.

NATURAL HISTORY: The habits of this bird are unknown; only 12 specimens have been collected. Stomach contents revealed insects (mostly beetles, grasshoppers and moths) and a frog had been taken as prey. One subspecies is described as having an clear and explosive 2-hoot call, repeated at 7- to 10-second intervals.

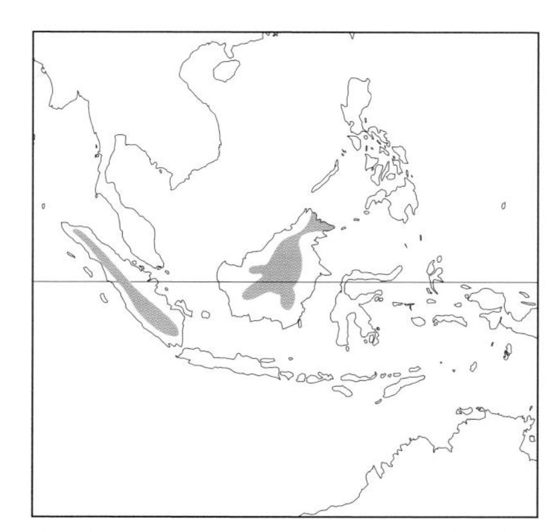

CONSERVATION STATUS: Not Globally Threatened (IUCN). Serious threats include deforestion in mountain regions. The Bornean form of this owl appears to be a separate species, but this is unsure as only 2 specimens have been collected; this population may already even be extinct.

25 Javan Scops Owl, *Otus angelinae*

DESCRIPTION: A relatively small owl (16 to 18 cm) with orange-yellow eyes. It has bold white "eyebrows" that extend up onto prominent ear tufts. Its facial disk is uniformly rufous. Plumage is dark rufous-brown, spotted with light and dark rufous coloring and light vermiculations. The underparts are light rufous, and along the sides of the breast there are conspicuous black herring-bone-like markings. The tarsus is fully feathered, even over the toes.

HABITAT: Montane primary forest and edges; mainly the middle and lower storys (1500 to 1600 m).

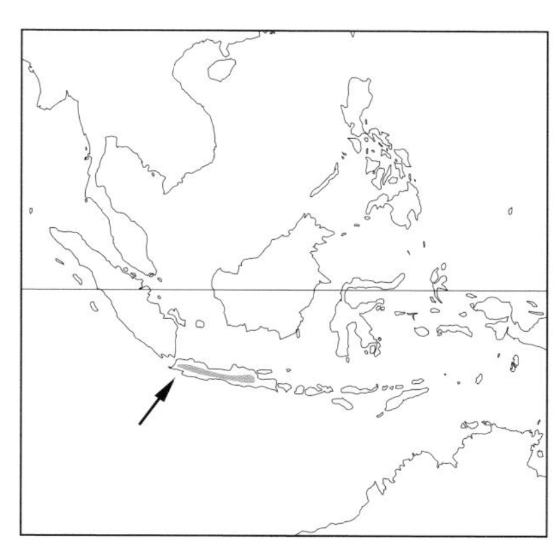

NATURAL HISTORY: Strictly a nocturnal bird, it roosts in low dense vegetation, usually on bare branches in an exposed position. It has also been observed roosting on small epiphytic bird's-nest ferns. When detected it assumes a tall-thin posture. Its diet comprises mainly larger insects such as beetles, grasshoppers, praying mantids, crickets and moths; sometimes it captures small lizards and snakes. Its prey is apparently captured with its feet from a branch, stem or leaf, or the ground, but probably not from the air. It is assumed to nest in tree holes and possibly in epiphytic bird's-nest ferns. Limited evidence suggests that full clutch size may only be 2 eggs. Apparently this species seldoms vocalizes. There seems to be no territorial call, and it vocalizes only under extreme stress or agitation. When stressed, it emits an explosive 2-note hoot, described as "poo-poo"!

CONSERVATION STATUS: Vulnerable (IUCN). Uncertain status due to its elusiveness; furthermore, because it is endemic to the montane forest of Java, it is threatened by habitat loss through deforestation.

26 Mentawai Scops Owl, *Otus mentawi*

DESCRIPTION: The Mentawai scops owl is the only scops owl found on the Mentawai Islands (off the west coast of Sumatra). It is a small owl (22 cm) with yellow or brown eyes. Its upperparts are dark blackish brown while the underparts are paler, peppered with small black streaks along the feather shaft. It has whitish "eyebrows" and ear tufts mottled with dark brown. There are 2 color morphs: rufous brown and blackish-brown.

HABITAT: Found in lowland rainforest and human settlements.

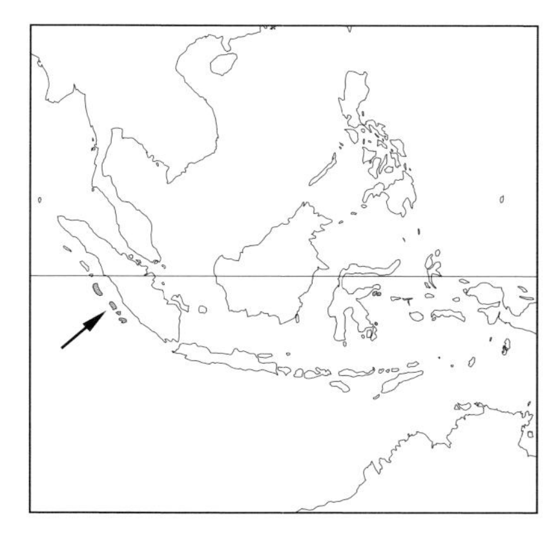

NATURAL HISTORY: A resident, nocturnal owl that responds to playback. Populations within the forest are elusive; those in populated areas are bolder and will perch on banana leaves or coconut fronds. Eats insects; no other information available on the habits of this owl. The male's call has been described as 3 to 4 rough barking notes in a row: "how-how-how."

CONSERVATION STATUS: Lower Risk—Near Threatened (IUCN). Locally common.

27 Indian Scops Owl, *Otus bakkamoena*

DESCRIPTION: The Indian scops owl is one of the largest *Otus* species in Asia (23 to 25 cm). It has large ear tufts and dark brown eyes. It is a sandy-brown owl with dark brown and black spots and a somewhat mottled pattern. Two basic color phases have been described. The gray-brown color phase has gray-buff underparts, whereas the rufous morph is more rufous-buff below. Their underparts also have a complex pattern created by shaft streaks, fine wavy transverse bars and vermiculations. This species has considerable geographic variation, resulting in much debate among owl taxonomists as to how many subspecies should be recognized. As early as 1874,

scientists proposed that it was closely related to North American *Otus* owls. Some authorities have suggested, based on the owl's physical characteristics, that many of these species should be combined into one. Most experts agree, however, that the currently recognized species are genetically distinct, a view supported by the owls' different vocalizations.

HABITAT: The Indian scops owl can survive in a variety of habitat types, including forests, desert vegetation, well-treed gardens and mango orchards in and around villages. At least some dense vegetation must be present, as it is known to seek out well-hidden day roosts. It has been found in lowland elevations and up to 2200 m.

NATURAL HISTORY: Its 3 to 4 whitish round eggs are laid in tree cavities. The ecological relationship of this owl with the smaller scops owls, which share its habitat and range, needs further study. Its diet is primarily insects, e.g., grasshoppers and beetles. Lizards, mice and small birds are sometimes taken. The call of the male (a regularly spaced "What? What? What?...") distinguishes it from the similar collared and Japanese scops owls. It is a resident, but may be partly migratory, especially where cooler winter temperatures may reduce insect populations.

CONSERVATION STATUS: Not Globally Threatened (IUCN). Widespread and locally common. Its ability to use landscapes heavily influenced by humans suggests that it is an adaptive generalist predator able to colonize newly created habitats.

28 Collared Scops Owl, *Otus lettia*

DESCRIPTION: This medium-sized species (23 to 25 cm) has 2 color phases. The rufous phase is similar to the Japanese scops owl, with white crescents between the eyes. It is overall browner and darker than the buff-gray morph. Its eyes are more golden- to chestnut-brown than the normally dark brown eyes of the buff-gray color phase. Both color phases have relatively short ear tufts compared to the Sunda scops owl.

HABITAT: This resourceful owl is able to use a variety of human altered-habitats such as managed forest and open country with patchy vegetation cover. It is found in the plains and foothills up to 2400 m in elevation.

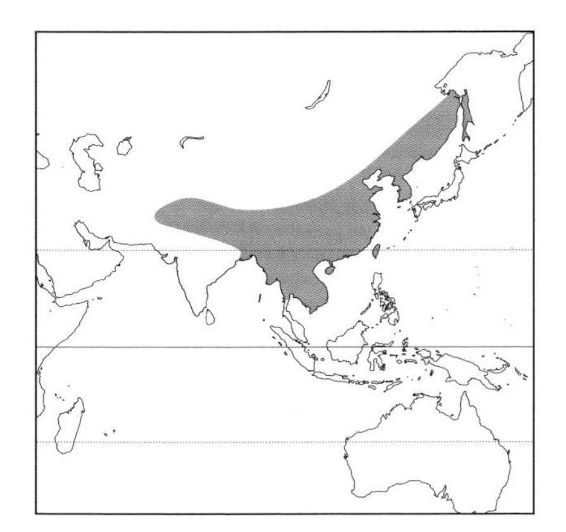

NATURAL HISTORY: Its diet is more consistently variable than the Sunda and Indian scops owls. It more regularly takes lizards, mice and small birds. The female lays 3 to 4, occasionally 5, eggs in tree cavities at heights from 2 to 5 m. Like most *Otus* owls, it is primarily detected at night when it is active and vocalizing. Males utter a single note "kwuo" at 15 to 20 second intervals, often in concert with the female's "kwiau" during pair formation. Day roosts are well concealed. It migrates south in winter to southern India and the Malay Peninsula.

CONSERVATION STATUS: Not Globally Threatened (IUCN). The collared scops owl has a patchy distribution in its widespread range. Where found, it is abundant.

29 Sunda Scops Owl, *Otus lempiji*

DESCRIPTION: The Sunda scops owl has plumage that varies considerably among individuals of some populations. Gray-brown, gray-buff and rufous morphs are present; these color patterns may also intergrade. Overall it is a medium-sized owl (20 to 23 cm) with sandy-brown coloration, rounded wings, large ear tufts and an ill-defined collar on its hind neck. Its eyes most commonly are deep brown, but some individuals have orange or bright yellow eyes. In Sumatran, this owl has been called "kuas cirit ayan," which means "fowl's excrement owl," so named because it hunts insects attracted to cow dung or poultry droppings!

HABITAT: This owl inhabits a wide variety of areas. It is commonly found in middle and lower stories of forest and secondary growth. It is also found in and attracted to areas of human habitation, such as plantations, treed gardens, villages and even larger towns containing suitable trees for cover, roosting and breeding. It seems to avoid undisturbed primary rainforest.

NATURAL HISTORY: A resident, nocturnal species, this owl is not often seen during the day. It roosts in densely foliaged trees, palms or bamboo groves. Prefers to call from a concealed area in densely foliaged trees. Both sexes share a musical, questioning call (although the female's is higher-pitched), and will sing together during the breeding season. Their calls can be heard year-round. They eat chiefly insects, including black dung beetles, grasshoppers, crickets, mantids, moths, cockroaches, and sometimes small birds. Usually 2, rarely 3, eggs are laid upon a nest built of vegetable fibers within natural holes in hollow trees. Nests may also be located between the straight, dead leaves of sugar, oil or coconut palms.

CONSERVATION STATUS: Not Globally Threatened (IUCN). Very common.

30 Japanese Scops Owl, *Otus semitorques*

DESCRIPTION: The Japanese scops owl (21 to 26 cm) has strikingly bright red or orange-yellow eyes. Its plumage is sandy, grayish-brown mottled with dark brown and black. The wings are pointed and the ear tufts are long. Its whitish "eyebrows" extend almost to the tip of the ear tufts. It has a gray collar on the hindneck and nape.

HABITAT: Found in treed lowland areas and mountainous habitats. Inhabits forests and frequents villages and suburbs, especially in winter.

NATURAL HISTORY: The species is both resident and migratory. It is strictly nocturnal, and during the day can be found roosting in tree cavities or thick foliage. The main call of the male is a low, sad-sounding "whoop." It nests in natural tree cavities, abandoned raptor nests and inside buildings. Clutch size is reportedly 4 eggs, but this and other aspects of its biology remain poorly studied.

CONSERVATION STATUS: Not Globally Threatened (IUCN). Widespread and locally common.

31 Palawan Scops Owl, *Otus fulginosus*

DESCRIPTION: The overall plumage of this small owl (19 to 20 cm) is a rich brown. A pale-colored collar is prominent on its hindneck. The face and underparts are rufous-brown, with the underparts streaked with dark brown. The eyes are orange-brown.

HABITAT: The species inhabits tropical rainforest and secondary forests in lowland areas. It reportedly can adapt to mixed cultivation and plantations.

NATURAL HISTORY: Its nesting habits are largely unknown. The male's call has been described as a deep croak, "krarr-kroarr," compared to a dry branch being cut with a handsaw. The Palawan scops owl preys on insects and probably nests in holes in trees.

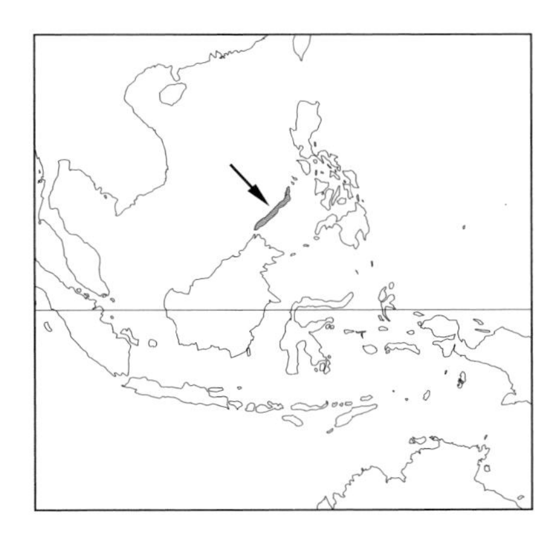

CONSERVATION STATUS: Lower Risk—Near Threatened (IUCN). Thought to be rare and endangered. Listed as Vulnerable by BirdLife International. Increased familiarity with its call has yielded new information that suggests it is more common and widespread than earlier believed.

32 Philippine Scops Owl, *Otus megalotus*

DESCRIPTION: For a scops owl , this species is relatively large (20 to 28 cm). The gray and rufous morphs are similar, with the latter more rufous above and below. The crown is blackish, and whitish "eyebrows" extend to the long ear tufts.

Both forms are heavily mottled with blackish vermiculations. The throat has a broad whitish-buff ruff of dark-tipped feathers. Dark arrowhead shaft streaks cover the underparts (with some cross markings). Discovered in 1893, it is remarkable that an obscure species such as this was ever found.

HABITAT: The species inhabits both tropical forests and secondary woodland. Seems to prefer dense forest at elevations from 300 to 1200 m where it is usually found. Quiet forest roads seem to be favored hunting grounds.

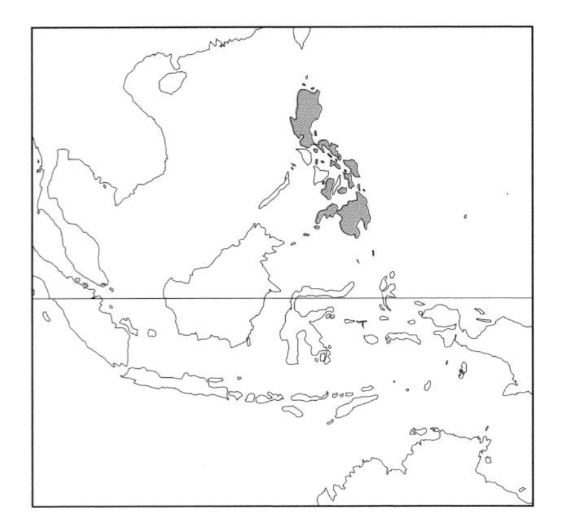

NATURAL HISTORY: This nocturnal species is nonmigratory. The female lays 3 to 4 eggs, most likely in tree cavities. The presence of a family group (two adults and 1 young) near the roots of upturned, dead trees suggested that it may have nested in tree roots. The breeding biology of this species needs more study. Its call has been described as peculiar and powerful, "oik-oik-oik-oohk," uttered soon after dusk.

CONSERVATION STATUS: Not Globally Threatened (IUCN). Some authors have described it as endangered but it is presumed by others to be locally common.

33 Wallace's Scops Owl, *Otus silvicola*

DESCRIPTION: A uniformly dull, pale owl with barring. Its upperparts are marked with black herringbone-like shaft streaks and reddish-brown vermiculations. Eyes are a dull orange-yellow. It is relatively large for a scops owl (23 to 27 cm) and has long ear tufts and powerful feet.

HABITAT: It can be found in tropical lowland forest (bamboo thickets) to submontane forests. Additional areas include secondary woodland, farms and towns. Has been found up to 1600 m in virgin montane rainforest. It occupies lowland areas up to elevations of 2000 m.

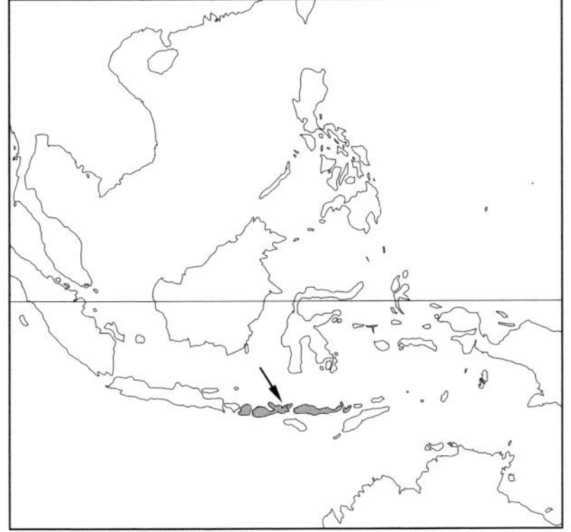

NATURAL HISTORY: This bird is difficult to observe, being strictly nocturnal and perching high in trees and under cover. The species is resident on the islands of Sumbawa and Flores. It preys mainly upon insects. A monotone, rough call consisting of a series of "rrow" notes repeated at 1-second intervals is its main call. More research is needed on this little-studied species.

CONSERVATION STATUS: Not Globally Threatened (IUCN). Wallace's scops owl's ability to tolerate and inhabit areas near and in human settlements may be beneficial to its future survival. However, deforestation may pose a threat to this species. While its status is somewhat uncertain, it is generally considered rare.

34 Mindanao Scops Owl, *Otus mirus*

DESCRIPTION: This small (19 cm) dark owl has rounded, short wings. Its facial disk is pale with light and dark concentric rings. The whitish "eyebrows" continue to the tips of small ear tufts. The dark plumage is heavily spotted with black, with rufous feather bases. Its underparts are noticeably pale and marked with distinct black streaks and crossbars.

HABITAT: The Mindanao scops owl is endemic to Mindanao, Philippines, where it inhabits higher elevation montane rainforest. It may be confined to elevations above 650 m, but more extensive surveys are needed.

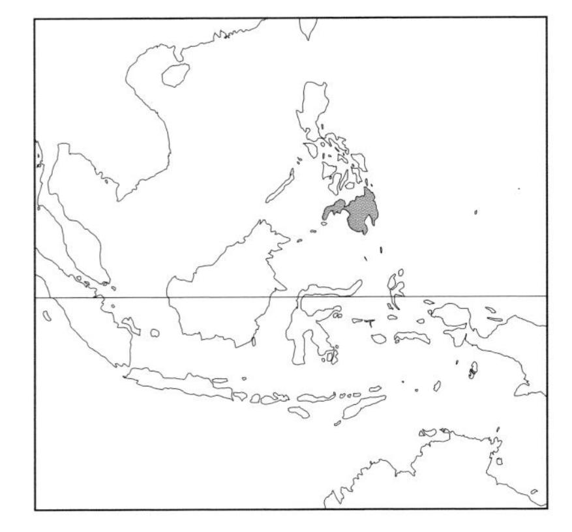

NATURAL HISTORY: Details of this bird's natural history are unknown. A nocturnal, resident species, it probably eats chiefly insects and other arthropods. Its call is described as a distinct whistle-like 10 to 15 second-long series of "pli-piooh" notes.

CONSERVATION STATUS: Lower Risk—Near Threatened (IUCN). This species seems to be rare within its small range. The ongoing loss of forest habitat may be a serious threat to this owl, which is listed as Vulnerable by BirdLife International.

35 Luzon Scops Owl, *Otus longicornis*

DESCRIPTION: A small owl (19 to 21 cm) with yellow eyes, whitish "eyebrows" and long ear tufts. Its upperparts are pale rufous with dark brown and black mottling and speckles. The underparts are white and rufous in color, and are heavily marked with black. The breast is rufous with the belly being whiter.

HABITAT: Endemic to Luzon, Philippines, where it inhabits closed-canopy forests

(moist and pine forest) extending up to at least 2200 m elevation.

NATURAL HISTORY: Its 2 to 3 eggs are laid in cavities in old trees. The species is nocturnal and eats mainly insects. The male's call is a descending, extended string of "wheehuw" notes. The call accelerates into a series of 2-note whistles in response to the call of neighboring or other owls intruding into its territory. More information is needed on this species' biology.

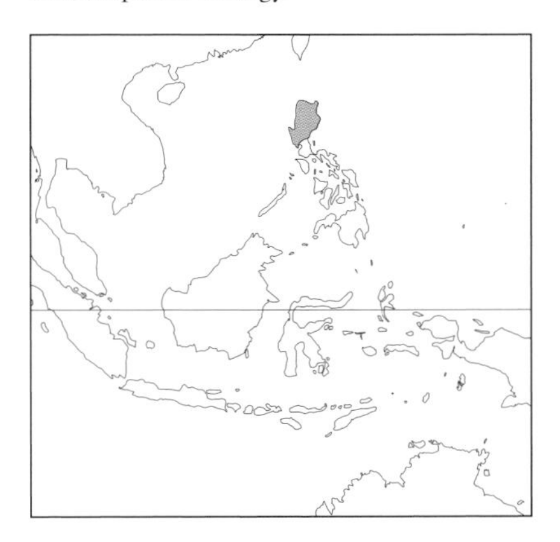

CONSERVATION STATUS: Lower Risk—Near Threatened (IUCN). The species is considered locally common, but has become rare in some areas due to habitat loss, especially in the lowlands. Listed as Vulnerable by BirdLife International.

36 Mindoro Scops Owl, *Otus mindorensis*

DESCRIPTION: The Mindoro scops owl is a small (18.5 cm), buff-brown owl with yellow eyes. Both upper and underparts are streaked, with more barring below. Its ear tufts are medium length. It has a buff, narrow, but ill-defined collar on the back of its neck.

HABITAT: It is endemic to Mindoro, Philippines, where it inhabits closed-canopy forest throughout the mountains in the centre of the island, usually at elevations above 870 m. More recently it has been commonly observed at elevations of 700 to 900 m in highly patchy lowland second-growth forest.

NATURAL HISTORY: A nocturnal, resident species that roosts in dense foliage or natural cavities; most likely nests in cavities in old trees. Preys mainly upon insects, as do most scops owls. Little else is known about this species. Its main call is a series of whistle-like "whoow" notes.

CONSERVATION STATUS: Lower Risk—Near Threatened (IUCN). Regarded as locally common, especially above 1000 m, but the species is threatened by habitat loss due to logging. It has a restricted range and is considered Vulnerable by BirdLife International.

37 Pallid Scops Owl, *Otus brucei*

DESCRIPTION: The pallid scops owl resembles the common scops owl in size (18 to 21 cm) and appearance. However, it is paler and more uniformly sandy-colored, lacks horizontal dark streaks and vermiculations. and the lesser wing coverts lack russet color: Also, its back and wing

coverts are distinctly streaked. Its plumage has been described as bark-like, and its color varies somewhat from sandy to sandy-gray.

HABITAT: This species is typically found in riverine forests of poplar, willow and tamarisk. In addition, it inhabits citrus and date palm plantations, orchards, vineyards, parks and gardens within villages and towns, deserts, roadside groves and cultivated farmlands and saxaul forest vegetation in semi-deserts. It is also found in arid foothills at elevations of 1500 m in Pakistan, and close to steep cliffs and rocky gorges scattered with small trees. Overall, it prefers lowlands and avoids mountain habitats in west-central Asia. During winter it can be found in cultivated and riverine woodlands.

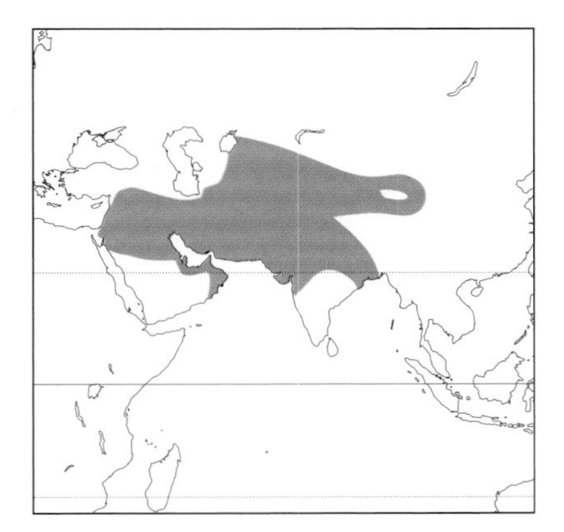

NATURAL HISTORY: The Pallid scops owl is resident in southeast Arabia and Iran. An owl that resides on the island of Socotra near Somalia may belong to this species, but is thought by some to be a unique species. Elsewhere, populations are mostly migratory, with wintering ranges including Pakistan, northwest India south to Bombay, Levant, northeast Egypt and Arabia, and probably throughout the Middle East. Apparently, migration routes do not exceed 2500 to 4000 km. The extent of these movements is unclear in some areas. Females usually lay 4 to 5 eggs, sometimes 2 to 6. Nests in cavities in trees 3 to 6.5 m above the ground, in holes made by woodpeckers, holes in river banks or stone walls, beneath roofs of houses, inside saxaoul bushes in semi-desert regions, and sometimes in old magpie nests. This species nests earlier than the common scops owl. The diet is mainly insects caught in flight, including beetles, mole crickets, grasshoppers, locusts and moths. Occasionally, it takes lizards, small birds and bats, and small rodents such as house mice, shrews and gerbils. Although mostly nocturnal, this bird has been observed hunting in early morning and late afternoon. A long series of dove-like "whookh" notes make up the male's call. Barking and rattling are uttered when it is alarmed.

Conservation Status: Not Globally Threatened (IUCN). While its general status is not well known, it is considered by some as fairly common over most of its range. Pesticide use may pose a threat in some areas.

38 African Scops Owl, *Otus senegalensis*

Description: The African scops owl is a very small owl (16 to 19 cm) with a slender build, small feet and narrow toes. Its head is angular with prominent ear tufts, which may be laid back flat, raised to resemble blunt lobes, or may appear as pointed horns. The facial disk is edged with a narrow black line. Its finely patterned plumage has delicate pencilling and bars of buff against gray and brown feathers; underparts are broadly streaked with deep brown across the center. These characteristics render the bird highly cryptic against favored trees. Overall, it is similar to the common scops owl, but has heavier markings and appears less "bark-like." There are 2 distinct color morphs, gray and rufous, each with much individual variation.

Habitat: The species inhabits dry, semi-open woodland without dense ground cover. It can be found in the bushveld of southern Africa, but avoids the treeless habitat, grassy upland plateau and mountains. It is absent from the dense forests of western Africa, as well as the arid, treeless habitat of the southwest and northeast. It prefers acacias and mopane (types of tree species) woodlands that characteristically contain old gnarled trees with many holes and cracked bark that provide roosting and nesting opportunities. In addition, it can be found in park-like habitats and gardens with mature trees. It is found at elevations from sea level to 2000 m.

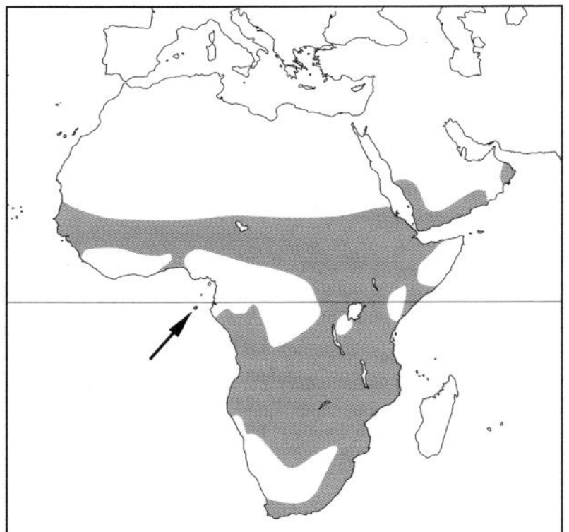

Natural History: The African scops owl is normally described as resident, but there is some evidence that owls from south of Limpopo River may travel north to Zimbabwe in winter. It is nocturnal, and roosts either in dense foliage or creepers, in a cavity or cleft in a gnarled tree, or against a branch or tree trunk. Its cryptic markings render it virtually impossible to detect except by its distinct call, a regularly repeated "kruup." Occasionally, pairs will roost together or within close proximity. Roost sites are frequently used regularly over a long time and droppings, pellets and bits of food accumulate under them. Often it will roost only 1 m above the ground. When disturbed, it adopts a "tall-thin" appearance, becoming slim and erect with eyes half-closed and ear tufts raised. The edges of the facial disk then become scarcely visible. Some have been so reluctant to fly that they allowed themselves to be picked up by hand. Not a powerful hunter, the species is basically insectivorous, preying upon beetles, mantids, crickets, grasshoppers, woodlice, moths and cockroaches. Will sometimes take spiders, frogs, scorpions, lizards, small rodents and passerine birds. Generally, the species hunts from the ground, but also from low rocks. Sometimes it will forage for prey along dead tree trunks, peering into cracks and pulling at bark. Insects are also captured in flight. The species is monogamous and solitary, but pairs may have close territories. Its 2 to 4 eggs are laid in natural cavities 2 to 10 m above the ground. It will also nest in old woodpecker holes, cavities in broken treetops or the end of a broken branch, and nest boxes. There is one record of a pair breeding in the side of an old stick nest of a lappet-faced vulture. Other birds, lizards, monkeys and snakes frequently prey on owlets in the nest.

Conservation Status: Not Globally Threatened (IUCN). Throughout its range the species is generally common; locally it is very common. More social than the common scops owl, as many as 12 individuals have been heard calling from the same location in wooded savannah habitat. While populations do not seem to be declining, and it will likely remain widespread throughout its range, better methods to assess populations and trends are needed. Possibly threatened locally by pesticide use.

39 Eurasian Scops Owl, *Otus scops*

Description: This small owl (16 to 21 cm) varies considerably in color in different parts of its range. Furthermore, it has distinct gray-brown and rufous-brown color phases or morphs. It can be overall silver-gray, gray-brown, rufous-brown or almost blackish with streaks and vermiculations. Its underparts are generally paler with shaft streaks, barring and white patches. The flight feathers and tail are barred dark and light. It has small ear tufts that are not always visible, especially when the plumage is relaxed. Its eyes are yellow, the bill and unfeathered toes are gray. Ecological differences, call characteristics and DNA evidence suggest that this species is distinct from its outwardly similar African and East Asian relatives.

Habitat: It occurs in open country and cultivated areas with scattered trees or small woods and avoids dense forests. Mountainous regions in warmer climates are also used, along with parks, Mediterranean scrub and rocky terrain.

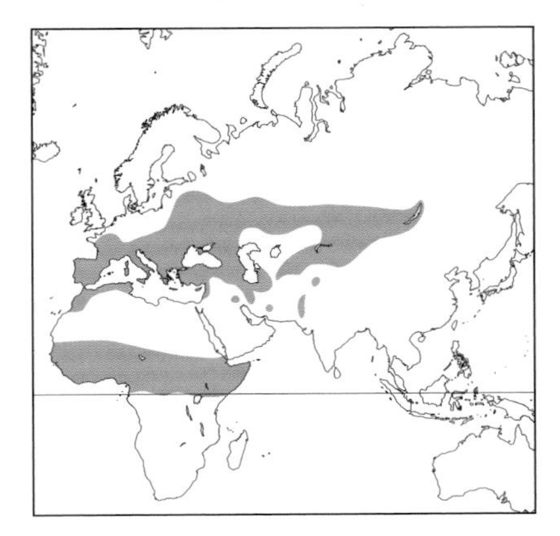

Natural History: This owl is active at night, and roosts by day in dense foliage or in holes in old trees, rocks or walls. Its eyes are kept almost fully shut, making it hard to find. Its call—starting at dusk—is a low, short whistled "tyeu" or "kyoot" repeated every 3 seconds for hours. Clutch size is 3 to 5, rarely 2 to 6, and incubation lasts about 25 days. Nest sites include cavities in trees, rocks, walls, steep ditch or sand pit-banks, and under roofs. Its diet is mainly insects, including crickets, moths, beetles, grasshoppers and caterpillars. It appears to only rarely take small mammals, lizards and birds. It is mainly migratory, overwintering in the treed savannas of Africa. It uses fat reserves to travel up to 8000 km in 2 months.

Conservation Status: Not Globally Threatened (IUCN). It is thought to be most common in the Mediterranean. Elsewhere, population decreases have been linked to agricultural pesticide use and habitat loss, while declines in some areas are attributed to increases in the tawny owl, a predator of smaller owls. An unknown number are shot for food during migration.

40 Oriental Scops Owl, *Otus sunia*

Description: A small owl (16 to 20 cm) with pale, streaked plumage, rounded wings and a short tail. Although similar in appearance to the common scops owl, it differs in that its underparts are more boldly streaked, its plumage is less bark-like, and its ear tufts are longer. Both gray-brown and rufous morphs are present.

Habitat: The species prefers deciduous and mixed-forests that are open to semi-open, particularly along river valley bottoms. Also found in evergreen forests, riparian woodland, savannahs with scattered trees, parks, orchards and tropical gardens in rural villages. Overwintering populations

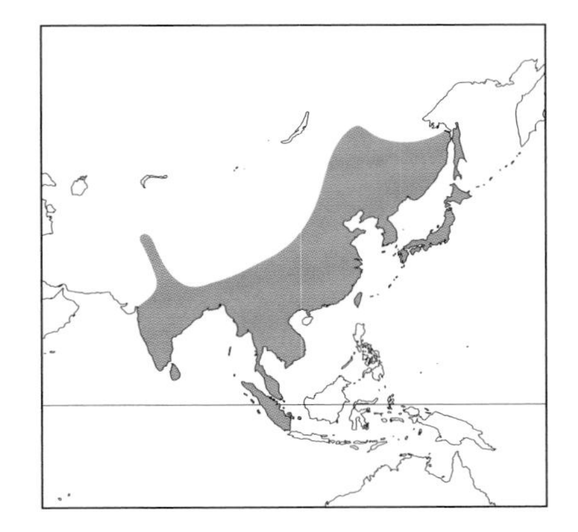

in India use deciduous and evergreen forests, orchards, gardens and cultivated land with groves of trees in and near settlements with dense foliage for roosting. In Sri Lanka it has been observed near streetlights in urban areas. In Japan taller trees near temples are apparently preferred areas. The species is usually found in lowlands and foothills to 1500 m, but up to 2300 m in the Himalayas.

Natural History: The southern populations of the Oriental scops owl are resident: birds from eastern Siberia and Japan migrate to India and China in winter. This nocturnal species can be found roosting in holes or against tree trunks, or in dense foliage. If spotted, it assumes a "tall-thin" posture, which involves a vertical stretch, compressed plumage, erect ear tufts and eyes almost closed. The 3 to 6 eggs, usually 3 to 4, are laid directly onto the floor in a hole in a tree or a stone wall, and also in nest boxes. Nest sites are frequently high above the ground. Although there is little information about its breeding behavior, its general biology is thought to differ from other scops owls. A wide variety of prey is taken, insects and spiders being preferred, including moths, beetles, cicadas, locusts, grasshoppers, mole crickets, caterpillars, small mice, shrews, and even rats and birds. When spiders are locally abundant in webs in bushes, large numbers may be taken. This owl hunts from perches and swoops down to catch prey on the ground, in flight or from the tree canopy. Preferred hunting habitat is along forest edges, in open country or in glades. The diagnostic call of the species is a monotonous 3-note song "kurook-took" or "wuk-tuk-tab," described as sounding similar to a pendulum or a pump engine. It can easily be heard over several hundred meters. Birds in Thailand and Sri Lanka have been reported to have a 4-note "toik" call, resembling the rhythm of "Here comes the bride."

CONSERVATION STATUS: Not Globally Threatened (IUCN). The species varies in abundance regionally. For example, it is scarce and very local in Pakistan, scarce but widespread in Sri Lanka, fairly common in most of the Indian subcontinent, and uncommon in Thailand and Japan; in southeastern Siberia it is reported as the most common owl. Installing nest boxes may increase local populations.

41 Flammulated Owl, *Otus flammeolus*

DESCRIPTION: A small owl (15 to 17 cm) with dark brown eyes and short, typically flattened ear tufts atop a square head. The facial disk is incomplete. The horn-colored bill has a lighter tip. It has a short tail and long, pointed gray-brown wings that are well suited for its long-distance migrations. Gray feathers with black shaft streaks and crossbars cover its body, with varying washes of rufescent resulting in either reddish or grayish mottled plumage.

HABITAT: Occurs predominantly in mature Ponderosa pine, but also dry Douglas fir at the northern reaches of its range. Food, shelter from predators and nest cavities are more abundant in mature forests. This owl appears to prefer forests with a variety of tree sizes with grass and shrubs growing beneath.

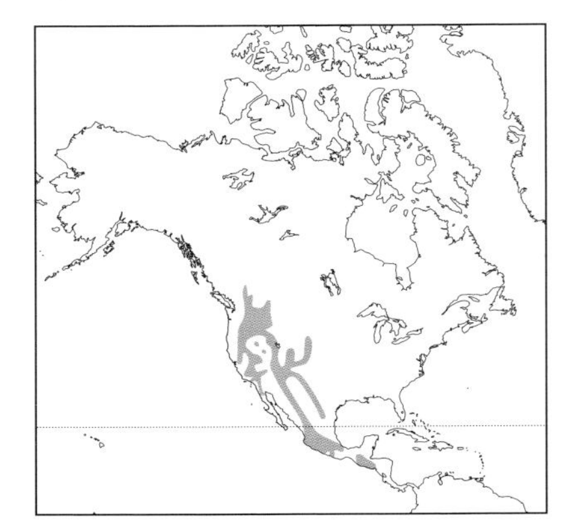

NATURAL HISTORY: The gray and red color phases of the flammulated owl may be adaptations related to the color of tree bark. In southern areas, where pine is most abundant, the reddish phase is more common. Where Douglas fir is dominant, the grayish phase occurs more often. A specialized throat structure produces hoarse, low frequency notes suggestive of a much larger owl. The main call is a flat, terse, ventriloquistic hoot, often repeated for hours on end early in the breeding season. It lays 2 to 3 eggs, rarely 4, in old cavities in trees built by pileated woodpeckers or northern flickers. Its breeding populations have a clumped distribution. It may be semi-colonial; in which case large areas of apparently suitable habitat remain unoccupied between nesting areas. Has a low clutch size and eats mainly insects, particularly moths and beetles. Northern populations migrate long distances at night to areas with ample prey populations in winter. The migratory status of populations in Mexico is uncertain and needs study. Almost no information is available on its winter diet, habitat or distribution.

CONSERVATION STATUS: Not Globally Threatened (IUCN). The flammulated owl was formerly considered rare until surveys used imitations of its call to solicit responses from territorial males. It is now considered the most common raptor of the montane pine forests of the western United States and Mexico. Nonetheless, it is thought to have been more common before widespread forest harvesting within its range. It has a low reproductive rate for a small species, and is restricted to commercially valuable forests.

42 Moluccan Scops Owl, *Otus magicus*

DESCRIPTION: This is a small to medium-sized scops owl (20 to 23 cm) that occurs in 3 morphs: buff gray-brown, rufous and yellow-brown. The upperparts are variably streaked and darkly barred. The crown is also darkly streaked. The underparts are much lighter with dark streaks and horizontal vermiculations. The belly is paler than the breast. The facial disk varies from whitish to brownish with a variable rufous to grayish tinge and an indistinct dark rim. The flight feathers are prominently barred with a whitish color.

HABITAT: Predominantly forested areas, often coastal swamp forest, secondary growth, heavily wooded limestone cliffs and plantations. Prefers large trees in

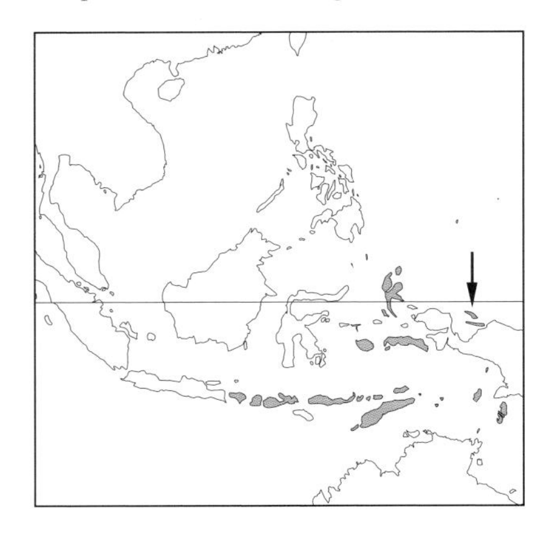

villages and also farming habitat. Usually found up to 900 m, locally to 1500 m.

NATURAL HISTORY: A resident species. Like other scops owls, it is nocturnal, hiding during the day. Nests are probably located in tree cavities; there is no other information on breeding biology. Eats primarily insects and other arthropods, and occasionally small vertebrates. Its call is a series of rough, deep frog-like "kwoark" notes spaced several seconds apart.

CONSERVATION STATUS: Not Globally Threatened (IUCN). In many areas the species is locally common, but elsewhere its status is unknown. Some forms observed in heavily degraded habitats. Forest destruction may be a threat over time.

43 Mantanani Scops Owl, *Otus mantananensis*

DESCRIPTION: This medium-sized owl (18 to 20 cm) has a pale facial disk with a distinct dark rim, a dark patch over its eyes and prominent dark-spotted ear tufts. The upperparts are dark brown and mottled with black, the underparts being whitish with black streaks and broken cross-lines. There are brownish and rufous morphs.

HABITAT: The species can be found in forest and wooded areas, lowlands, foothills, coconut groves and casuarina trees. Hunting habitat includes forest edges, clearings and areas with secondary growth.

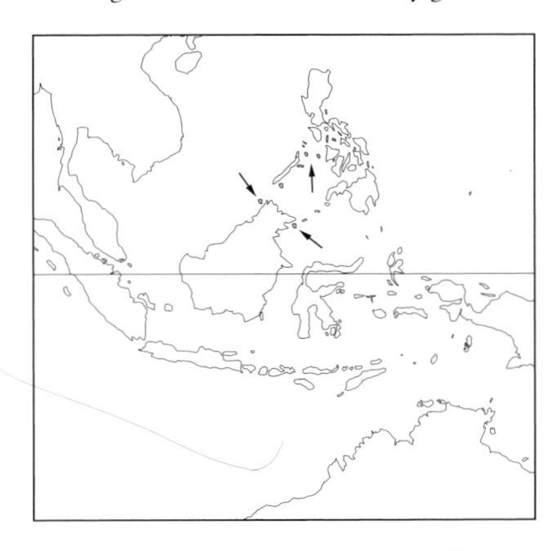

NATURAL HISTORY: All aspects of its natural history are in need of study. It is nonmigratory throughout its range. A nocturnal species, it roosts in dense foliage, coconut groves and cavities, or against tree trunks. Nests in tree cavities, and possibly other cavities as well. Insects make up most of the diet, but it probably takes small vertebrates as well. Its call is a sequence of goose-like "kwoank" notes spaced about 5 to 6 seconds apart.

CONSERVATION STATUS: Lower Risk—Near Threatened (IUCN). Locally common. Confined to the islands of the central Philippines and those off northeast Borneo, the cumulative area it occupies is small. Noted as common on Mantanani Island, but information on its abundance is lacking for other areas. The species may be threatened by habitat loss, but some reports suggest that it may avoid undisturbed forest. It is also found near farmland areas, so it may tolerate human-altered habitats.

44 Ryukyu Scops Owl, *Otus elegans*

DESCRIPTION: A medium-sized owl (20 cm) with long, pointed wings and long ear tufts. Its upperparts are a drab brown or dark reddish-brown, and are finely mottled and vermiculated. Its silvery-gray facial disk has a rufous-colored rim. The underparts are paler with an intricate black pattern. Two morphs occur, buff-grayish and rufous.

HABITAT: Inhabits subtropical, dense evergreen forests from sea level to 550 m, and likely higher. The species was originally

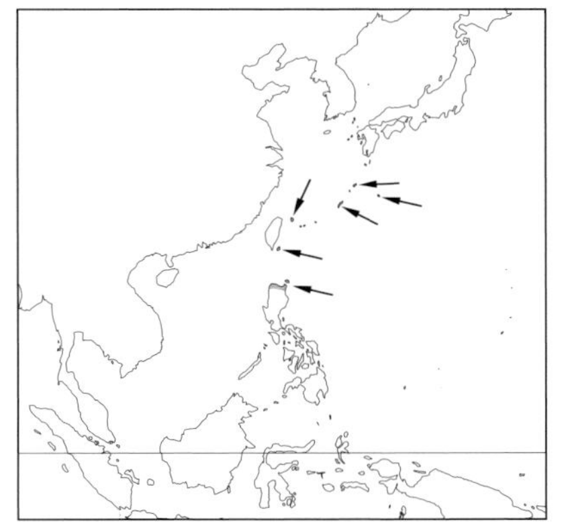

found in mature forests, but has adapted locally to habitats that have been human-altered, and it can be found in or near villages.

NATURAL HISTORY: Northern populations are apparently partly migratory, southern ones resident. Northern populations may disperse south following breeding. The species is nocturnal. Its clutch size is usually 3 to 4 eggs (sometimes up to 5) in holes in trees, usually those made by woodpeckers. Sometimes nests in the axils of coconut palms. Has been observed attempting to re-nest after failure of the first nest. Eats chiefly insects, such as beetles, moths, grasshoppers, crickets; also spiders, small birds and mammals. The male's call is a regularly repeated hoarse note, "kew-guruk," and likely different from that of the female's.

CONSERVATION STATUS: Lower Risk—Near Threatened (IUCN). The species is reported to be common on Nansei-shoto and throughout Ryukyus where suitable habitat is present. Populations are widespread on Amani, northern Okinawa and Iriomote. The Lanyu race was recently estimated at 1000 individuals (compared to 200 in the mid-1980s), but these owls are targeted by hunters and their habitat must be protected to ensure survival. Said to be fairly common in forest and forest edges on Batan, Calayan and Sabtang, where ranges are very small. Deforestation eliminates potential nest sites, thus posing a serious threat to local populations. Providing nest boxes may be beneficial in these cases. According to studies, breeding may be unsuccessful in disturbed areas, and rural owls may live longer than suburban ones.

45 Sulawesi Scops Owl, *Otus manadensis*

DESCRIPTION: This small owl (19 to 20 cm) occurs in brown and rufous morphs, with the rufous morph uncommon. Its plumage is extremely variable. It has orange-yellow to yellow eyes, ear tufts of moderate length and distinct white "eyebrows" curving around the bill. Its dark brown cheeks are surrounded by lighter areas. The upper tail and tertial are typically less barred.

HABITAT: The Sulawesi scops owl is found in humid forests and lowlands to 2500 m. It is also found in clearings and along forest edges.

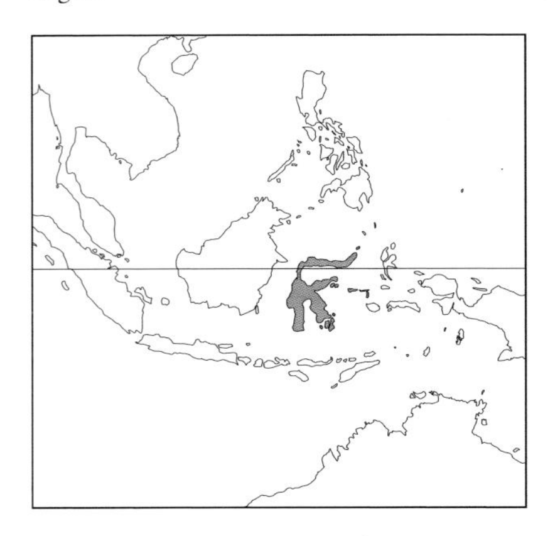

NATURAL HISTORY: Resident throughout its range. The species is nocturnal, probably nests in tree cavities, and is believed to prey mainly on insects and other arthropods, possibly small vertebrates. No other information is available, hence requires study in all areas of its life history. One version of this owl's call ("ploo-ek-oo-ek-oo-ek") is said to be a good omen, while another: ("kiek-kok-kok") is said to be a bad omen for traveling or house building.

CONSERVATION STATUS: Not Globally Threatened (IUCN). On Sulawesi Island the species is considered relatively common. No other information is available on populations in most areas; all are poorly known and are likely to be uncommon. Deforestation likely poses a major threat to populations.

46 Sangihe Scops Owl, *Otus collari*

DESCRIPTION: A relatively small scops owl (19 to 20 cm) with drab brownish plumage and indistinct markings. The upperparts have dark shaft streaks with buff spots, with lighter underparts marked with bold, black shaft streaks and fine vermiculations. It has moderate-length ear tufts, and a rather long tail and long narrow wings. The species is endemic to Sangihe Island.

HABITAT: The Sangihe scops owl is found in forest, secondary growth, mixed plantations and agricultural areas containing bushes and trees. Occurs in lowlands and mountain slopes from sea level up to 315 m.

NATURAL HISTORY: Extremely little is known about this species. It is presumably resident on the island. It is nocturnal and probably nests in tree cavities and eats insects. Its call, an irregularly spaced "peeyuuwit," is decribed as more drawn out than that of the Sulawesi scops owl. The call note follows an initial downward inflection "peeyuu..." followed by an upward ending "...wit."

CONSERVATION STATUS: Not Globally Threatened (IUCN). Although there are no data on population numbers, the species appears to be common and widespread on Sangihe Island. It can tolerate human-altered habitats, and has been found in plantations that have been in cultivation for many decades. It does not seem to have any obvious threats.

47 Biak Scops Owl, *Otus beccarii*

DESCRIPTION: This owl is unique from all other *Otus* owls in that it lacks streaking on both its upperparts and underparts. Its overall plumage is densely barred (excluding the mantle) with light and dark. It occurs in a dark rufous-brown and a dark brown or "black" morph. It measures 20 to 25 cm, and has moderate length ear tufts and yellow eyes.

HABITAT: Found in heavily forested and wooded lowland habitat; also along forest edges and locally near villages.

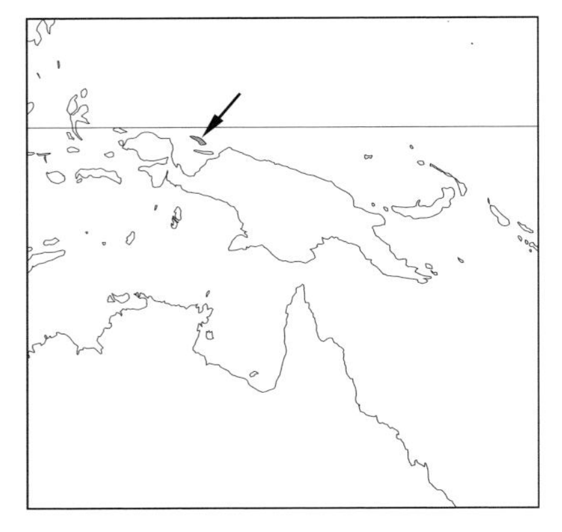

NATURAL HISTORY: The species is presumed to be resident on Biak Island. It is strictly nocturnal, preys upon insects and spiders (likely small vertebrates and invertebrates as well), and nests in tree cavities. Its call is a hoarse, raspy corvid-like croak repeated in series. No other information is known.

CONSERVATION STATUS: Endangered (IUCN). A very poorly known species; known only from 3 specimens and a few sight records. Only 1 pair was found in 1973 after extensive surveys in the last piece of undisturbed forest. Because of its small range and loss of habitat (most of the tall lowland habitat has been removed or degraded), the long-term survival of the species could be threatened. The forests upon which it depends do not regenerate readily. It is thought to occur in the Biak-Utara and Pulau Supiori Nature Reserves. Surveys using taped playback of its calls are needed to better estimate its population size (now thought to be decreasing to between 2500 and 10,000 individuals), distribution and habitat use. Previously thought to be a geographic race of the more widespread and commoner Moluccan scops owl.

48 Seychelles Scops Owl, *Otus insularis*

DESCRIPTION: This is a small (20 cm) scops owl with buff plumage, minute ear tufts and unfeathered legs with rather long toes. The crown and nape are heavily spotted and streaked with black. The underparts are very lightly barred with white and broadly streaked with black. The species is probably endemic to Mahe Island in the Seychelles, but possibly there are some pairs on Praslin. It is the only small owl in the Seychelles.

HABITAT: Currently, this owl occupies secondary forest in uplands and valleys (between 250 and 600 m), usually with nearby water. It is thought that it once inhabited both lowland and highland forests over the entire Mahe Island before the forests were destroyed. The extent of its preferred upland forest habitat may have increased over the last 40 years with the decline in forest-based industry and subsequent forest regeneration.

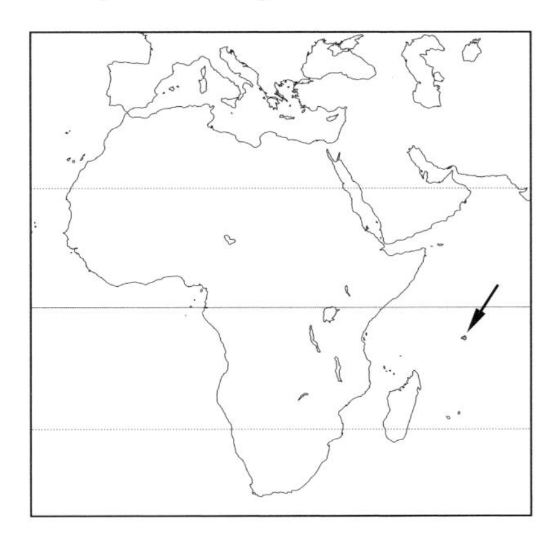

NATURAL HISTORY: The species is nocturnal, and readily responds to playback of its song. It roosts in bushes or trees, and is nonmigratory. Although no nests have been found, it possibly uses cavities in old trees, clefts in rocks or perhaps even ground cavities. It is assumed that only 1 egg is laid. New information suggests that individuals use the same nest area year after year. Insects, tree frogs and lizards (geckos) are apparently the most predominant prey, but other prey items retrieved from stomach contents include beetles, grasshoppers and vegetation. The owl is frequently on the ground in areas where there are many boulders. The call is described as a "deep saw-like croak," ("kroahk-kraa, kroahk-kraa"), interspersed with knocking sounds ("tok tok").

CONSERVATION STATUS: Critically Endangered (IUCN). Discovered in 1880, the species was not observed again for such a long time that it was designated as extinct in 1958. It was rediscovered again on Mahe Island in 1959. In 1975–76, research showed that at least 12 pairs were present. The pairs were spaced at 1000 m apart. Based on this distribution and available habitat, the total population was estimated at 80 pairs in the early 1990s, but numbers have likely been reduced since then. Other current estimates range from 180 to 360 owls. The species is still thought to be extremely rare and endangered. Thought to be present on Praslin Island and Felicite Island, but these reports are unconfirmed. New forestry extraction techniques may revitalize the forest industry. Such techniques have rendered areas previously inaccessible (e.g., upper Grand Bios Valley) open to logging. This constitutes a major threat as forests are removed for housing development and timber exploitation. The effects of increased ecotourism (trails and disturbance) in upland habitat are unknown. Introduced cats and rats, and the barn owl, also pose a threat as predators of nests or adults. Although the Morne Seychellois National Park has been established to provide habitat and to help protect the survival of the species, it seems that elsewhere it may disappear. In order for conservation efforts to be successful, the biology and ecology of this critically endangered species must be better understood. Surveys are currently underway, as well as efforts to determine if other suitable islands exist in case future translocations need to be considered.

49 Simeulue Scops Owl, *Otus umbra*

DESCRIPTION: A very small scops owl (16 to 18 cm) with a dark reddish-brown plumage and a paler rufous facial disk with an indistinct rim. The ear tufts are short, and its wings are rounded. The crown has some thin dark streaking, and the upperparts have fine black vermiculations; underparts are somewhat paler with whitish vermiculations and rufous barring. The eyes are yellow to greenish-yellow.

HABITAT: It is found in forests, forest edges and areas where forests have been cleared and planted with cloves. Remnant forests are associated with steep slopes, especially coastal forests.

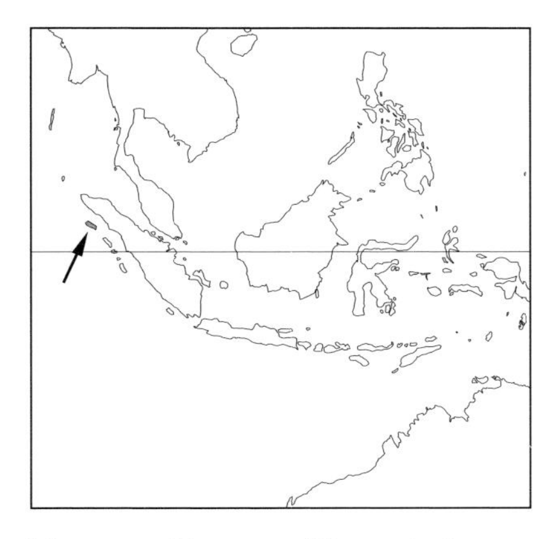

NATURAL HISTORY: The species is nocturnal and resident on Simeulue Island. Presumably nests in tree cavities. Insects apparently make up the main diet. The male's call is a distinct "took took tutook" uttered in a short, jerky series. No other information is known.

CONSERVATION STATUS: Lower Risk—Near Threatened (IUCN). The species is endemic to Simeulue Island, off the northwest coast of Sumatra. Because it is confined to a single island, habitat loss is likely the main threat. Some report that primary forest habitat loss in the Sundaic lowlands of Indonesia has been so extensive that this species may disappear as early as 2010. Conversely, it may not be under immediate threat because it seems to tolerate secondary growth and edge habitats. There are variable reports on its status, ranging from rare to common. There is no information regarding population numbers.

50 Enggano Scops Owl, *Otus enganensis*

DESCRIPTION: A small (16 to 20 cm) brown scops owl similar to the Simeulue scops owl. It differs in that it has more distinct dark markings, especially on the crown and wings. Its upperparts are also duller and darker, and dorsal coloration can be chestnut to olive-brown. A pale nuchal collar divides the nape from the mantle. The "eyebrows" are whitish.

HABITAT: Presumably forest and wooded areas, and forest edges. Poorly studied.

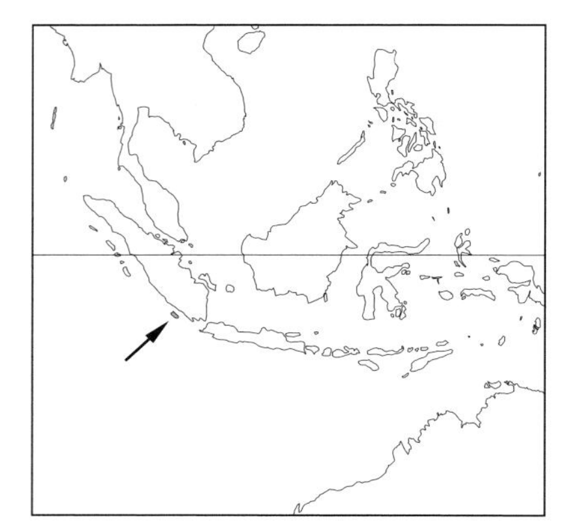

NATURAL HISTORY: Presumably resident. It is nocturnal, but more detailed habits have not been reported. Likely nests in tree cavities, and apparently eats mainly insects, spiders and other arthropods, but there is no information available. Its calls are unknown, which precludes using surveys to assess its status and distribution.

CONSERVATION STATUS: Lower Risk—Near Threatened (IUCN). This species is found only on Enggano Island, off southwestern Sumatra, Indonesia. Habitats have remained undisturbed on the island, but recent agricultural development proposals may result in habitat loss and potentially threaten this species. The status of the owl is uncertain, and reports have been contradictory; reported as very rare to common.

51 Nicobar Scops Owl, *Otus alius*

DESCRIPTION: A small scops owl (19 to 20 cm) with warm-brown plumage that is finely and distinctly barred all over, including the rounded, medium-length ear tufts. The back and mantle lack streaking, and the underparts have reduced streaking and heavy tri-colored barring. The facial disk is somewhat pale, lightly barred at the lower edge, and has an indistinct rim. Eye color is unknown, but presumed to be yellow. The legs are sparsely feathered, with the rear edge and lower part being bare. It is apparently endemic to Great Nicobar Island.

HABITAT: Two specimens were collected in 1966 and 1977 in coastal forest close to sea-level, otherwise unknown.

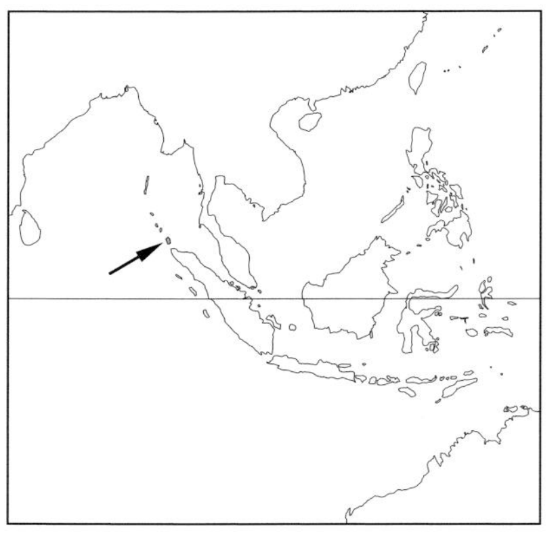

NATURAL HISTORY: Presumably resident, and probably nocturnal. There is no information on breeding biology. Very little information is available on its diet; 1 owl is known to have eaten a beetle and a spider, another specimen had a gecko in its stomach. The male's call is unknown, but females repeatedly utter a moaning, 2-second (or longer) "oun" note for up to half an hour.

CONSERVATION STATUS: Data Deficient (IUCN). Likely rare and/or endangered. The status is unknown as only 2 specimens have been collected from the same site, Campbell Bay on Great Nicobar Island; no other records are available. It is possible that the species occurs on other Nicobar islands, like Little Nicobar, which is only a few kilometers north. General threats include the loss of coastal forests to agriculture, road and industrial development, and expanding human settlements.

52 Pemba Scops Owl, *Otus pembaensis*

DESCRIPTION: A small scops owl (16 to 20 cm) with a rufous and brown morph, both rather uniform. It has small ear tufts and pale "eyebrows" in varying degrees of prominence. The primaries are barred, as are the outer tail feathers, the central ones being plain rufous. The eyes are yellow. The only other owl on Pemba Island is the larger and very different barn owl. It is said that the native inhabitants of the island identify the owl with witchcraft, and believed it gave birth to already hatched young.

HABITAT: The Pemba scops owl is found only on the island of Pemba, a lush and hilly island near Zanzibar, 50 km off the coast of Tanzania. It occurs in wooded areas or semi-open habitat with densely foliaged trees. It is frequently observed in overgrown clove and mango plantations.

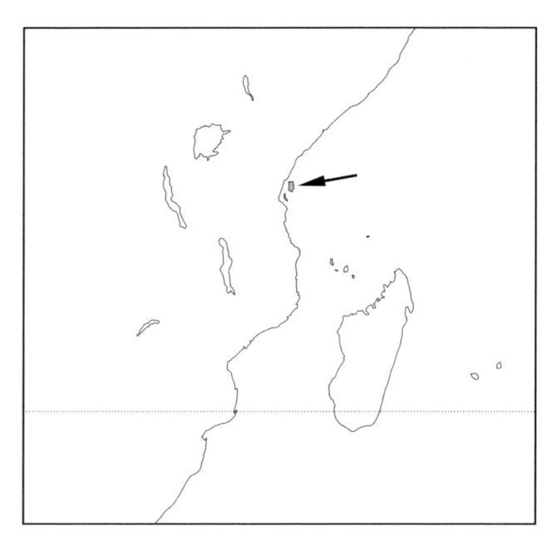

NATURAL HISTORY: A nocturnal bird it roosts in thick foliage or dense undergrowth about 2 m above the ground. It is rarely observed during the day and prefers to sit tight; even when approached it will not move and can even be picked up. It has a habit of diving and flying low above the ground and then swooping up as it moves between perches. Probably nests in tree cavities, but virtually nothing has been recorded of its breeding biology. Hunts insects from a perch or in flight, and also gleans them from foliage. Reported as resident. Its call is a long series of monotone, hollow notes: "hoo hoo hoo..."

CONSERVATION STATUS: Not Globally Threatened (IUCN). Apparently common and widespread (particularly in clove plantations), but there is no information on its population. It may be persecuted by humans. The fact that it is range-restricted renders the species vulnerable. Continuing forest degradation and clearance for agriculture may result in population declines.

53 Comoro Scops Owl, *Otus pauliani*

DESCRIPTION: This small scops owl (15 to 20 cm) was first discovered in 1958. Light and dark morphs are present, and the plumage is remarkably uniform. The ear tufts are very small and often not visible. The crown is finely barred and the underparts are densely vermiculated without many distinct black shaft streaks. The dark grayish-brown plumage is finely barred or with dark vermiculations and some lighter spots. The gray or brown facial disk has several concentric rings. The eyes are yellow to dark brown, and the lower tarsus is bare.

HABITAT: The Comoro scops owl is thought to be present along a forest-high mountain heath transition zone, a habitat that extends around Mt. Karthala. It is found in both primary or degraded mountain forest and along forest edge on upland slopes (460 to 1900 m).

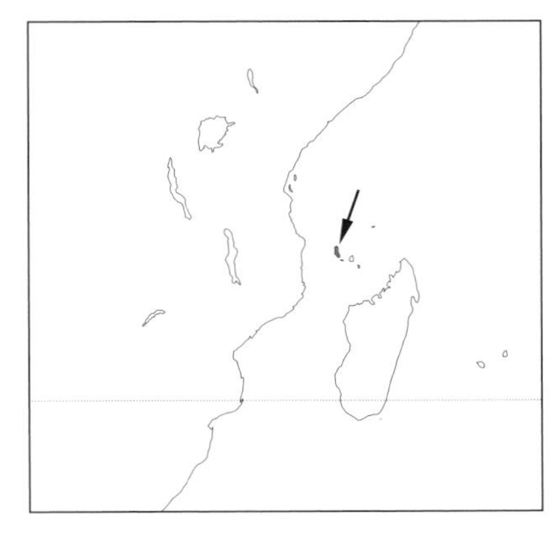

NATURAL HISTORY: The species is presumably resident. It is nocturnal, becoming active at dusk. It is very vocal year-round, and extremely territorial, and will fly close when its call is imitated. Territory size reported as about 5 ha. Nests in cavities, but other breeding biology information is unknown. The stomach of the only specimen collected contained a few beetle remains. Weak talons imply it primarily eats invertebrates. Its call consists of a series of "gluk" notes are uttered at a rate of 2 per second.

CONSERVATION STATUS: Critically Endangered (IUCN). Its population is estimated to be about 1000 to 2000 pairs within the estimated 100 square km of suitable habitat left on Mt. Karthala, an active volcano on Grand Comoro Island. For now, the species appears stable, but its range and numbers will decline as the island's human population continues to increase. Habitat fragmentation is a serious threat to the species as deforestation slowly works its way up the mountain. Habitat loss is estimated at over 25% since 1983. Areas are being burned for grazing land for cattle, making the habitat more and more suitable for the introduced common myna to invade and compete for nest sites. Other exotic invasive species, such as the strawberry guava (*Psidium cattleianum*) threaten its habitat while introduced rats eat eggs or nestlings. Plans to build a road to Mt. Karthala's crater would be an added threat. Immediate action is needed to establish a nature reserve on Mt. Karthala in order to ensure the survival of the species.

54 Anjouan Scops Owl, *Otus capnodes*

DESCRIPTION: A small owl (20 to 22 cm) with minute, almost nonexistent ear tufts. The light morph has a whitish-cream facial disk with dark, narrow concentric rings and a dark rim. The upperparts are grayish-brown and mottled in appearance. The underparts are finely barred and have dark shaft streaks. The rufous-brown morph is similar, but less speckled, and the dark morph is a dark chocolate-brown to earth or gray-brown with fine buff (not white) speckles. The eyes are yellowish-green.

HABITAT: The species is confined to primary forest on the mountain slopes of Anjouan Island above 550 m, but usually above 800 m.

NATURAL HISTORY: Although movements are unknown, it is presumably sedentary and movements are local. It frequently perches in dense vegetation 3 to 15 m above the ground. A nocturnal species, it leaves its roost at dusk to hunt. Most likely nests in holes in large old trees.

No data on its diet, but it probably eats insects. Its main call is a plover-like sequence of 3 to 5 long whistles, "peeooee," uttered at 10-second intervals. A 1.5 second–long screeching call is also attributed to this species.

CONSERVATION STATUS: Critically Endangered (IUCN). Thought once to be extinct (no sightings since 1886), the species was rediscovered in 1992. Apparently, the species was overlooked because its vocalizations are unlike typical owl calls. The local people, however, have always been familiar with them. Fairly common on steep forested slopes, very local elsewhere. Based on sightings and preferred habitat, the population is estimated at 100 to 400 pairs. Continued population declines are inferred due to ongoing habitat loss. Due to deforestation of primary habitat, only 10 to 20 square km of suitable habitat is left. This owl is also hunted for food, and introduced rats may affect breeding success; these factors pose major threats to the current population. Human overpopulation and natural resource degradation on Anjouan are severe. Severe winds are a significant threat to the small patches of remaining forested areas. Given the above circumstances, any proposed conservation efforts will be complicated; however, urgent action is recommended to ensure the continued survival of the species. A translocation of this species to the neighboring island of Mohéli was proposed until the discovery of the previously unknown Mohéli scops owl, now also Critically Endangered. Establishing a captive population may be the best guarantee against the loss of this species.

55 Mohéli Scops Owl, *Otus moheliensis*

DESCRIPTION: A medium-sized scops owl (22 cm) with plumage rather uniform overall, very small ear tufts and an indistinct facial disk. The 2 morphs, brown and rufous, are similar but the latter is more rufous-cinnamon all over. The rufescent-brown plumage has blackish spots and bars on the upperparts and crown, particularly

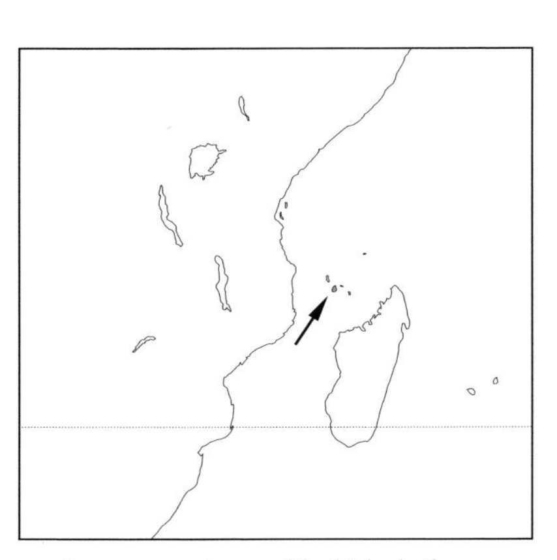

on the nape, and some blackish shaft streaks. The underparts and facial disk are rusty-cinnamon. The lower third of the tarsus is bare, and the eyes are greenish-yellow. The species is endemic to Mohéli Island in the Comoro Islands, and was discovered only in 1998.

HABITAT: It is found on the dense humid forested slopes and summit of Mohéli Island from 450 to 790 m.

NATURAL HISTORY: The species is presumably resident. Strictly nocturnal, it becomes active after sunset. There is no information on its breeding biology. It likely eats mainly insects and spiders. Its call is poorly known, but has been described as a series of 1 to 5 hissing whistle notes. An understanding of its basic ecology is needed to help form effective conservation plans.

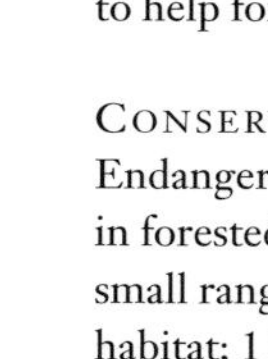

CONSERVATION STATUS: Critically Endangered (IUCN). Seems to be common in forested uplands within its extremely small range (1 bird/5 ha in undisturbed habitat; 1 bird/10 ha in degraded habitat). It is estimated that there are 400 individuals on the island. Habitat loss is a threat to the species, as forests continue to be destroyed for subsistence agriculture. Between 1955 and 1968 the species' habitat was estimated at 5000 to 6000 ha (30% of the island's surface area). In 1995, it was reduced to 1070 (5%). Invasive exotic plants also threaten its remaining forest habitat. Introduced rats are numerous and eat eggs or nestlings and may compete for food. Remaining habitat needs protection and management if this species is to survive.

56 Madagascar Scops Owl, *Otus rutilus*

DESCRIPTION: A small (19 to 23 cm) cryptically colored scops owl with relatively small ear tufts and yellow eyes, and feathered legs. Gray, brown and rufous morphs are present, with much variation. The brown morph has a pale facial disk and a dark brown rim. The upperparts are pale brown with ochre and whitish markings and black shaft streaks. Rufous and gray morphs are similar but with ground colors more rufous or gray respectively. The rufous morph sometimes lacks black markings. Western birds more frequently are gray, while eastern birds are more often rufous.

HABITAT: The species can be found in several different habitats, including primary and secondary forest to thickets, humid brushy regions, dry forest (also dry areas with shrubs, brush and scattered trees) in the west and southwest. Also found in urban parks and in trees near villages, and remnants of partially exploited forests. Plantations are avoided. Sea level to 2000 m.

NATURAL HISTORY: Apparently the species is resident. A nocturnal bird, it roosts in thick foliage, between branches or against tree trunks. Its 4 to 5 eggs are usually laid in tree cavities, but also in nests on the ground or sometimes in abandoned stick nests of other birds. Insects, particularly moths and beetles, are the preferred prey. Small vertebrates are also possibly part of its diet. Captures prey either on the ground or in flight. A series of 5 to 9 distinct "oot" notes form its main call. A vocally unique population is thought to occur in drier portions of western Madagascar.

CONSERVATION STATUS: Not Globally Threatened (IUCN). Locally rather common. Has been reported as common or even abundant in some areas (in suitable habitat between Betsileo and Tanala, Ankafana and Zahamena). Scattered records suggest a wide but sparse distribution. It is also present in protected areas such as Ranomafana National Park, Perinet Special Reserve and Berenty Private Reserve.

57 São Tomé Scops Owl, *Otus hartlaubi*

DESCRIPTION: This small scops owl (16 to 19 cm) is known only from São Tomé Island in the Gulf of Guinea. The facial disk is a light rufous-brown, with a white chin and "eyebrows" and very small ear tufts. The upperparts, which are almost solid dark brown, have indistinct rufous and dark markings. The underparts are lighter with fine rufous and brown barring and blackish shaft streaks. The tail is practically unbanded, but has narrow buff bars. A gray-brown morph occurs, but is rare.

HABITAT: The species occurs in humid, densely foliaged primary forest, dense secondary forest bordering the Io Grande River, and plantations. This owl is present in high densities in clove and mango plantations (400 m) and secondary forest from sea level to 1300 m.

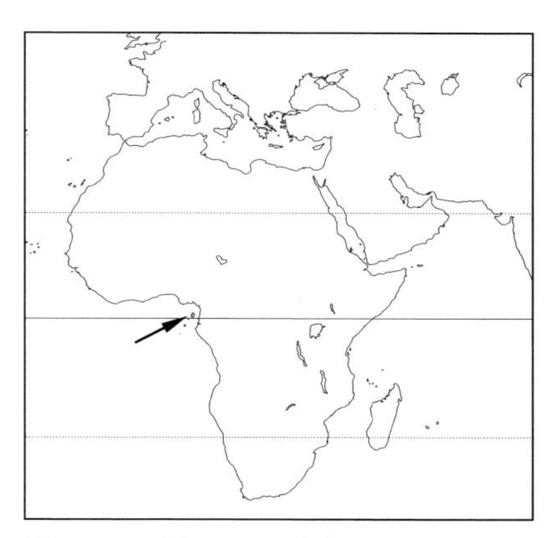

NATURAL HISTORY: The species is presumably resident on São Tomé Island. Its call is a low-pitched, rough "kwow" given at 15- to 20-second intervals. A nocturnal bird, it roosts in dense foliage and tree cavities, or against tree trunks. Probably nests in tree cavities, but it may also nest on the ground. Its diet is predominantly insects (grasshoppers, beetles and moths), but also small lizards. Hunts in dense foliage at low levels, occasionally on the ground. More information on its ecological needs is required to conserve this rare species.

CONSERVATION STATUS: Vulnerable (IUCN). Reportedly common in some areas, but with a highly fragmented distribution. Apparently it can tolerate some changes to its habitat on the edge of its range, and will move into areas abandoned after cultivation. Historically, the establishment of coffee and cocoa plantations, which it avoids, resulted in the loss of large forested areas. Recent road development on the east and west coasts has increased access to previously remote habitats. Forest loss continues and is accelerating as small farms proliferate in newly accessible areas. Avoids coffee and cocoa plantations. Potential threats to remaining populations include competition from common barn owls, and predation by cats. Heavy pesticide use from 1963–73 likely had a negative impact on the species. The current small but stable population (250 to 1000) estimate is simply a guess based on the small area of suitable habitat within its single island range.

58 Western Screech Owl, *Otus kennicottii*

DESCRIPTION: This small (19 to 25.5 cm) owl with feathered ear tufts was formerly considered a subspecies of the similar eastern screech owl. A blackish bill, parallel streaks on upper body feathers, and ventral feathers with robust shaft streaks and finer, perpendicular, horizontal bars distinguish it from its close relative. Its eyes are lemon-yellow. There is considerable variation in size and color over its range—from brown to gray-brown in the Pacific Northwest, gray in southern desert habitat, and some reddish-brown individuals in coastal populations.

HABITAT: Most frequently found in riparian deciduous woodlands, it also occurs in a wide range of forest and treed habitats. It avoids higher elevations, especially at the northern part of its range.

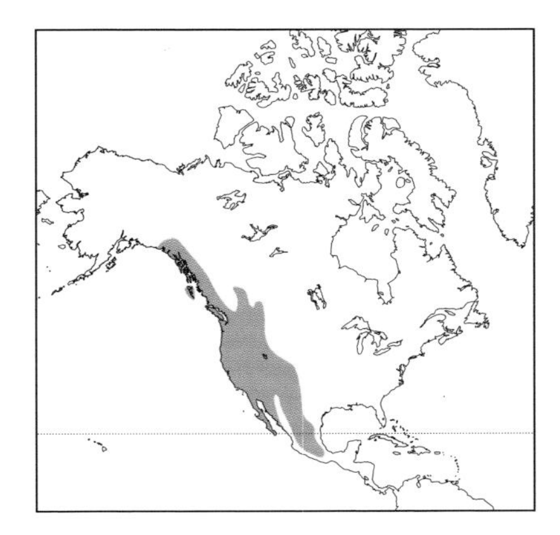

NATURAL HISTORY: The western screech owl captures mainly small rodents, but also takes worms, snails, slugs, insects, fish, amphibians, reptiles and birds. It has a distinct call, a series of 5 to 15 hollow, whistled hoots that speed up, like a bouncing ball coming to rest. This resident owl stays on its territory year-round. Its 2 to 7 eggs, usually 3 to 5, are laid in tree cavities in shaded areas. It readily uses suitable nest boxes. The male feeds its mate during incubation and while she broods their young. The young are self-sufficient and disperse from their parents' breeding territory by mid-summer, moving up to 300 km.

CONSERVATION STATUS: Not Globally Threatened (IUCN). Housing development and clear-cut forest management have eliminated habitat, thus causing local population declines. Conversely, it appears to be slowly expanding its range eastward along restored or established treed riparian habitat in Texas, Colorado and the western Great Plains north to Saskatchewan. Otherwise, this common owl is thought to have stable populations over most of its range. The barred owl has only recently expanded into the Pacific Northwest range of the western screech owl. Some fear that the barred owl, a significant predator of smaller owls, will ultimately cause a regional decline of the western screech owl. Eggshell thinning has been linked to pesticide (DDE) residues in some eggs.

59 Balsas Screech Owl, *Otus seductus*

DESCRIPTION: A rather large screech owl (24 to 26 cm) with grayish-brown plumage overlaid with pinkish coloring. It has short ear tufts, brown eyes (rarely golden-brown) and powerful talons with big bristled toes. The upperparts are streaked with dark vermiculations and the shoulders have a whitish band. The lighter underparts have thin, dark shaft streaks and delicate vermiculations. The neck and upper breast are irregularly spotted with deep chestnut or dark brown broad shaft streaks. Only 1 morph is known.

HISTORY: This owl inhabits arid open to semi-open regions, thorn woodland and deciduous woodland, mesquite, tall cactus and secondary growth, from 600 to 1500 m.

NATURAL HISTORY: Very little information is available on this owl's biology and ecology. It is resident in its southwest Mexican range. No nests have been recorded, but it probably nests in tree cavities. Its diet includes insects and small vertebrates, though no details are known. It habits are likely similar to those of other screech owls. Its call consists of "bookh" notes that accelerate to a "bobobrrr" trill, also described as a bouncing-ball rhythm.

CONSERVATION STATUS: Not Globally Threatened (IUCN). It is described as fairly common to common. Habitat loss may be more significant than previously thought and it is being considered for entry into the Red Data Book.

60 Pacific Screech Owl, *Otus cooperi*

DESCRIPTION: A relatively large screech owl (20 to 25.5 cm) with dark barring on

the crown and ear tufts. The facial disk is pale gray with a prominent dark border. The upperparts are tawny-gray with dusky, delicate vermiculations and black shaft streaks. The legs are completely feathered and the toes are bristled.

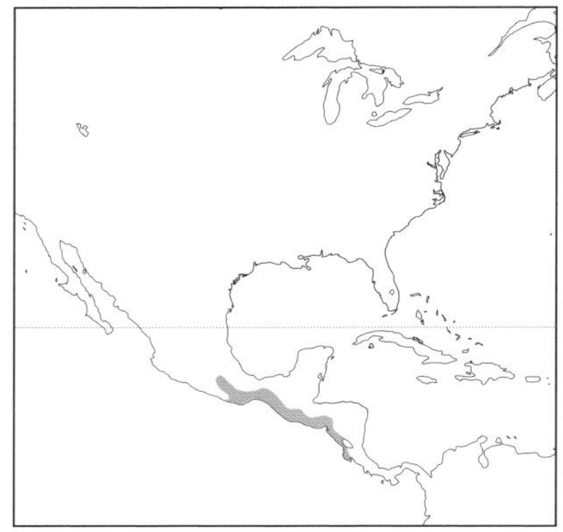

HISTORY: Inhabits arid to semi-arid forest, partly open areas with some trees, palms, shrubs, and giant cacti, secondary growth and lakeside fields. The species can also be found in low scrub, mangroves and swamp forests from sea level to 1000 m.

NATURAL HISTORY: Appears to be resident. Little is known about its natural history, but it is probably like that of other screech owls. The 3 to 5 eggs are laid in a cavity or old woodpecker hole. Preys upon large insects such as katydids, beetles and moths; as well, it takes scorpions and probably some small vertebrates. Both sexes appear to use a single "woof" note to keep in contact. A fast trill, slowing to a staccato "gurrrgogogogogo" has also been described. The male's primary courtship call has yet to be discovered and recorded.

CONSERVATION STATUS: Not Globally Threatened (IUCN). Thought to be fairly common to common, but there is no information on population levels. Found in protected areas in Costa Rica: Palo Verde, Santa Rosa and Barra Honda National Parks.

61 Eastern Screech Owl, *Otus asio*

DESCRIPTION: This small owl (16 to 25 cm) owl with prominent ear tufts comes in 1 of 2 colors—gray or bright cinnamon-rufous—but some intermediate brownish individuals have been noted. Its pale to lemon-yellow eyes apparently darken with age. A yellowish to light-green upper bill, upper body feathers with streaks that flare transversely, and ventral feathers with robust shaft streaks and equally thick horizontal bars arranged in an anchor pattern distinguish it from the similar western screech owl. There is considerable variation in the proportion of color morphs over its range—patterns thought to be related to temperature and humidity.

HABITAT: This owl is found in a great variety of natural and human-altered treed habitats, typically below 1500 m, including northern and tropical deciduous and mixed forest, riparian deciduous and southern pine forest, southwestern oak-juniper and subtropical thorn woodland, and rarely in boreal and montane coniferous forest. It also occurs in a variety of forest ages (early successional to mature) and in lowland, mountain and river valley landforms.

NATURAL HISTORY: This resident owl stays on its territory year-round. It captures mainly terrestrial and aquatic invertebrates, e.g., insects, crayfish and earthworms, but also takes small mammals, birds, reptiles, amphibians and fish. Its 2 to 6 eggs, usually 3 to 4, are laid in tree cavities in wooded environments. It readily uses nest boxes. Adults incubate for 30 days, and feed the nestlings for about as long. Adults are very aggressive toward animals that approach their recently fledged young, even attacking humans in urban areas—hence such colorful nicknames as "the feathered cat." A descending or monotone toad-like trill call helps identify it, but it also gives various barks, hoots, chuckles, rasps and screeches. Another call is reminiscent of a horse's whinny. The young are self-sufficient and disperse when 12 to 14 weeks old up to 17 km from their parents' breeding territory.

CONSERVATION STATUS: Not Globally Threatened (IUCN). This species appears to thrive in treed urban and suburban habitat. Population cycles (4 and 9 to 10 year) have likely been confused with population declines in some areas. This emphasizes the need for long-term studies of population density and regulation in order to assess the status of this and other owl species.

62 Whiskered Screech Owl, *Otus trichopsis*

DESCRIPTION: This small owl (16 to 19 cm) has yellow eyes and only moderately developed ear tufts. It has feathered legs, small feet and bristles on top of relatively short toes. It occurs in a gray and red phase.

Its basic color ranges from light gray in northern populations to reddish brown at the southern limits of its range. The herringbone pattern of streaks on its upper and lower parts is darker and more coarsely patterned than that of other similar screech owls. Relatively long and numerous, black hair-like rictal bristles ("whiskers") surround its pale grayish-yellow to dull greenish bill and its face, especially the upper facial disk.

HABITAT: Occupies dense, closed groves of primarily mixed oak, pine and sycamore woodland on mountain slopes, in canyons and along riparian areas. It is reportedly common in coffee plantations (El Salvador). Occurs at elevations from 1500 to 2500 m, above that inhabited by the western screech owl, but below that of the flammulated owl.

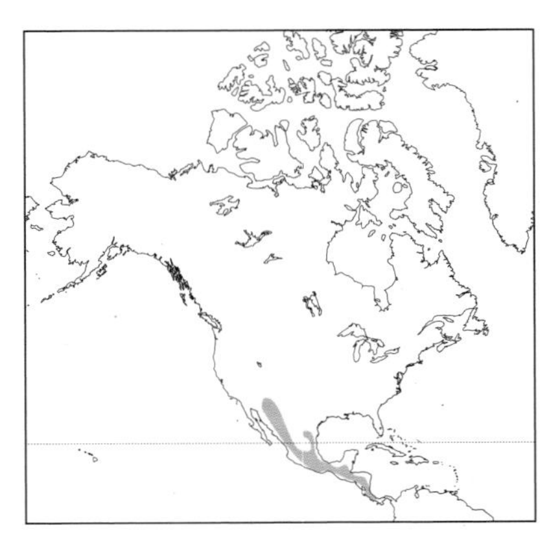

NATURAL HISTORY: This presumably resident owl is strictly nocturnal. Its call is Morse code–like: a series of short, monotonous and evenly spaced whistles of variable duration, given relatively slowly compared to the western screech owl. It is always reluctant to leave its daytime roost on the rare event when its effective camouflage fails it. Flying insects caught on the wing make up most of its diet, which includes small mammals, birds, reptiles and amphibians. It rarely pounces on prey on the forest floor or hunts more than 10 m above the ground. An aggressive owl, it excludes other birds and small owls from its well-defended nesting and hunting territories. Its 3 to 4 eggs are laid in natural tree cavities and unoccupied woodpecker holes. There is no information on other aspects of its breeding biology, such as the incubation and fledging periods.

CONSERVATION STATUS: Not Globally Threatened (IUCN). Despite historic and current intensive human-altering land use of pine-oak woodland, it continues to be locally abundant. There is no information available on population size or trends. Conservation of dense montane forests should ensure its long-term survival.

63 Tropical Screech Owl, *Otus choliba*

DESCRIPTION: This owl (20 to 24 cm) is the typical small owl of South America. It is found from Costa Rica through to South America where it is common and widespread. It is also present on Trinidad. It is the most adaptable and successful screech owl in South America. Most birds are of the gray-brown morph, and are characterized by ear tufts, light gray facial disk outlined prominently in black and a distinct pattern of dusky streaks that are herring-bone-like on the underparts. Brown and rufous morphs also occur, with the latter being rare.

HABITAT: The species is found in a wide variety of habitats: timbered savannah, open to semi-open forests, forest edges, farmlands or pasture with suitable trees, openings in rainforest, dry forest, second-growth thickets, riverine forest, plantations and urban parks. It does not occur in dense primary forest, deserts, treeless areas, or temperate cloud or mist forest. Usually from sea level to 1500 m, but has been recorded at 3000 m.

NATURAL HISTORY: The movements of this species are unknown, but likely it is resident. A nocturnal owl, it roosts in the dense foliage of bushes and trees, preferring thorny shrubs or epiphytes (orchids, mosses and climbing cactus) on tree trunks. Males and females often roost together. Usually bold and can be approached at close range. The 1 to 3 eggs (sometimes 6) are usually laid in natural cavities, abandoned bird nests, old woodpecker holes, nest boxes or even rotted-out fence posts. Insects, including moths, grasshoppers, cicadas, beetles, mantids, scarabids, bumblebees, crickets, scorpions and spiders, form the main part of the diet. Sometimes snakes, bats and small mammals are taken. Forages from a perch at low level, and captures prey on branches, on the ground or on the wing. Will also hunt along the road perched on trees or telephone wires. Its main call is a terse, trill-like call, ending in 2 or more sharp "gurrrrku-kuk" notes. It also utters a bubbling "bubububububu" during courtship and a gentle "woog" to keep in contact with its mate and/or young. A descending laughing "hahahaha..." call is given when alarmed or surprised. Although the species is common throughout its range, little is known of its biology.

CONSERVATION STATUS: Not Globally Threatened (IUCN). Widespread and common throughout South America, but there is no information on population numbers. It may be affected by pesticide use and local human persecution. Traffic mortality can be high as it often hunts along roadsides. Deforestation has been beneficial to the species by providing clearings and secondary growth, thus creating suitable habitat.

64 Koepcke's Screech Owl, *Otus koepckeae*

DESCRIPTION: This screech owl (24 cm) is dark gray and has prominent ear tufts with dark shaft streaks. It has a dark crown, and almost no whitish fringe is visible around the hindneck. The whitish facial disk is mottled and has a black rim. The eyes are yellow, with dark eyelids, and the bill is whitish-blue. The upperparts are brown to dark gray-brown with dark streaks and bars, and brown-white spots. The underparts are grayish-white with bold, broad, dark streaks. The legs are yellowish with fine brown spots.

HABITAT: The owl is present on the Andean slopes from 2500 to 4500 m. Here it occupies wooded areas and arid forest patches.

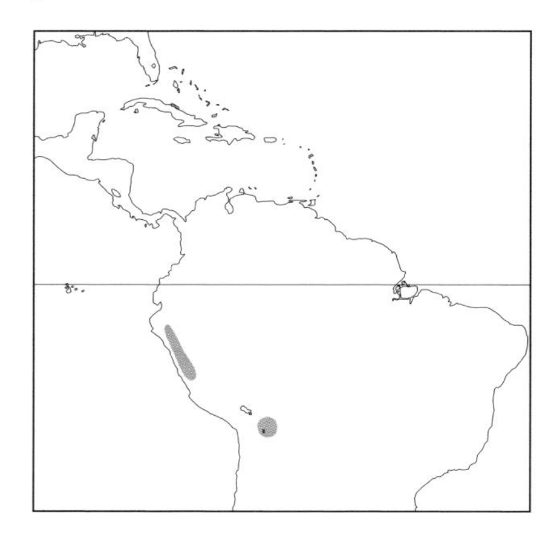

NATURAL HISTORY: Active at night, it roosts during the day in bushes or trees with dense foliage. It probably nests in cavities and eats primarily insects, but there is no data to support this. A very poorly known species. Its call is described as a rising series of 8 to 10 low-pitched "uk" notes, but the last note is descending.

CONSERVATION STATUS: Not Globally Threatened (IUCN). Some consider it to be not uncommon locally. Its status is more likely Data-Deficient as there is little information on population status, ecology or biology.

65 Peruvian Screech Owl, *Otus roboratus*

DESCRIPTION: The Peruvian screech owl (20 to 22 cm) is found only in northwest Peru and southwest Ecuador. The gray-brown morph has a grayish-white facial disk with a black rim, yellow to golden eyes, small ear tufts and a dark crown. The white "eyebrows" contrast strongly with the dark crown. The gray-brown upperparts have dark barring and black shaft streaks, with whitish underparts lightly vermiculated with fine dark shaft streaks. The middle of the belly is frequently clear white. The rufous morph has a more rufous plumage and dark brown markings. This bird, little known to the outside world, remains a mystery.

HABITAT: Inhabits hilly lands and mountain slopes (500 to 1200 m, sometimes higher). Here, the typical habitat is dry, deciduous woodland interspersed with bushes. The *pacificus* race is found in arid coastal plains and foothills of the Andean slopes (below 500 m) and open country with cacti; prefers mesquite woodland and acacia shrub in arid hillsides or plains.

NATURAL HISTORY: Presumably a resident species. Active at night, it roosts in dense foliage and bushes. Responds to playback. Nests in tree cavities, and possibly in abandoned mud nests of pale-legged hornero. Breeding biology is otherwise unknown. Prey items include caterpillars, crickets, grasshoppers, beetles and their larvae. Its main call is a purring trill, ascending then fading away. Two forms are distinguished by variations in the trill call, and they also have different inflections in their yelping, aggressive "kew" or "kyui" calls.

CONSERVATION STATUS: Not Globally Threatened (IUCN). Very common locally, but throughout its range it is rare, possibly vulnerable. Has experienced habitat loss since the early 1970s due to grazing by goats and tree cutting for firewood and charcoal. Providing nest boxes may mitigate local habitat loss, enabling the species to breed in areas with few or no trees.

66 Bare-shanked Screech Owl, *Otus clarkii*

DESCRIPTION: This is a relatively large screech owl (20 to 25 cm) with small ear tufts and a large head. The facial disk is cinnamon to tawny-brown in color with an ill-defined rim. The lower third of the legs are bare (other *Otus* owls in the same range have completely feathered legs). The upperparts are reddish-brown, and the underparts have broken dark streaks and bars, with whitish square markings on the sides of each shaft streak, which gives it a scalloped look. Only this species of screech owl occurs in the mountain forests of Panama.

HABITAT: This owl can be found predominantly in thickly wooded montane forests and also along the forest edge (900 to 2350 m, sometimes to 3300 m), but also in thinned forests.

NATURAL HISTORY: The species is presumably resident. Active at dusk and during the night, it roosts during the day within the dense foliage of trees or epiphytes. Nests in natural cavities in trees (only 1 nest found, in a natural cavity in an oak tree). Apparently the owls tend to be social during the breeding season when they are frequently observed close together in "family size" groups. Large insects, such as grasshoppers, crickets and beetles, form its main diet. Other prey include spiders, shrews, small rodents and perhaps other small terrestrial vertebrates. Prey is usually captured on the ground or from branches. The main call is a repetitive and deep 3-note "woohg-woohg-woohg" call. Another rhythmic call, "bubu booh-booh-booh" is often uttered by breeding pairs.

CONSERVATION STATUS: Not Globally Threatened (IUCN). The status is uncertain (population levels unknown), but seems to be rare or uncommon; locally frequent. In Costa Rica, the owl is found in some protected areas: Volcáno Poás National Park, Moneverde Biological Reserve, the Panama Highlands and Darién Highlands Endemic Bird Areas. Because it is restricted to the mountains of Middle America, and its habitat is negatively impacted by the dairy industry and the destruction of dense cloud forest, the species may be at risk.

67 Bearded Screech Owl, *Otus barbarus*

DESCRIPTION: This is a small screech owl (16 to 20 cm), with relatively dark plumage that is heavily and crisply marked, with scallop-shaped markings on the underparts. Its facial disk is pale with a narrow dark rim and whitish "eyebrows." The ear tufts are very small, and the wings reach beyond the short tail. Both gray and red morphs are present.

HABITAT: Inhabits humid pine-oak forests (also in open areas) in highlands, at 1400 to 2500 m, but more frequently above 1800 m.

NATURAL HISTORY: The species is resident throughout its range. It is an elusive, nocturnal owl. Little is known of its biology and ecology, but both are probably similar to other screech owls. Presumably nests in tree cavities or old woodpecker holes. It lays 4 to 5 eggs. Insects and other arthropods are probably its main prey. Its main call is a cricket-like "teerrrrrrrrrt," which is repeated at several-second intervals.

CONSERVATION STATUS: Lower Risk—Near Threatened (IUCN). Its status is uncertain, but it is currently considered rare and likely to be endangered. Noted as fairly common but very local by others. Occurs in North-Central American Highlands Endemic Bird Area. Potential threats include habitat fragmentation in Chiapas. Although population levels are unknown, it is suspected that numbers may be decreasing.

68 Rufescent Screech Owl, *Otus ingens*

DESCRIPTION: The rufescent screech owl is a large screech owl (24 to 28 cm) that lives in the eastern Andes, along a narrow strip from Venezuela to Peru and northern Bolivia. It has fairly uniform plumage, small ear tufts, large brown eyes and a long tail. The buff-brown facial disk has ill-defined rings and a weakly defined rim; the "eyebrows" are pale. In the brown morph, the upperparts are tawny-olive to light gray-brown (sandy rufous in the rufous morph) with fine, dark vermiculations. There is a distinct, whitish scapular line. The underparts are light with fine barring of light brown. The tarsus is densely and completely feathered, and the talons are powerful.

HABITAT: The species is found on steep Andean slopes, from 1200 to 2250 m. The habitat is described as dense, humid forest with epiphytes; also scrub.

NATURAL HISTORY: The owl is a permanent resident throughout its range. Strictly nocturnal, it becomes active at dusk, and roosts by day on branches among epiphytes or against tree trunks. It probably

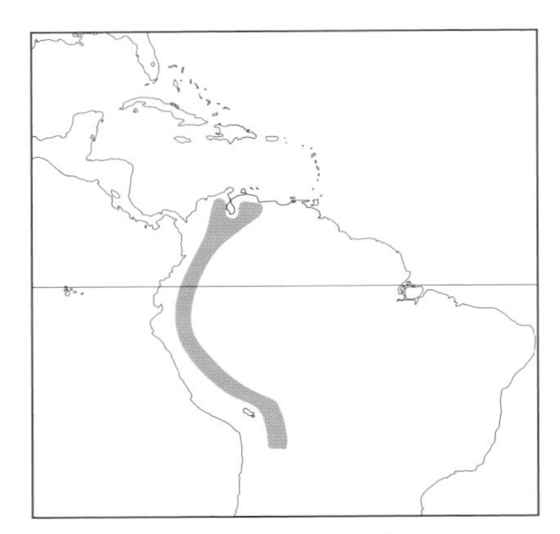

nests in tree cavities; it eats mainly insects, spiders and small vertebrates, which it apparently hunts in middle story and canopy. Its primary call is a 10-second-long series of 50 or so "ut" notes.

CONSERVATION STATUS: Not Globally Threatened (IUCN). The species is present in 2 protected areas in Colombia: La Planada and Rio Nambi National Reserves. Habitat loss due to deforestation is probably a threat to populations. The status of this owl is uncertain, but likely it is rare.

69 Colombian Screech Owl, *Otus colombianus*

DESCRIPTION: A large screech owl (26 to 28 cm) with uniform rufescent-brown or grayish-brown plumage, with tawny under-feathering. It has short ear tufts, a long half-feathered tarsus, powerful talons and brown eyes. The scapular stripe is indistinct and more buff than white; the flight feathers and underparts lack white coloring.

HABITAT: Inhabits dense undergrowth in cloud forests. Found from 1300–2300 m elevation.

NATURAL HISTORY: A resident, nocturnal bird. Probably nests in tree cavities, and eats large insects and small vertebrates. Its main call is a sequence of 10 or more flute-like "bu" notes. There is no other reliable information on this species, which requires study.

CONSERVATION STATUS: Lower Risk—Near Threatened (IUCN). Can be found in

the Chocó Endemic Bird Area. Its status is uncertain. Habitat loss due to deforestation is a threat to this restricted-range species.

70 Cinnamon Screech Owl, *Otus petersoni*

DESCRIPTION: This owl, first discovered in 1976, was named in honor of the late Roger Tory Peterson. It is relatively small (21 cm), and distinguished by its uniform warm buff-brown to cinnamon-brown plumage. It has medium-length ear tufts, dark brown eyes, a thin, buff nuchal collar, and almost fully feathered tarsi. Unlike most *Otus* owls, there are no white markings on the wings or body.

HABITAT: The species is found in cloud forest that is very moist, with dense undergrowth, mosses and epiphytic plants. Occurs between 1690 and 2450 m. While locally sympatric with the rufescent screech owl, it prefers higher elevation, though in some areas they live side by side at the same elevation.

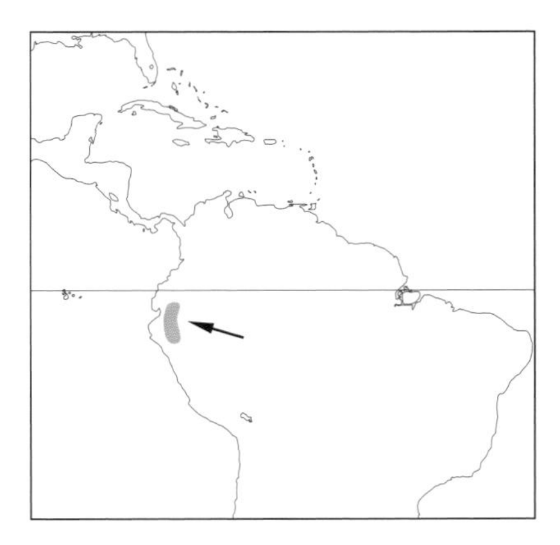

NATURAL HISTORY: A resident, nocturnal bird. No nests have been recorded, but it probably nests in tree cavities. Its breeding biology is probably similar to that of other screech owls. Eats mainly insects, perhaps occasionally takes small vertebrates. Its main call is a fast series of "bu" notes that starts at a low volume and ends suddenly.

CONSERVATION STATUS: Not Globally Threatened (IUCN). This is a restricted-range species. Occurs in Ecuador-Peru East Andes Endemic Bird Area. The status is uncertain, but it is probably rare, and doubtless threatened by deforestation. Its status needs to be reassessed.

71 Cloud Forest Screech Owl, *Otus marshalli*

DESCRIPTION: The cloud forest screech owl, discovered in 1967, was 1 of 3 new owl species discoveries over a few years. It generally resembles other screech owls, but has transverse white spots separated by black and rufous bars and streaks over the entire underparts. The upperparts are chestnut-colored with black bars but no streaks or spots. A dark, prominent rim borders the facial disk. The short ear tufts have dark outer edges, buff with brown bars on the inner. The eyes are dark brown, and its legs are feathered to the base of the toes. The lower half of the belly is off-white and darkly streaked with thin crossbars.

HABITAT: Generally found at 1900 to 2500 m, this owl is 1 of 4 screech owls occurring at varying altitudes along a slope from the valley floor to the cordillera crest. These are, from lowest to highest elevation: tawny-bellied, rufescent, cloud forest and white-throated, the latter in the highest montane forest. The habitat of this species is persistently cold and misty to wet. The cloud forest is characterized by dense undergrowth, rich in mosses and epiphytes, climbing bamboo, tree ferns and orchids. Trees grow to a height of 40 m.

NATURAL HISTORY: A resident, nocturnal bird, with habits likely the same as other screech owls. Virtually no information

available on breeding biology; probably nests in tree cavities. Likely insectivorous. Requires study in all areas of its natural history. A monotone, fast series of "eeu" notes is reported as its main call, but more research on its vocalizations is needed.

CONSERVATION STATUS: Lower Risk—Near Threatened (IUCN). It is a restricted-range species, and occurs in the Peruvian East Andean Foothills Endemic Bird Area. A little-known bird, its status is uncertain, but it is locally abundant (e.g., Cordillera de Vilcabamba at 2130 to 2190 m). Deforestation is a long-term threat.

72 Tawny-bellied Screech Owl, *Otus watsonii*

DESCRIPTION: This medium-sized screech owl (19 to 24 cm) is elusive and poorly known. The upperparts are dark grayish-brown with small light spots, giving the bird a dusty appearance. The upper breast is dusky brown, with dark and light mottling. The underparts are sandy-brown with thin dark shaft streaks, crossbars and fine vermiculations. The eyes are brownish-orange to yellow.

HABITAT: This owl is found in the interior of dense rainforests, where it prefers the lower story and is often less than 30 m above the ground. Less frequently, it occurs in clearings or edges, and remnant forests along rivers. It shows a preference for old-growth primary forest. It is usually at 600 m above sea level, but has been recorded at 2100 m in the Perijá Mountains and northeast of the Orinoco River.

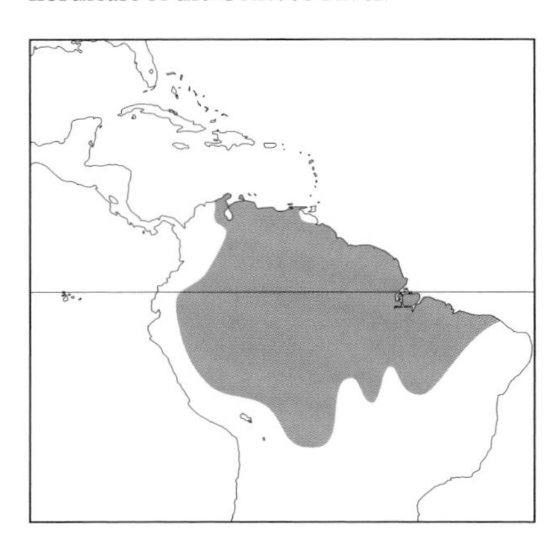

NATURAL HISTORY: A resident, nocturnal species. No information is available on its breeding biology. Forages in lower levels for insects, and possibly small vertebrates on occasion. Its main call is a staccato series of "bu" notes similar to that of the cinnamon screech owl, but is relatively faster and lasts longer.

CONSERVATION STATUS: Not Globally Threatened (IUCN). Nominate races occur in these protected areas: Amacayacu National Park in Leticia, Colombia, and Imataca Forest Reserve in Bolívar, Venezuela. The status is uncertain, but it is probably threatened by deforestation.

73 Guatemalan Screech Owl, *Otus guatemalae*

DESCRIPTION: This medium-sized screech owl (20 to 23 cm) occurs in brown and rufous morphs. It most closely resembles the vermiculated screech owl, but unlike that species its legs are feathered

to the base of the toes. The brown morph has a light-brown facial disk with a narrow, dark border. White "eyebrows" contrast with short, dark ear tufts. The upperparts range from dark gray-brown to blackish-brown with darker streaks and vermiculations. Underparts are pale with blackish shaft streaks and vermiculations, especially on the breast.

HABITAT: The species occupies humid lowlands, semi-arid evergreen forest, thorn forest, semi-deciduous forest, dense secondary growth or scrubby woodland and in plantations, from sea level to 1500 m.

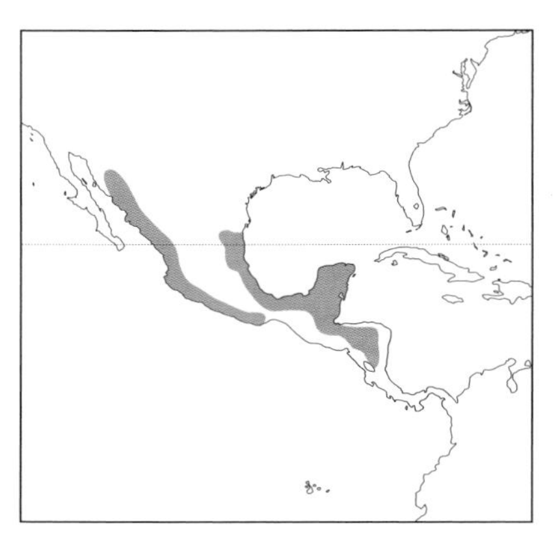

NATURAL HISTORY: This nocturnal bird, which roosts in thick foliage or cavities, is resident throughout its range. Not often seen, it can be lured closer with playback. Its main call is an extended trill of "u" notes lasting up to 20 seconds. Its 2 to 5 eggs are laid in natural tree cavities, frequently woodpecker holes and knotholes, but it has also been observed using an old trogon nest. Primarily insectivorous, it eats beetles, katydids, orthopterans, mantids and phasmids. Other prey include frogs, fish, reptiles and sometimes rodents. Often hunts in openings or along the edge of the forest. Catches prey by swooping down on it on the ground or on branches, or in flight.

CONSERVATION STATUS: Not Globally Threatened (IUCN). The status of this bird is uncertain, but it seems to be locally common. Habitat loss through deforestation poses a threat over the long term.

74 Vermiculated Screech Owl, *Otus vermiculatus*

DESCRIPTION: This rather small screech owl (20 to 23 cm) occurs in a grayish-brown morph and a rufous morph. It has short ear tufts, yellow eyes and an indistinctly rimmed light rufous-brown facial disk. Light "eyebrows" are barely visible, sometimes lacking altogether. There is a prominent white scapular stripe. The underparts have delicate dark shaft streaks, with dense light and dark vermiculations.

HABITAT: Found up to 1200 m in humid tropical forest, lowlands and foothills.

NATURAL HISTORY: Generally shares the same habits as other screech owls. It is resident and nocturnal, and responds readily to playback. The main call is a trill, reminiscent of a purring cat, that starts and ends with relatively quiet "u" notes. Its 3 eggs are laid in natural cavities in trees, old woodpecker holes or knotholes. Primarily insectivorous, but possibly also takes small vertebrates.

CONSERVATION STATUS: Not Globally Threatened (IUCN). The species requires almost solid forest, hence it is somewhat restricted in its habitat use. It appears to be locally common, but deforestation poses a threat over the long term.

75 Hoy's Screech Owl, *Otus hoyi*

DESCRIPTION: This is a medium-sized owl (23 to 24 cm) with small ear tufts and yellow eyes. Its gray-brown facial disk is well defined with a black border. The back of the head has a small semi-circular pattern of white feathers. The upperparts are marked with dark vermiculations and streaks, and the scapulars have a long row of large white spots. The light underparts are strongly streaked with black, and each side of the upper breast and flanks has strong black shaft streaks that are spade-shaped. Three morphs are present: gray, brown and rufous.

HABITAT: Occupies montane forest and cloud forest of the eastern Andean slopes, usually from 1000 to 2800 m. Normally occurs in forest areas where there are many epiphytes, as well as dense shrubby undergrowth, climbing bamboo and creepers. In southeast Salta it has been recorded close to dry bush and closed forest.

NATURAL HISTORY: It may migrate to lower altitudes in winter. The species is nocturnal and usually roosts in the dense foliage of the crown of a tree or among epiphytes on branches. Its eggs are laid directly onto the floor of natural cavities in trees, most often old woodpecker holes. One study revealed that Hoy's screech owl is present only in forests with old woodpecker cavities. It is possible that breeding may coincide with the onset of heavier rainfall. The male broadcasts a loud series of staccato "bu" notes, ending with a "bop." This owl responds readily to playback, and up to 6 birds (male and female) have approached their taped calls, suggesting high densities in some areas. Territories appear small, and loose colonies are sometimes formed. The species is primarily insectivorous, but has been observed capturing spiders, locusts and moths. Forages in the upper story, undergrowth, forest edge and on the ground.

CONSERVATION STATUS: Not Globally Threatened (IUCN). Occurs in some protected areas in Argentina, including El Rey National Park and Salta. Locally quite common. Parts of the range of this species are inaccessible, which may prove to be important to its survival as more accessible areas are being logged or overgrazed.

76 Variable Screech Owl, *Otus atricapillus*

DESCRIPTION: This medium-sized screech owl (23 to 24 cm) has a dark blackish crown, prominent ear tufts and dark brown eyes in the brown morph, and golden-yellow to yellow-brown eyes in the gray and rufous morph. Coloration and plumage patterns are extremely variable, as its name suggests.

HABITAT: The species inhabits primary and secondary forests within extensive rainforests. This habitat is characterized by dense undergrowth, epiphytes and climbing bamboo. Old trees with cavities are an important feature of its habitat. The variable screech owl is also found near forest edges, open woodland, and near human settlements and busy roads. In the northern part of its range it is found to at least 600 m, and up to 250 m in the south.

NATURAL HISTORY: Movements are unknown, but it is probably a resident species. It is nocturnal, roosting in tree cavities or among dense foliage during the day. During the breeding season, pairs tend to nest in close proximity to conspecifics, suggesting they are somewhat semi-colonial. The birds are bold and can be approached closely, and they respond readily to playback. Its main call is a rapid, high-pitched trill up to 20 seconds long. Clutch size is probably 2 to 3 eggs, which are laid in natural cavities or old woodpecker holes. Preys mainly on locusts, moths, beetles, cicadas, spiders and perhaps occasionally on small vertebrates. Hunts from a perch, often in low undergrowth.

CONSERVATION STATUS: Not Globally Threatened (IUCN). There are no data on populations, but it is considered to be common locally. Deforestation poses a threat to the species, as it seems to require large tracts of forest in order to breed in colonies to keep populations stable; remnant forests may be unsuitable.

77 Long-tufted Screech Owl, *Otus sanctaecatarinae*

DESCRIPTION: It is a relatively large (24 to 27 cm) and heavy screech owl. Its ear tufts are long and wide, its body is bulky and the talons are strong. In the more common brown morph, the facial disk is ochre-brown with a distinct dark rim. The upperparts are brown with a touch of ochre and without vermiculations. The underparts are coarsely marked, but again without vermiculations. Gray and red morphs are similar. The eyes are light-yellow to orange-yellow, rarely light brown.

HABITAT: It can be found in predominantly semi-open forests and open pasture with scattered trees. It avoids extensive, dense forests. It is also found along forest edges next to farmland, in secondary growth, near human settlements with trees and in upland moors. Normally at elevations of 300 to 1000 m.

NATURAL HISTORY: Presumably resident. A nocturnal bird, it roosts in the dense tree foliage. Not as bold as some of the other screech owls, it responds to playback but does not approach too closely.

The main call is initiated with a quiet grunting sound, which then grows in volume to a gruff, fast trill. It nests in natural cavities such as knotholes or old woodpecker holes. One nest found in a knothole was 5 m above the ground. It eats mainly insects, especially grasshoppers, mantids, moths, beetles and cicadas; also spiders and small vertebrates. It is a skillful hunter that can launch from a perch and swoop down to capture prey on leaves, branches or the ground.

CONSERVATION STATUS: Not Globally Threatened (IUCN). It is locally common in some areas (Rio Grande do Sul and Santa Catarina), but not in others (north Misiones in northeast Argentina). The greatest threat to the species is habitat loss as a result of overgrazing, logging and burning. It is sometimes overlooked because of its similarity to the variable screech owl.

78 Puerto Rican Screech Owl, *Otus nudipes*

DESCRIPTION: A medium-sized owl (20 to 25 cm) with a round head, no ear tufts and orange-brown eyes. The lower half of the yellow legs is bare. The darkish, ruddy facial disk lacks a distinct rim. Brown and rufous morphs are present. The upperparts have light spots and narrow dark bars while the pale underparts are covered with many dark vermiculations and shaft streaks. The belly is whitish and not as patterned.

HABITAT: This bird can be found in dense forests, coastal thickets, caves, coffee plantations and patches of trees in human settlements.

NATURAL HISTORY: A resident bird on the islands in its range. It is nocturnal, and roosts in caves, thickets or dense foliage in trees. Its 1 to 4 eggs, normally 2, are laid in a variety of nest sites such as caves, crevices in limestone cliffs, natural cavities, knotholes, old woodpecker holes or even under the eaves of houses. Primarily insectivorous (grasshoppers, crickets), but sometimes also takes small vertebrates. The bird's call, a loud "coo coo" (similar to that of the burrowing owl), has led to the local name: "cuckoo bird."

CONSERVATION STATUS: Not Globally Threatened (IUCN). The species is endemic on Puerto Rico and the Virgin Islands. It can be found in Puerto Rico and Virgin Islands Endemic Bird Areas. Considered to be common on Puerto Rico, with no current threats. It is possibly extinct on Vieques. There have been no confirmed reports on St. Croix since the 19th century, and no 20th-century records from St. Thomas and St. John. It is considered to be critically endangered on the Virgin Islands because recent surveys (1995) produced no owls. Habitat loss and egg predation have contributed to its decline.

79 White-throated Screech Owl, *Otus albogularis*

DESCRIPTION: This relatively large screech owl (28 cm) has a white chin and throat that contrasts sharply with its overall dark plumage. Its complete lack of a white scapular stripe, large fluffy-feathered head with almost no ear tufts and large size make this bird easily identifiable. It is also the only screech owl in the higher Andes with such a bold white throat. The eyes are orange-yellow.

HABITAT: The species inhabits the humid montane rain and cloud forests from 1300 to 3600 m, but most often occurs at 2000 to 3000 m. It prefers stunted alpine forest, open forest and edges, glades and semi-open regions with tree patches. Bamboo thickets and epiphytes distinguish its habitat.

NATURAL HISTORY: Presumably resident, but possibly makes seasonal altitudinal movements. It is nocturnal. The male's call is a series of descending, mellow "bu" notes uttered at a rate of 4 to 5 notes per second. Little is known of its breeding biology, but it probably nests on the ground among grass and ferns, in abandoned bird nests in trees or bushes, and in tree cavities. One nest was found in a deserted cup-shaped nest just above the ground. This owl forages in the canopy where it preys primarily on insects. Apparently some small vertebrates are also taken.

CONSERVATION STATUS: Not Globally Threatened (IUCN). The status of this species is uncertain, but it is probably common. More information on its status is needed.

80 Palau Owl, *Pyrroglaux podarginus*

DESCRIPTION: This relatively small (22 cm) owl occurs on the remote Palau Islands, east of the Philippines. It is a dark reddish-brown owl with a big rounded head and barely visible ear tufts. The slightly lighter facial disk has thin rufous-brown rings with whitish lores (area between the eye and the base of the bill) and "eyebrows." The upperparts are lightly streaked, and the lighter underparts have pale spots and bars. The ear tufts are either brown or orange-yellow. The featherless legs are whitish. Somewhat distinct, the female has darker brown upperparts with fine black vermiculations, and pale to dark rufous underparts with white and brown bars or spots.

HABITAT: The species inhabits all forest types, including mangrove swamps, rainforest, woodlands, ravines and steep forested ridges.

NATURAL HISTORY: The species is resident on the Palau Islands. A nocturnal bird, it is frequently found hunting in and near villages. Its call is a series of "kwuk" notes spaced about 1 second apart, becoming more rapid when disturbed. Pairs and family groups remain together year-round in territories reported as 100 to 200 m diameter. Nests in cavities or hollow trees, where 3 to 4 eggs are laid. Preys on a variety of invertebrates including

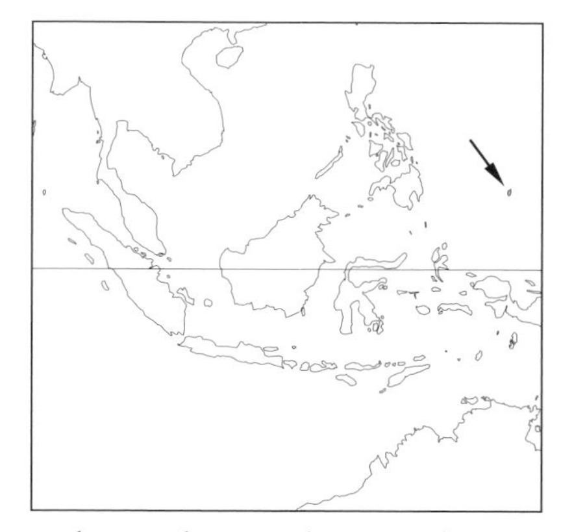

earthworms, large grasshoppers and centipedes.

CONSERVATION STATUS: Not Globally Threatened (IUCN). It occurs in the Palau Endemic Bird Area. Although its current population status is unknown, surveys in the mid-1940s found 33 pairs on Koror and 4 on Peleliu, and it was thought to be abundant in its small, restricted range. It is now considered by some to be highly endangered.

81 Cuban Screech Owl, *Gymnoglaux lawrencii*

DESCRIPTION: A rather small owl (20 to 23 cm) similar in appearance to the burrowing owl, but even smaller. It has a rounded head without ear tufts, prominent white "eyebrows" and long, bare yellow legs. The brown upperparts are spotted with brown, and the underparts are creamy.

HABITAT: Inhabits tropical semi-deciduous forest, thickets, plantations and semi-open limestone areas with caves and eroded gullies.

NATURAL HISTORY: Resident on Cuba and the Isle of Pines. A nocturnal bird, it roosts in the dense foliage of trees, rock crevices, caves or thickets. Its 2 eggs are laid in tree cavities, old woodpecker holes, caves or crevices in cliffs. Primarily insectivorous, but also preys on frogs, snakes and birds (rarely). It is usually seen on the ground where it hunts, but it will also hunt from a perch. It is locally called "cucu" after its call, a quiet ascending "coooo-cooo-coo-

cu-cu-cu."

CONSERVATION STATUS: Not Globally Threatened (IUCN). Thought to be somewhat common and widely distributed throughout its range. Considered rare on Cayo Coco, Cayo Romano, Guajaba and Sabinal Peninsula.

82 Northern White-faced Owl, *Ptilopsis leucotis*

DESCRIPTION: Quite striking in appearance, this is a relatively large, stocky scops owl (19 to 25 cm) with orange eyes, a whitish face with a prominent black rim, and black bands along the long ear tufts. Its overall plumage is pale grayish-brown, with dark streaks and delicate vermiculations. Paler upperparts distinguish it from the southern white-faced owl.

HABITAT: The species occupies a variety of woodland habitats, with a preference for dry savannas with thorn trees. It is also found along wooded desert waterways, dense closed-canopy woodlands, clearings, forest edges, near human settlements, and suburban gardens and towns. Thick tropical rainforests and deserts are avoided. Sea level to 1700 m.

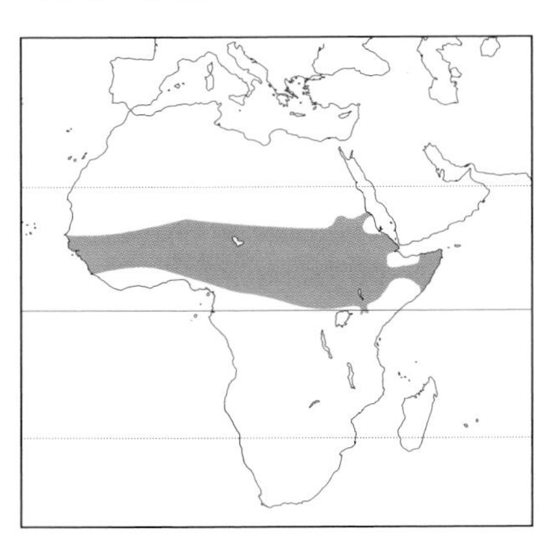

NATURAL HISTORY: The species is resident, but with some small dispersal movements related to prey abundance and rainfall. A nocturnal bird, it roosts in either relatively unconcealed sites in open canopy or against branches or tree trunks. Its bubbling and stuttering 6- to 8-note "po-prool" call is distinctive. When roosting it will adopt a "tall-thin" posture, with its body elongated, eyes half-closed and ear tufts erect. Its 2 to 3 (sometimes 1 to 4) eggs are laid on either small, unstable nests of other birds (e.g., pigeons), larger stick nests (e.g., eagles), the tops of bulky nests (e.g., buffalo weavers) or in tree cavities or on crotches (usually 2 to 8 m above the ground). Males sometimes incubate for up to 30 minutes during the night. Nests have been recorded only 200 m apart, suggesting small territories, but they are still defended by the pair. Eats primarily small mammals, but also insects and sometimes birds.

CONSERVATION STATUS: Not Globally Threatened (IUCN). It is locally common to uncommon throughout its distribution, but rare in Somalia. Present in many protected areas: Abuko Nature Reserve (Gambia), west National Park (Niger), Ouadi Rimé Reserve (Chad and Bamingui-Bangoran) and Manovo-Gounda-Saint Floris National Parks (Central African Republic). May be vulnerable to pesticide use and habitat loss.

83 Southern White-faced Owl, *Ptilopsis granti*

DESCRIPTION: This rather large scops owl (20 to 24 cm) is very similar to the northern white-faced owl. However, its plumage is grayer, and the upperparts and crown are darker gray with black shaft streaks. The facial disk is whiter and the black rim broader and bolder. Adults have orange-red to red eyes.

HABITAT: The species occupies savannah scattered with trees and thorny shrubs, dry open forest (e.g., mopane woodlands), forested regions beside rivers, the edge of forests and openings.

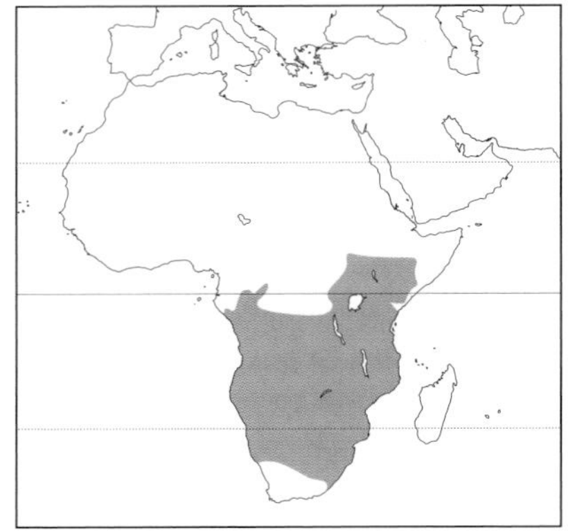

NATURAL HISTORY: Although the species is generally resident, it is described as nomadic in dry areas, with irruptions occurring after rains or as a result of prey abundance fluctuations. Apparently it will fly long distances to grass fires in order to prey on flushed rodents and insects. It is nocturnal, and its call is a series of staccato "whhhh" trills ending with a higher pitched "hooh." Its 2 to 3 eggs are laid in tree cavities or abandoned stick nests of other birds. Eats mainly large insects, spiders and scorpions, but also small mammals, birds and reptiles. Its powerful talons enable it to capture bush squirrels and doves.

CONSERVATION STATUS: Not Globally Threatened (IUCN). Thought to be common in most of its range. It is scarce in Kenya but rare in northern Tanzania and at the northern limits of its distribution. Present in many protected areas: Lilonge Nature Sanctuary (Malawi); Hwange

(Wankie); Victoria Falls, Gona-re-Zhou and Rhodes Matopos National Parks (Zimbabwe); Moremi National Park (Botswana); Etosha National Park (Namibia); Gorongosa Reserve (Mozambique); and Kalahari Gemsbok and Kruger National Parks (South Africa). Endangered in areas of pesticide use.

84 Giant Scops Owl, *Mimizuku gurneyi*

DESCRIPTION: The largest of the scops owls (30 to 35 cm), its brown to dark rufous plumage is spotted and streaked with black.

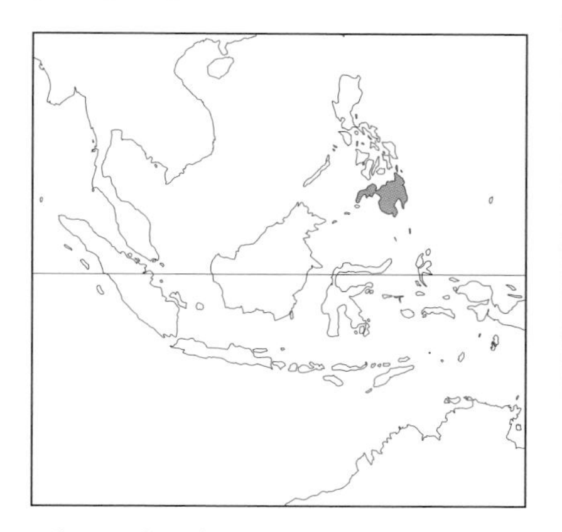

It has a pale rufous facial disk narrowly bordered in black, broad white "eyebrows" and long ear tufts. The white underparts and rufous-buff breast have broad, drop-shaped streaks, giving it a unique appearance among the scops owls.

HABITAT: Found in lowland rainforest and secondary growth. Sea-level to 1200 m, with a questionable report up to 3000 m.

NATURAL HISTORY: The species is resident in its southern Philippines range. It is strictly nocturnal, and its call is a coarse snarling "wuaah" call note repeated at 10 to 20 second intervals. There is no other information available on its biology or ecology. It probably eats insects and small vertebrates.

CONSERVATION STATUS: Vulnerable to Endangered (IUCN). It is rare throughout most of its restricted range. Thought to occur naturally at low densities. Habitat loss is a serious threat, and is the cause of population declines. Despite this, it is believed most montane species may not be in extreme danger. Recently, it has been found at higher altitudes (Mt. Apo and Mt. Katanglad) with suitable remaining habitat. Additional surveys of suitable habitat on surrounding islands may document new occurrences.

85 Great Horned Owl, *Bubo virginianus*

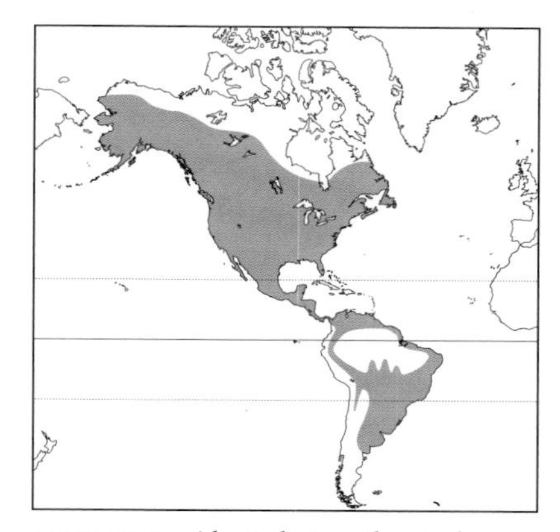

DESCRIPTION: The great horned owl is a large (46 to 63 cm) and heavy (up to 2.5 kg) long-lived owl with ear tufts, a white bib and large yellow eyes. It has exceptionally strong feet and talons. It is generally a brown, buff and black-lined owl with a medium-sized facial disk. Its color varies with regional humidity, ranging from the darkest populations found along humid coastal regions to the palest individuals found in deserts and subarctic prairies.

HABITAT: This owl can survive in deserts, grasslands, human-altered landscapes and forested habitats. Consequently, it has the most extensive distribution of any American owl. It appears unable to survive in arctic-alpine regions.

NATURAL HISTORY: The great horned owl hunts mostly at night, but may be seen pursuing prey in broad daylight. It can feed on a great variety of prey, from small rodents to birds as large as geese and herons. The majority of its prey are mammals, including rabbits, hares and small rodents. Most great horned owls are permanent residents, but southeasterly irruptions from northern regions occur in response to population crashes of snowshoe hare in boreal forest and aspen parkland. One winter, an owl repeatedly returned to feed on a large prey item over several days, and sat on it daily for up to an hour to thaw the portion it consumed. Typical territorial call is a deep and powerful 3 to 6 notes: "who-ho-o-o, whoo-hoo-o-o, whoo," or variations thereof. Nests in stick nests in trees, but also uses tree cavities, snags, cliffs, deserted buildings and artificial platforms; has even nested on the ground. It lays 2 to 3 eggs, rarely as many as 5.

CONSERVATION STATUS: Not Globally Threatened (IUCN). This owl is considered widespread, locally abundant, secure and not in need of management, but information on long-term population trends is lacking. Starvation of young owls is the most significant mortality factor affecting populations. Locally, secondary poisoning by accumulation of pesticides such as anticoagulant rodenticides or strychnine from contaminated prey may cause death. Control or removal of individual great horned owls that prey on species at risk, such as the roseate tern, is sometimes necessary. The effect of this powerful predator on other raptor species, such as the spotted owl, needs study.

86 Magellanic Horned Owl, *Bubo magellanicus*

DESCRIPTION: This species is very similar to the great horned owl, but with some notable differences. It is smaller in

size (45 cm), and also has a smaller bill and ear tufts. The border around the facial disk is more prominent, and its talons are weaker. The barring on the underparts is narrower and denser, and its vocalizations (see below) are completely different. Overall plumage ranges from gray-brown to dark brown.

HABITAT: Occupies rocky mountainous regions with pasture, semi-open forests (especially those dense with mosses and lichens), semi-desert, shrubby ravines and adjacent grasslands. Also found in human settlements and parks. Found locally at sea level, but at 2500 to 4500 m in the Andes.

NATURAL HISTORY: Considered to be resident, with southern birds moving north in winter and juveniles wandering in autumn. Both nocturnal and crepuscular, with occasional activity before sunset in its southern range. The male's call ("bu-hoohworrrr") is 2 deep hoots plus a final deep but quiet purr. Birds roost against tree trunks, concealed by lichens and foliage; cliff crevices; ledges or cave entrances. Its 2 to 3 eggs are laid on the ground (in a shallow depression scratched out by the male), often at the base of a tree or under a branch. It will also use abandoned stick nests of other birds, or a crevice or cavity on cliffs and between rocks or beneath rocky overhangs. Both sexes are aggressive at the nest. Eats mainly mammals, but also a wide variety of prey including birds, reptiles and invertebrates.

CONSERVATION STATUS: Not Globally Threatened (IUCN). Common in Chile, Patagonia and Tierra del Fuego. Threats include human persecution, road casualties and possibly poisoned fox bait in Patagonia.

87 Eurasian Eagle Owl, *Bubo bubo*

DESCRIPTION: Among the largest living owls (58 to 72 cm) in the world with a body weight of 2400 to 3000 g. Its upper body is golden-brown to tawny-buff with black feather tips; the feathers on the nape and underparts are broadly streaked in black. Considerable geographic color variation; intergrades from darkest forms in Scandinavia to grayer forms in northern Africa and those with an ochraceous wash in parts of Asia and Iran. Its eyes are red-orange to bright golden-yellow, and its bill is black. The tarsi and toes are densely feathered to the black claws. Prominent ear tufts crown an indistinctly rimmed facial disk.

HABITAT: This versatile owl has been documented in a great variety of habitats and climate zones, including wooded and

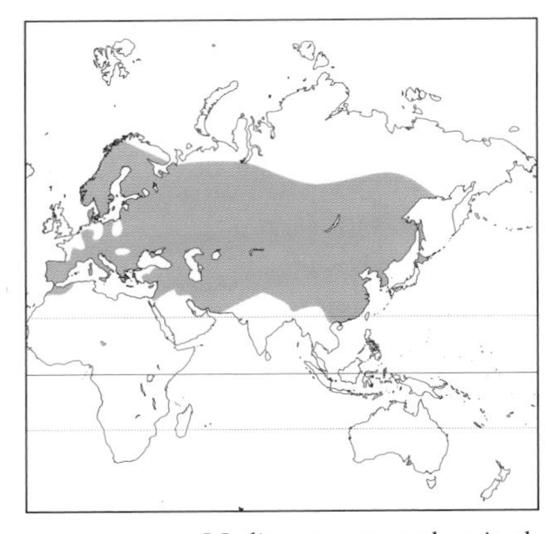

grassy steppes, Mediterranean scrub, mixed deciduous and boreal coniferous forests. It also inhabits deserts, providing there are rocky outcrops or other dark hiding places such as cliff faces. Recorded nesting at elevations of 2100 m (Alps) and reportedly seen up to 4700 m (east Tibetan highlands).

NATURAL HISTORY: This owl is sedentary and highly territorial. Marked young typically remain within 100 km of their natal home range, only rarely dispersing farther than 200 km or more. Some southward irruptive flights of starving individuals from northern parts of its range occur, presumably in response to prey scarcity. Its best known call is a resonant "oohoo" or "oohu," sometimes a simple "hoo." It also utters a barking sound if alarmed, while exchanges between paired male and female include rattling, growling and laugh-like sounds. Its 1 to 2, rarely 3 to 4, eggs are laid in nest scrapes (a shallow, saucer-like depression) on the ground or in holes, crevices and fissures of rocky slopes or cliff faces protected by overhanging rocks, bushes or trees, such that few mammalian predators can easily access them. Rarely uses stick nests of large raptors. Young leave the nest when 5 to 6 weeks old and can fly at about 7 to 8 weeks. The young become independent as late as October. While largely nocturnal, it may hunt during the day, especially in northern areas with limited sunless periods during summer. It preys opportunistically on a great variety of species—over 250 species recorded to date—from small rodents and shrews to species as large as half-grown wild sheep and their lambs and roe deer fawns. Preys heavily on other owl species and on diurnal raptors and their nestlings; perhaps these species are taken simply as prey, but they may also be killed to eliminate possible competitors for food.

CONSERVATION STATUS: Not Globally Threatened (IUCN). Conservation efforts to reintroduce the Eurasian Eagle Owl in parts of Germany, Belgium and Switzerland have resulted in the unfortunate local extirpation of the peregrine falcon, another species of conservation concern. While some populations of this owl have been eliminated or reduced by direct persecution by people, others have benefited from hunting habitat created by deforestation activities. Many owls are killed by direct collision with overhead wires and by electrocution on high power lines.

88 Rock Eagle Owl, *Bubo bengalensis*

DESCRIPTION: A large brown owl (50 to 56 cm) with an oval-shaped facial disk, long ear tufts and orange to deep orange-red eyes. Its legs are fully feathered, and it has very strong talons for its size. Both light and dark morphs occur, with the latter being darker overall and with fewer light spots above.

HABITAT: Occurs in a variety of habitats, such as rocky country with hills and bushes, forests with scrub and ravines, old mango plantations near human settlements and semi-desert containing thorny brush and rocks. Humid evergreen forest and pure desert are avoided.

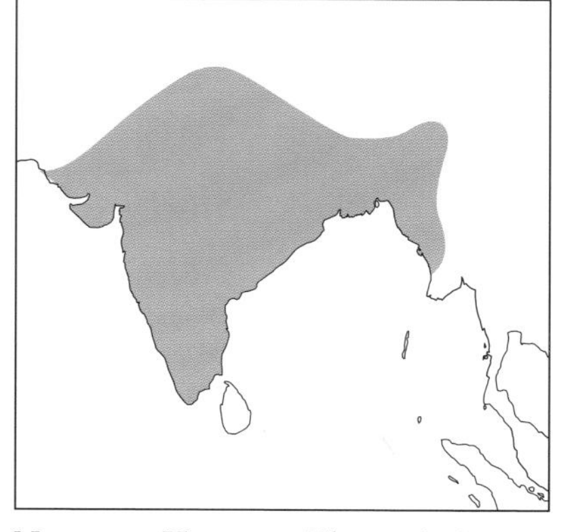

NATURAL HISTORY: The species is resident. Predominantly nocturnal, but can be observed before sunset and after sunrise. The male has a low 2-note "bu-whuoh" call that is repeated every few seconds. Roost sites include branches with dense foliage, rock crevices, sheltered cliff ledges or vacant buildings. Its 4 eggs (2 to 5) are laid directly onto the substrate of the ground, on a sheltered rock ledge, between rocks on a slope or in a recess in a cliff. Both sexes will scratch out the nest scrape. It eats mostly mammals (rodents), but also birds, frogs, reptiles, crabs and insects. It has a deep double-hoot call, with emphasis on the second note; also hisses, coughs and growls.

CONSERVATION STATUS: Not Globally Threatened (IUCN). Uncommon, but may be more common in northern and central India. More information is needed to assess population levels.

89 Pharaoh Eagle Owl, *Bubo ascalaphus*

DESCRIPTION: This relatively large eagle owl (46 to 50 cm) is similar to the Eurasian eagle owl, but is smaller in size, with shorter ear tufts and paler plumage that is more mottled and less streaked. Its sandy-ochre plumage has black streaks and blotches, and its upper chest has drop-shaped black streaks, with the lower breast finely vermiculated.

HABITAT: The species is found in rocky deserts and semi-desert, outcrops of oases, temporary watercourses with cliffs,

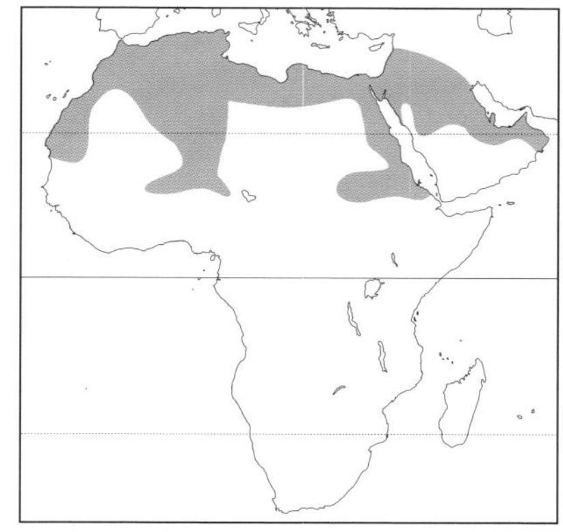

mountainous regions, and arid rocky mountain slopes with trees or bushes. In the south, it occasionally occupies dry savannas.

NATURAL HISTORY: A resident species; young birds are likely to wander and disperse. Nocturnal, it roosts between rocks, at cave entrances, on the ground, high on steep cliffs or in trees. Its courtship behavior is similar to that of the Eurasian eagle owl, and a "hu-huhooh" call has been associated with pair formation. The male's main call is a terse, descending "buo." Adults are monogamous, pair for life, and use the same territories for years. Its 2 eggs (sometimes up to 4) are laid directly onto a scrape among rocks, in a crevice, on the ground, in tree cavities or in old nests of larger birds. It has even nested deep within a dark water well. Hunts from a perch. Eats primarily mammals (e.g., gerbils, hares, bats, fox, hedgehogs), but also takes birds, reptiles, insects and scorpions.

CONSERVATION STATUS: Not Globally Threatened (IUCN). The status is uncertain, but it is likely common throughout most of its range. Human persecution may cause local populations to become endangered. Population levels are poorly studied.

90 Cape Eagle Owl, *Bubo capensis*

DESCRIPTION: This owl is large (46 to 61 cm) and very powerful. Three subspecies are distinguished by size. Overall plumage is dark brown with the sides of the upper breast heavily marked with blackish

blotches, and heavily mottled and barred on the wings, tail and back. The ear tufts are long and pointed, the eyes yellow to orange-yellow, and the legs and toes densely feathered.

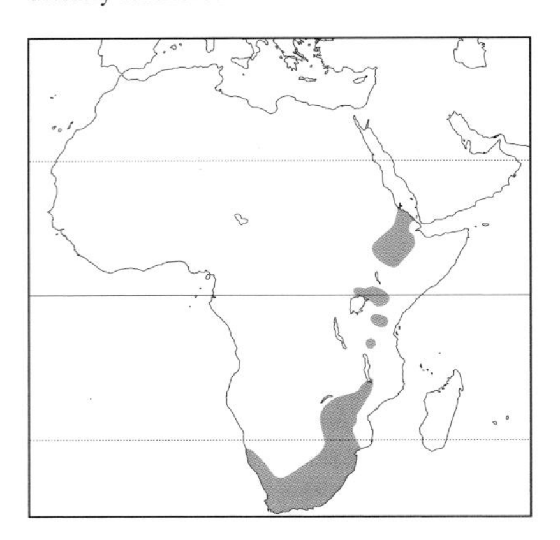

HABITAT: The species occurs predominantly in rugged mountains (from 2000 to 4200 m) where it prefers wooded gullies, rocky areas and gorges usually associated with running water. Also can be found in flat, dry, open country. Farther south it is found in open savannah, and locally in towns.

NATURAL HISTORY: Although generally resident, juveniles may wander, and adults are sometimes found beyond their normal range outside the breeding season. It is nocturnal and crepuscular. The male's loud, deep and startling "boowh" hoot is followed by a quieter "hu" note. Roost sites are secluded: on the ground under bushes, between rocks, in crevices or holes in cliffs, beneath overhanging rock ledges, in dense foliage of trees or locally on buildings in towns. This owl is reluctant to fly when approached. Nest sites are usually discovered by the presence of thick white droppings or the accumulated remains of prey. Two eggs (1 to 3) are laid directly on a shallow scrape on the ground, in a sheltered rock ledge, at a cave entrance, between rocks or in abandoned stick nests of large birds. Does not always breed every year. Forages opportunistically. Eats mainly mammals, especially hares and hyraxes, but also young antelope, bats, small rodents, mongoose, genet and civet. Also takes birds, reptiles, amphibians, crabs, scorpions and insects.

CONSERVATION STATUS: Not Globally Threatened (IUCN). Status varies from locally rare or absent in some areas to common in others (Mau Plateau in Kenya). Main threats are nest predation, vehicle traffic, hitting power lines and barbed wire, pesticide use and being taken for pets, usually resulting in death or ill health due to improper diet. Despite its wide but discontinuous distribution, the species is elusive and difficult to find.

91 Spotted Eagle Owl, *Bubo africanus*

DESCRIPTION: This is the most common and familiar owl in southern Africa. It lacks the size (45 cm) and strength of the other African eagle owls, and has rather small feet. The upperparts, crown and neck are dusky-brown with whitish spots. Below, it is grayish-brown to creamy-white with dark bars. The ear tufts are long, and the gray facial disk is rimmed in black. The eyes are yellow. Despite its name, it is more strongly barred than spotted. The rare chestnut-brown morph occurs in more arid regions and is browner above and buff below, with orange eyes.

HABITAT: Occurs in a variety of habitats, from open to semi-open woodland with bushes, savannah with thorny shrubs, riverine woods adjacent to rocky hills or stony slopes, desert and semi-desert and cultivated parklands. Favors low hilly areas with scrub and grassland. Dense rainforest is avoided. Sea level to 2100 m.

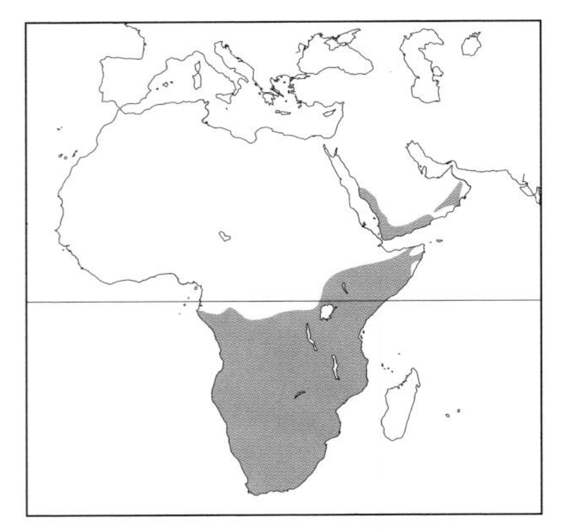

NATURAL HISTORY: Resident, but in Malawi moves to higher elevations in summer. Predominantly nocturnal, but sometimes hunts before sunset and in some areas will hunt during the day. The 3-note "wuhuhu" call of the male is followed by a lower "whooh." Roosts against tree trunks where it will do a "tall-thin" display if approached; also roosts on rock ledges, on the ground between rocks, under bushes, in tall grass, in cave entrances and in the burrows of larger mammals. Pairs mate for life, and nest sites may be used for decades. Its 2 to 4 eggs, usually 2, are laid in a wide variety of places on the ground or on rocky ledges, but also steep banks, amongst boulders, on buildings (e.g., window ledges, drainpipes, sloping roofs, window boxes), in old raptor stick nests, atop hamerkop stork or sociable weaver nests, haystacks or, less frequently, in tree cavities. An opportunistic hunter, it eats mainly arthropods (e.g., beetles, spiders, scorpions), mammals (e.g., moles, hares, bats, hedgehogs, gerbils) and birds (up to tern and falcon size). Additional prey taken include reptiles, snails, amphibians and crabs, occasionally fish and carrion. Its feeding and breeding habits explain its widespread and common distribution in southern Africa. Frequently observed because of its habit of hunting along roads and near streetlights.

CONSERVATION STATUS: Not Globally Threatened (IUCN). Widespread and common. May be endangered locally due to pesticide use. Other threats are road-kill casualties, collision with fences and overhead wires, bushfires and human persecution.

92 Greyish Eagle Owl, *Bubo cinerascens*

DESCRIPTION: A relatively small eagle owl (43 cm) with dark brown eyes and reddish-fleshy eyelids. The overall plumage is grayish-brown, and upperparts are finely

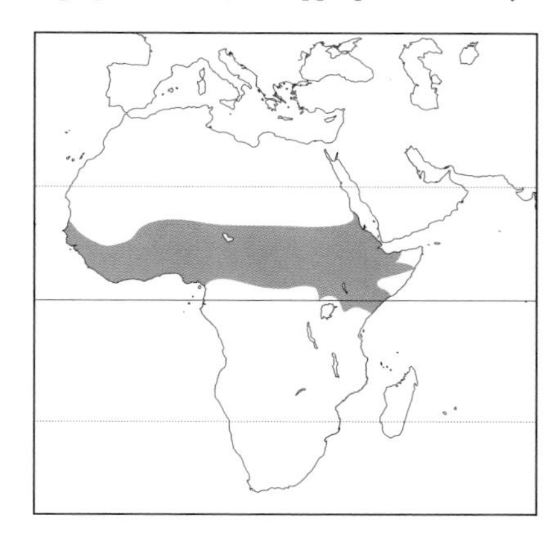

vermiculated with dark brown and a few lighter and darker spots. The underparts are also densely vermiculated, and the top half of the breast has some dark spots.

HABITAT: Occupies dry, rocky semi-desert, savannah with trees and thorn bushes, rocky mountain slopes, open hilly country, low woodland in Somalia, and towns and suburbs in the east. Rainforests are avoided.

NATURAL HISTORY: Nocturnal and crepuscular, with altitudinal movements in summer. The main call is a loud "kuo" followed by an extended low "wooh." It roosts in crevices, between rocks, holes in earth banks or walls, in trees, or on the ground under bushes, or between rocks. Its 1 to 3 eggs are laid in a scrape on the ground, between rocks or on cliff ledge, occasionally in abandoned stick nests of larger birds. Prey includes insects and other arthropods, birds, small mammals, bats, frogs and reptiles. Hunts from a perch or captures prey in flight.

CONSERVATION STATUS: Not Globally Threatened (IUCN). Status varies locally from common to rare.

93 Fraser's Eagle Owl, *Bubo poensis*

DESCRIPTION: A relatively small eagle owl (39 to 45 cm) with wide and large ear tufts, dark brown eyes with pale blue eyelids and a distinctly dark-rimmed, pale rufous facial disk. The fully feathered tarsus meets the bare, pale blue-gray toes. Overall plumage is rufous and buff with dark barring or blotches.

HABITAT: The species occupies predominantly lowland primary evergreen forest. Also present in secondary forest, clearings, forest edge and plantations. Sea level to 1600 m.

NATURAL HISTORY: A resident, nocturnal and crepuscular species. Its calls include a course staccato "kororororor" and a 2-note "twowooht" that is uttered at 3- to 4-second intervals. Roosts by day in tree foliage (up to 40 m above the ground). Details on its breeding biology are unknown, but nests in tree cavities and possibly on the ground. Young are in full

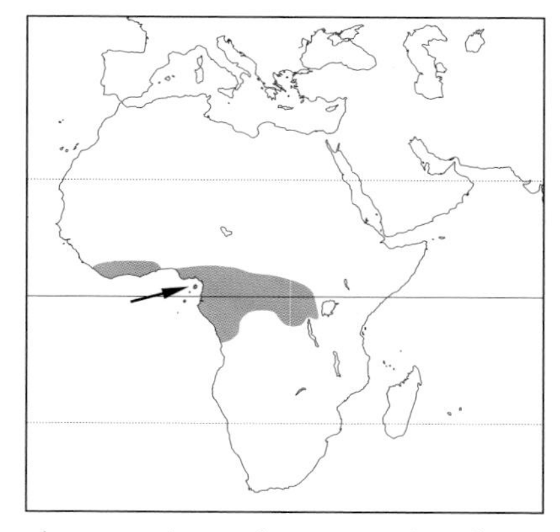

plumage at 1 year of age, suggesting a long post-fledging dependence. Eats mammals (e.g., mice, squirrels, tree hyraxes, bats and galagos) and insects (e.g., beetles, millipedes, grasshoppers); frogs and small birds are also taken. Questionable reports of it occasionally eating fruit.

CONSERVATION STATUS: Not Globally Threatened (IUCN). Uncertain status; appears to be common in Cameroon and Liberia, but uncommon in Sierra Leone. Occurs in some protected areas: Gola Forest Reserve (Sierra Leone), Dzanga-Ndoki National Park and Dzanga-Sangha Rainforest Reserve (Central African Republic) and Impenetrable (Bwindi) Forest National Park (Uganda). Appears to tolerate habitat disturbance and will use logged forest and secondary habitats.

94 Usambara Eagle Owl, *Bubo vosseleri*

DESCRIPTION: A relatively large eagle owl (48 cm) very similar in appearance to

Fraser's eagle owl, but larger, with browner upperparts, irregular barring on a whiter underside and a more densely marked breast. Its eyes are dull yellow-orange or orange-brown.

HABITAT: Occupies montane evergreen forest (900 to 1500 m) and forest edges of tea plantations (200 m).

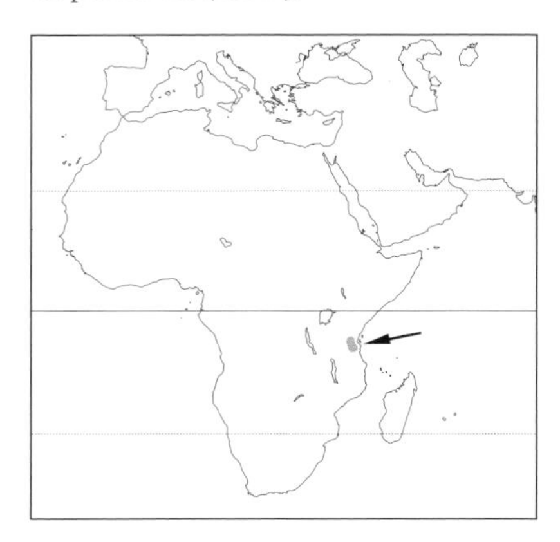

NATURAL HISTORY: Its call is described as a descending series of deep, quiet "po-a" notes lasting 5 to 7 seconds and repeated up to 4 times every 30 to 60 seconds. Virtually no other information is available on the natural history of this owl. It is thought to be a nocturnal and secretive bird.

CONSERVATION STATUS: Vulnerable (IUCN). A restricted range species, it occurs in the Tanzania-Malawi Mountains Endemic Bird Area. First discovered in 1908, but not seen again until 1962. Population is estimated at 500 pairs (2 pairs/square km). It is unlikely that all known breeding pairs are in a single subpopulation, hence its global population may be greater. Threats include forest destruction for subsistence farming, and tea and cardamon plantations. Cardomon is grown under an intact canopy, and young birds have been located in these areas, which suggests that undisturbed forest may not be critical to populations. Conservation efforts are currently underway to maintain known populations of this species.

95 Forest Eagle Owl, *Bubo nipalensis*

DESCRIPTION: A large and powerful eagle owl (51 to 64 cm) with remarkably long, outspread ear tufts of variably sized feathers. Overall plumage is brown, mottled and greatly barred buff-white. The breast is covered with dark brown V-shaped spots. The legs are completely feathered, and the eyes are brown to hazel-brown. It has been described as a "feathered tiger" because of its large prey and its aggressiveness while defending its nest or young.

HABITAT: Occupies dense evergreen and deciduous tropical lowland forest. In the north, including in the Himalayan foothills, it is found in hilly regions mixed with forest. In the south, it uses montane evergreen wet temperate forest. Generally from lowlands to 1500 m, in the Himalayas from 900 to 2100 m (locally to 3000 m), and at 1800 m in Sri Lanka.

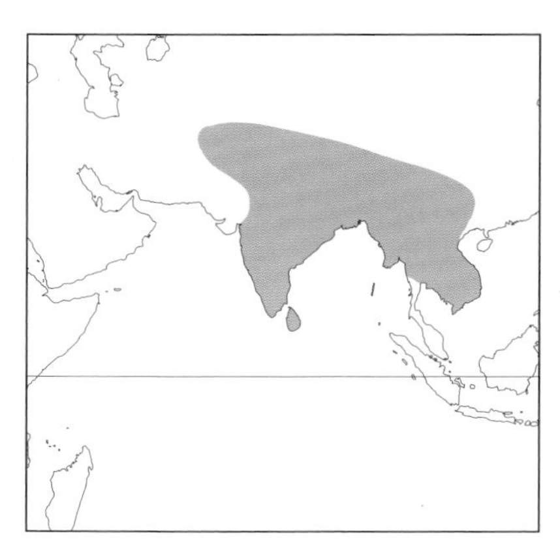

NATURAL HISTORY: The species is resident and mostly nocturnal, roosting in dense cover. After a courtship characteristically accompanied by dreadful shrieks and strangled cries, its single egg is laid in a hollow in an old tree, in an abandoned stick nest of a large bird, in a cave, or on a horizontal surface of a rock wall. Its regular call is a low 2-note "hoo-hoo" call uttered at 2-second intervals. A powerful hunter, the main diet consists of large birds such as jungle fowl, pheasants and peafowl. It also eats hares, jackals and young barking deer; also lizards, snakes and fish, as well as carrion. It hunts from a perch and often takes roosting birds.

CONSERVATION STATUS: Not Globally Threatened (IUCN). Some consider it near threatened. Local and rare in the Indian Subcontinent and Sri Lanka, uncommon in Thailand and rare elsewhere. Occurs at low densities throughout its range. Tropical lowlands and hill forest habitat is threatened and degraded areas may not support this species.

96 Barred Eagle Owl, *Bubo sumatranus*

DESCRIPTION: The barred eagle owl is a relatively small eagle owl (40 to 46 cm) with large brown eyes, and long, horizontally slanted, finely barred ear tufts. The blackish-brown upperparts are barred rufous-buff, and the flight feathers and tail are widely barred. Lighter underparts are also barred with brown. The breast is noticeably darker than the belly.

HABITAT: The species is found in evergreen and semi-evergreen forest with pools and streams, clearing, forest edge, secondary growth, old plantations, and forested gardens (e.g., Botanical Gardens, Bogor, Java), also near human settlements. Sea level to 1000 m, but in some areas up to 1600 m.

NATURAL HISTORY: It is resident throughout its range. A nocturnal or crepuscular bird, it roosts singly or in pairs in densely foliaged trees. It is assumed to pair for life, and the same nest site may be used for years. Its 1 egg is laid in holes of large trees or atop bird's-nest ferns. Eats large insects (e.g., grasshoppers, beetles), small mammals (rodents), birds and snakes. A low, descending "hoo" or "hoo-hoo" is its main call.

CONSERVATION STATUS: Not Globally Threatened (IUCN). Status varies across its range; uncommon in Thailand, uncommon to fairly common in Peninsular Malaysia, low densities in Bornea and locally fairly common in other areas. No serious current threats; ability to tolerate disturbed habitat is beneficial to the species.

97 Shelly's Eagle Owl, *Bubo shelleyi*

DESCRIPTION: A large and powerful eagle owl (53 to 61 cm) with wide shoulders and a large head with seemingly expressionless dark brown eyes. Overall plumage is dark with heavy barring. This is the largest and most heavily barred of the 3 eagle owls in the west African lowland forest. Fewer than 20 specimens are known.

HABITAT: Occupies primary lowland rainforest, also forest edges.

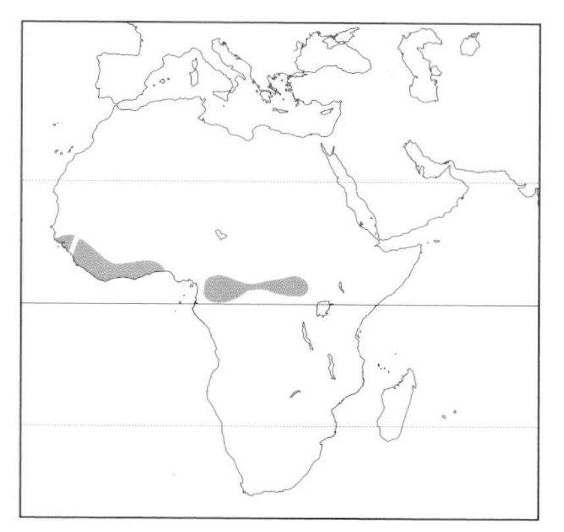

NATURAL HISTORY: Resident and nocturnal, it roosts in foliage, occasionally low above the ground. A wavering "kooouw" uttered irregularly is its main call. Virtually no other information is available regarding its breeding or food habits. Its size and powerful talons indicate it is capable of securing mid-sized to large prey; there is 1 record of it eating a large flying squirrel.

CONSERVATION STATUS: Not Globally Threatened (IUCN). Some consider it to be rare and endangered throughout its range. One of the least studied African owls, more information is needed to more accurately define its status. Habitat loss is the main threat to the species.

98 Verreaux's Eagle Owl, *Bubo lacteus*

DESCRIPTION: This is the largest owl (60 to 65 cm) in tropical Africa and one of the heaviest (3 kg) in the world. It is pale gray overall with fine darker gray vermiculations above and below. The whitish face is boldly bracketed with black and topped with stumpy ear tufts. Prominent black bristles surround a large pale-blue bill. Pink eyelids highlight its large dark eyes.

HABITAT: The Verreaux's eagle owl prefers savannah woodlands, especially riverine acacia forests. It occupies semi-arid regions, but is absent from arid regions within its range.

NATURAL HISTORY: It lays 1 to 2 eggs, rarely 3, at up to 7-day intervals, in stick nests built by other birds such as the Hamerkop, Wahlberg's eagle, bateleur and buffalo weavers, but also in tree hollows, and even on flat surfaces (roofs). The eggs are large, white and round with small nodules on the surface. The incubation period is estimated at 32 to 34 days, but also reportedly as long as 38 to 39 days, hence more study is needed. Females incubating eggs appear to lie as flat as possible to avoid detection. Only 1 young is raised; apparently it is often present during its parents' subsequent breeding efforts, and is sometimes fed by the male. Its variable diet includes termites, snakes, lizards, bats,

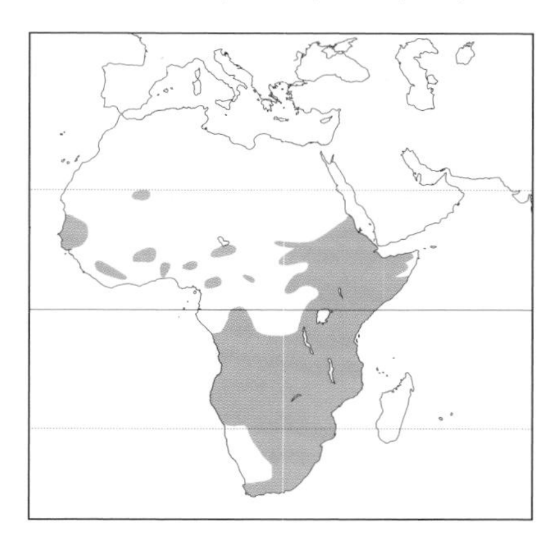

rodents, hares and hedgehogs; it peels the latter, leaving skins on the ground below nest sites. Evidently some hedgehog skin can be ingested without harm, for spines are sometimes found in pellets. It can catch insects with its feet while flying, yet its large powerful feet enable it to catch larger prey including monkeys, warthog piglets, mongoose and birds as large as secretary birds, eaglets and nestling vultures. It hunts mostly at night, but opportunistically takes prey during the day from shaded roosts on large branches. A quantitative analysis of prey taken is needed. The male's call is a series of deep nasal grunting "gwonk" notes uttered at irregular intervals. It is a year-round resident. A captive owl was observed to sunbathe while lying flat on the ground with its wings spread. The Verreaux's eagle owl often bathes in shallow water.

CONSERVATION STATUS: Not Globally Threatened (IUCN). Widespread, but can be locally rare because of direct human persecution and mortality from pesticide use. No population trend data exists, and more study is needed of its basic biology.

99 Dusky Eagle Owl, *Bubo coromandus*

DESCRIPTION: This is a large eagle owl (48 to 53 cm) with grayish or sooty plumage. Its ear tufts are narrow and prominent, often standing erect and close together. Dark brown shaft streaks that mark the plumage are wide and indistinct above, thinner and more defined below. The underparts are more whitish. The legs are fully feathered, and the eyes are yellow.

HABITAT: Occurs in flat, forested country near water. Also occupies old mango plantations, densely foliaged groves (especially tamarinds), roadsides, riparian forest and irrigated forest plantations. Absent in desert regions.

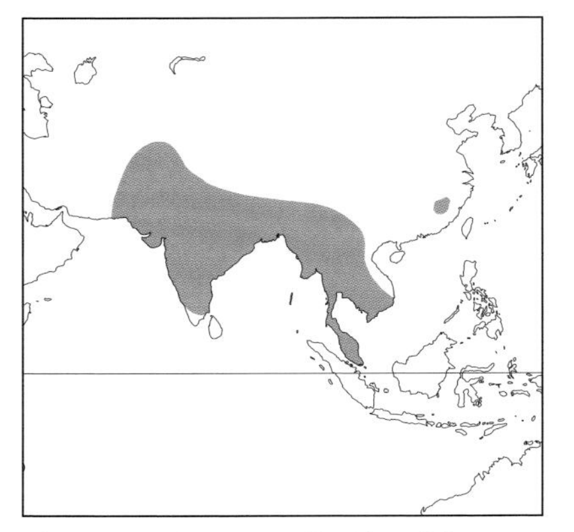

NATURAL HISTORY: Resident. Not strictly nocturnal, it will hunt during daylight hours, especially on cloudy, rainy days. An increasingly fast staccato "kro-kro-kro-krohohohohoh" is reportedly its main call. Roosts in densely foliaged trees. Pairs have been known to occupy the same grove for many years. Its 1 to 3 (normally 2) eggs are laid in abandoned raptor stick nests, quite high and preferably in or near water; ground nesting has been reported in Bombay. Reports of the species building its own nests or adding greenery to an old one are doubtful. Both sexes incubate. Diet includes birds (e.g., crows, doves, pigeons, parakeets, coucals, coraciiforms, hawks, coots and pond herons), bird's eggs, mammals (e.g., rats, hares, squirrels, porcupines), also reptiles, amphibians, fish and insects (especially water beetles).

CONSERVATION STATUS: Not Globally Threatened (IUCN). Variable reports on its status, e.g., common to uncommon, even rare in India and Bangladesh, more frequent in Pakistan, local and rare in Nepal, suggest more study is needed. Very rare in the east and in Thailand. Last recorded from Peninsular Malaysia in 1915. Chinese populations are unknown and require study. Current status in Myanmar unknown, but previously thought to be fairly common.

100 Akun Eagle Owl, *Bubo leucostictus*

DESCRIPTION: This relatively small eagle owl (40 to 46 cm) is the only west African eagle owl that has yellow eyes. Its overall plumage is brown to rufous-brown. The light rufous facial disk has rufous concentric rings and a dark rim, and the ear tufts are prominent. The bill and small feet are weak.

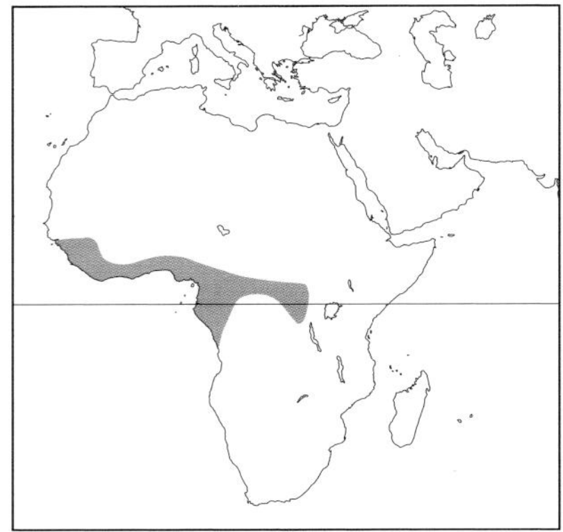

HABITAT: Occupies primary and old secondary lowland rainforest, preferring edges and clearings, and riverside forests and forested islands. Observed in swamps within primary forests and farmland with tall trees (Liberia).

NATURAL HISTORY: The species is resident. A nocturnal bird, it roosts high and alone (sometimes pairs roost together) within the foliage of large trees. Breeding biology is poorly known; apparently it nests on the ground in a shallow scrape. Primarily insectivorous (cicadas, beetles, locusts and cockroaches), its small feet indicate it cannot secure large birds or mammals. It hunts from a perch or on the wing. Owls heard interacting utter a series of "kok" notes, and more study is needed to determine if this is also its main territorial or courtship call.

CONSERVATION STATUS: Not Globally Threatened (IUCN). The status is uncertain, but it is generally thought to be uncommon in its patchy distribution in West Africa. Said to be uncommon in Sierra Leone, rare in Nigeria, and common in Liberia (1 pair/square km in the northwest). More information is needed to more accurately assess its status; because it is often overlooked, it may be more common than available records indicate. Habitat loss through logging may be a potential threat.

101 Philippine Eagle Owl, *Bubo philippensis*

Description: This is a small to medium-sized eagle owl (40 to 43 cm) with yellow eyes and fairly long, horizontal ear tufts. The upperparts are tawny-rufous and marked with wide dark brown stripes. The upper area of the throat is buff and unmarked, while the lower breast and belly are buff-white with dark brown streaks. The facial disk is rufous-buff, and the head and breast are tawny-rufous in color. The legs are feathered but the toes are bare.

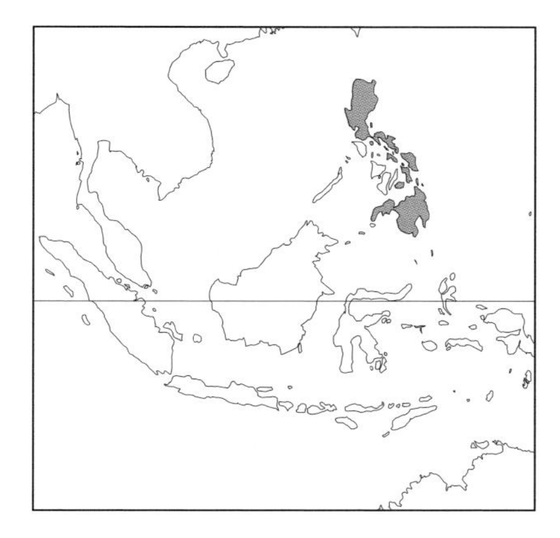

Habitat: The species is found at low elevations in forests close to rivers and lakes.

Natural History: The male's call is an extended series of 5 "bu" notes, spaced at 4-second intervals. Virtually no other information is available on this species. It is resident, and probably eats small mammals and birds.

Conservation Status: Vulnerable (IUCN). This is a rare species; the few recent records have come only from Luzon (one from Bohol in 1994). Population declines are a result of the widespread destruction of lowland habitat, and possibly human persecution. Immediate protection is necessary for its survival.

102 Blakiston's Eagle Owl, *Bubo blakistoni*

Description: The Blakiston's eagle owl is one of the largest of all the owls (60 to 72 cm), with a 180 to 190 cm wingspan. The ear tufts are wide, long and horizontal. The buff-brown upperparts are broadly streaked. The whitish tail has dark barring, and the throat is white. The feathers below are a pale buff-brown, marked with long thin streaks.

Habitat: It occupies lowland riverine forest (broadleaf or mixed broadleaf/coniferous), especially along fast-flowing rivers or streams that are not completely frozen during winter. In Kurils, it is found in dense coniferous forests (fir and spruce) along lakes, rivers and seacoasts. To the north, it hunts along rocky coasts.

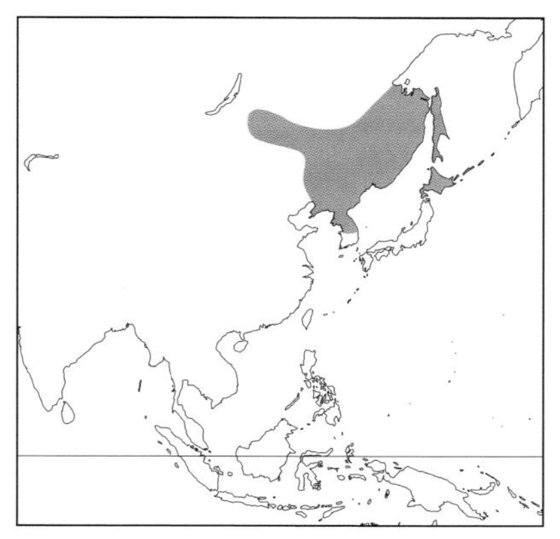

Natural History: It is generally considered to be a resident species, but it is thought that many birds migrate from the Bikin River valley, Ussuriland, during winter (unknown destination). Birds are active at night and during the day. Pairs are thought to mate for life, and are highly territorial. A resonate "boo-boo uoo" is its main call. Normally 2 eggs (1 to 3) are laid in large holes of hollow trees (12 to 18 m above the ground), in nest boxes, on fallen trees or on the ground. Breeds in alternate years (possibly in relation to prey availability and conditions). Eats mainly fish, including large species such as Amur pike, catfish, burbot, trout and salmon. Also eats crabs, crayfish, frogs and birds, and sometimes captures bats in flight. In winter, hares, marten, cats and small dogs are captured. It hunts from low perches overlooking rivers, and will even walk on the ground, jumping directly into the water feet first and pouncing on prey. It also swoops down over water, and up to 6 birds will wait for prey beside openings in the ice.

Conservation Status: Endangered (IUCN). This is one of the world's rarest owls. The Russian population was estimated at 300 to 400 pairs in 1984. In Japan, it is now restricted to the east and central areas with an estimated 80 to 100 birds in the late 1980s, and only approximately 20 pairs breeding yearly. Extremely local and rare in northeast China, possibly even extirpated. Since the 1950s, populations have declined drastically. Currently, only 15 pairs (or fewer) in Ussuriland (26 pairs in mid-1970s). Main threats include riverside habitat loss due to development (e.g., lower Bikin River), logging of taiga forests in the Far East and Japan, and fish stock depletion. Some owls are killed by hunters and anglers at fishing holes, accidentally killed in mink traps set by fur-trappers and hunted for food in Siberia. This owl has used nest boxes since 1983 on Kunashiri Island and Hokkaido, but captive breeding attempts have failed.

103 Brown Fish Owl, *Ketupa zeylonensis*

DESCRIPTION: A large owl (48 to 57 cm) with bright golden-yellow eyes set in a poorly defined tawny facial disk. The typically horizontal feathered ear tufts are black-tipped with a white base. It has a prominent white patch on the throat and foreneck. The tarsi and toes are bare. Its body is rufous-brown above with black or blackish-brown streaks above; underparts are whitish with black streaks and narrow wavy brown bars.

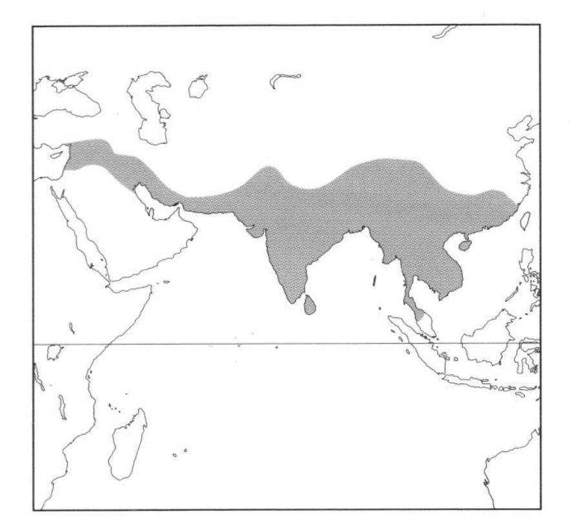

HABITAT: It occurs near water in dense lowland forest, but also in open woodland habitat. It is tolerant of settled areas such as older mango groves and plantations, dense treed areas along canals, and roads with water-filled ditches.

NATURAL HISTORY: Its typical call is either a 3-note "tu-whoo-hu" or a moaning "oomp-ooo-oo"; also recorded is an explosive "boom-o-boom" or "boom-boom." It frequently departs its day roosts in large trees to hunt well before the sun sets. Thus it has often been seen bathing in shallow water. It hunts mainly fish, frogs and crabs, but also takes rodents, birds, reptiles and invertebrates such as large beetles. Fish are grabbed from the water surface either while it is flying over streams, or after spotting them from low perches. A pair was observed feeding on a putrefied crocodile carcass. Its 2 eggs (range 1 to 3) are laid in a variety of nest sites near water, including old stick nests on cliffs or in trees, hollows in the cradle of large forked branches, crevices in rocky banks and the ruins of abandoned buildings. It is thought to be nonmigratory.

CONSERVATION STATUS: Not Globally Threatened (IUCN). Considered common in lowland areas but rare in forested hills and montane habitat to 2000 m. Others describe it as uncommon, but perhaps the most common owl of Sri Lanka. More information is needed on its basic biology, especially its behavior and vocalizations. Drainage of habitat, pollution and mass poisoning of rodents with thallium sulfate (consequently poisoning surface waters and associated prey species) have led to local extirpation from many parts of its range.

104 Tawny Fish Owl, *Ketupa flavipes*

DESCRIPTION: A large owl (48 to 58 cm), it is similar to the brown fish owl, except its tarsus is feathered along half its length, and its beak is larger and more powerful. Like other fish owls, it lacks the comb-like fringes on the leading flight feathers of the wing—therefore, its flight is probably not silent, an unimportant trait for a species that preys on fish. It has yellow eyes, an orange-rufous facial disk with an undefined dark border and a well-defined white throat patch. Its upperparts are tawny or orange-rufous with broad, black shaft streaks; brighter rufous-orange below.

HABITAT: Whereas the brown fish owl is associated with slow, sluggish water courses,

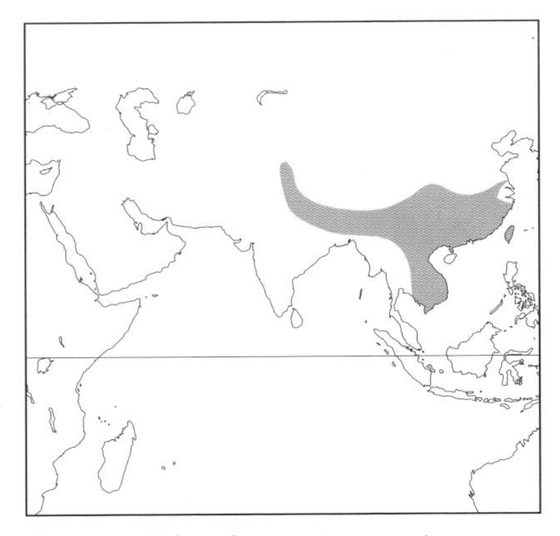

the tawny fish owl seems to occur in areas with running or fast-flowing streams. It occurs in deep wooded ravines, and dense broadleaf forests near streambanks and pools, up to 1500 m elevation, and rarely to 2450 m in India.

NATURAL HISTORY: While it is thought to be partly diurnal, at least crepuscular, little information is available on its behavior. Its 1 to 2 eggs are laid in caves in cliffs, cavities in river banks, hollows in tree branch forks, but less frequently in abandoned stick nests. Like other fish owls, it is a sedentary, resident species. Its call is a resonate "whoo-hoo." The tawny fish owl swoops down to take fish just below the water's surface. Eats mainly fish, but also wades in shallow water to catch crayfish and crabs. Rodents (bamboo rats), lizards, large insects and birds as big as pheasants are also taken.

CONSERVATION STATUS: Not Globally Threatened (IUCN). Given the lack of available information on this species, its conservation status is highly questionable and essentially unknown. Therefore, it should be a high priority for additional surveys and in-depth study.

105 Buffy Fish Owl, *Ketupa ketupa*

DESCRIPTION: This large owl (38 to 48 cm) is the smallest of the 3 fish owls in the genus *Ketupa*. Its upperparts are yellow-brown with buff variegations with pale rufous-edged and dark-streaked feathers. The warm buff underparts are streaked with dark brown. It has relatively long, unfeathered tarsi and long, curved talons. A blackish bill and yellow eyes are set in a pale buff-colored facial disk, with white "eyebrows"; the ear tufts are held outward, almost horizontally. Its warmer and darker buff color and a lack of thin wavy bars on the underparts distinguish it from the larger brown fish owl.

HABITAT: It occurs primarily in lowlands and plains but ventures locally up to elevations of 1100 m; in Sumatra it ranges to greater than 1600 m. It occupies forests and mangroves adjacent to lakes, rivers, streams, rice fields and fish ponds. It tolerates humans, and hence is often found in mature parks, plantations and larger gardens with wetlands.

NATURAL HISTORY: It is generally a resident species, but vagrants are occasionally found over 1000 km from its principal range. Breeding pairs are very vocal, with characteristic calls including a loud "kootoo kookootook....", a musical "to-whee to-whee," and a ringing "pof pof pof..." Its single egg may be laid in a variety of nest sites, including cavities in large trees, atop the compressed fronds of the bird's-nest fern, in old raptor nests, and infrequently on rock ledges. On rare occasions when 2 eggs are laid, only 1 chick survives. It takes mainly fish, hunting primarily at night. Fish are detected visually from perches and seized from just below the water's surface. It also captures other prey (e.g., invertebrates, amphibians, reptiles, birds and small mammals) while standing beside or in shallow water, sometimes actively searching by walking or wading about in the water.

CONSERVATION STATUS: Not Globally Threatened (IUCN). Its status is uncertain, and more rigorous inventories are needed. Local direct persecution reportedly continues, especially near fish ponds. Its ability to occupy a wide range of habitat, including human-altered environments, suggests that it is generally not at risk.

106 Snowy Owl, *Nyctea scandiaca*

DESCRIPTION: The snowy owl is a large owl (53 to 66 cm) with indistinct ear tufts atop a relatively small rounded head. Its facial disk is small and appears incomplete. The black bill, toes and talons are partly concealed by thick feathers. Its golden eyes are relatively small. The snowy owl is somewhat unique in that adult males are nearly pure white, whereas adult and juvenile females are more heavily marked with dusky-brown bars and spots. Juvenile males are similarly, but less heavily, marked, though not on the crown or neck. We can only guess that it was the beauty of this majestic owl that inspired ancient artists to so accurately record its image on the walls of prehistoric caves such as Les Trois Freres, a cave in the French Pyrenees.

HABITAT: Circumpolar, breeds in Arctic tundra habitat from ocean shores to near the tree line. Winters regularly south to the northern Great Plains of North America

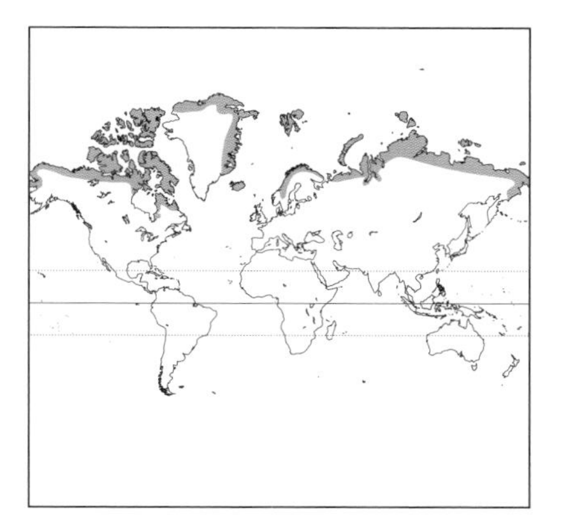

and similarly to central Europe, central Russia, northern China and Japan. It nests on wind-blown hummocks and other exposed sites.

NATURAL HISTORY: This species' nomadic movements are thought to be related to annual changes in numbers of lemmings, upon which it feeds extensively. In times of plenty it can produce large clutches, raising up to a dozen young. In contrast, it may not breed for a year or 2 when prey is scarce. Adapted to nesting in northern areas with perpetual summer daylight, it is largely diurnal. Even in winter on southern prairies it hunts during brief daylight hours, capturing a wide variety of prey. One was observed catching a rock dove in flight. A study of winter prey identified 27 meadow vole skulls in 1 pellet. Breeding pairs are vocal and aggressive, attacking humans and wolves that approach too close to their nest. Researchers have concluded that snowy owl pairs are typically monogamous, exhibiting a strict division of labor (female defends the nest while the male hunts). This behavior ensures successful reproduction in unpredictable and often harsh and open environment.

The male utters 2 to 6 gruff "krooh" notes reminiscent of a baying hound.

CONSERVATION STATUS: Not Globally Threatened (IUCN). Northern European populations thought to be declining. Overwinter mortality can be high, especially among juveniles. Persistent organic pollutants transported to Arctic regions from industrial nations may be of long-term concern to its survival. There is a need to better monitor long-term shifts in the numbers and distribution of this species.

107 Pel's Fishing Owl, *Scotopelia peli*

DESCRIPTION: Pel's fishing owl is more widely distributed and relatively better known than the other 2 fishing owls found in Africa. Owing to its restricted distribution and specific habitat, few have seen this elusive owl. It is a large owl (51 to 61 cm), weighing up to 2.3 kg. Its lower legs and toes are bare and straw-colored. It has long talons, long legs and osprey-like feet with soles with spicules to help grasp slippery fish. It has large blackish-brown eyes and a black bill set in a rounded head. It is generally rufous-colored with barred markings above, and black-spotted and streaked below—effective camouflage in dappled light reflected off water. There is much individual variation in markings and color, including some with prominent melanistic and albinistic feathers.

HABITAT: It occupies forests with large trees along slow-moving rivers, lakes and

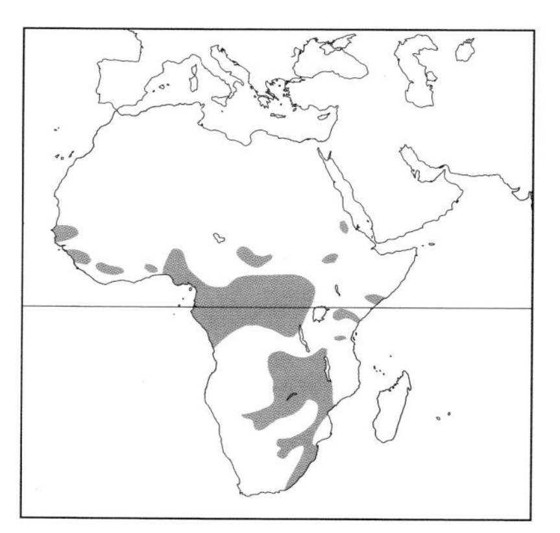

swamps with abundant fish populations from sea level to elevations of 1700 m. Ideal habitat consists of numerous small islands covered with thick riverine forests amongst dense reed beds.

NATURAL HISTORY: This owl prefers to hunt at night from low branches over calm or stagnant water where it catches fish up to 2 kg in weight. Fish and other prey are located visually when they make ripples on the surface. The owl then dives down and captures its prey with legs extended, only rarely getting substantially wet. Fish heads are eaten first, and sometimes the body is left untouched. It also eats frogs, crabs and young crocodiles. It is reported to eat birds, but a captive owl refused them even when hungry. It catches mussels by wading into shallow water near sandbanks. Unlike most owls, its heavy flapping flight is noisy, and it has a rudimentary facial disk, characteristics perhaps related to its decreased dependence on hearing to locate prey. It roosts by day in shady perches in the same general area. It is vocal from midnight to dawn, and pairs can be heard calling to each other year-round. It has a sonorous hoot "hooommmmm-hut" preceded by a series of low grunts "uh-uh-uhu" and followed by a couple of grunts. Nesting is initiated during seasonal flooding, after which water levels drop, making fishing easier at a time when the food demands of rapidly growing young are greatest. It lays 1 to 2 round white eggs in the hollow at the fork of a tree or in a hole in the main trunk, typically 7 m above the ground. Rarely uses abandoned stick nests built by other species. Only 1 chick is reared after a 33-day incubation period. Like many owls, the female has a dramatic distraction display given when intruders approach the nest or recently fledged young. It is typically a long-term resident, but may journey to other areas in search of food when water levels change dramatically. One unusually marked individual used the same territory for at least 8 years. There is much need for additional study of its behavior, calls and genetic relationship with the rufous and the vermiculated fishing owls.

CONSERVATION STATUS: Not Globally Threatened (IUCN). Water pollution may threaten local populations.

108 Rufous Fishing Owl, *Scotopelia ussheri*

DESCRIPTION: A large owl (46 to 51 cm) with a tawny head without a defined facial disk or ear tufts. It has dark rufous upperparts and light rufous underparts. The breast has narrow rufous streaks. The eyes are dark brown and the bill is blackish-gray. Its yellowish tarsi and toes are bare.

HABITAT: It occurs in coastal mangroves and primary forests along large rivers, lagoons and lakes. Secondary forests and degraded habitats, including plantations, are also used.

NATURAL HISTORY: Like Pel's fishing owl, it is probably resident. Its call is a deep moaning and dove-like "whoo," repeated at 1-minute intervals. Its biology is poorly known. There is scant information on its diet; a catfish was found in the stomach of 1 of 7 specimens collected in the rainforest

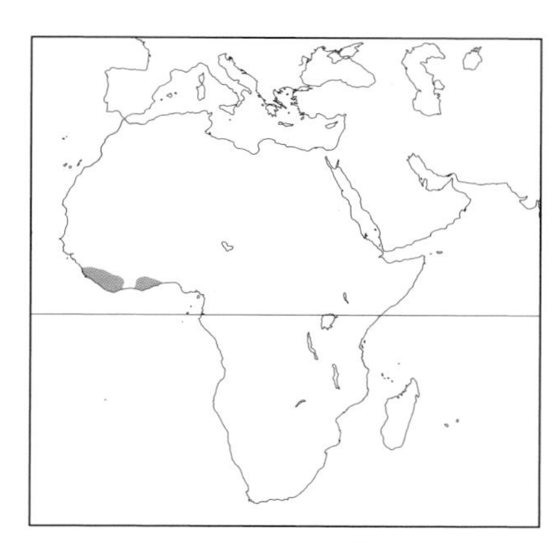

region of western Africa. It likely eats mainly fish, hunting mostly at night from low branches over water. There is no information on its breeding biology (e.g., nest sites, clutch size or nestling period). No more than 1 young has ever been found out of the nest with its parents.

CONSERVATION STATUS: Endangered (IUCN). It has a restricted range and is presumed to be rare. Threats include habitat destruction, human persecution (including the capture of young owls for pets) and water pollution from mining operations. Known protected areas include the Gola Forest Reserve (Sierra Leone), the Tai Forest National Park (Ivory Coast) and Sapo National Park (Liberia). This poorly known bird should be a priority for future research, surveys and monitoring.

109 Vermiculated Fishing Owl, *Scotopelia bouvieri*

DESCRIPTION: A large owl (46 to 51 cm) with dark rufous-cinnamon upperparts with fine deep brown vermiculations; underparts are light rufous to creamy with heavy dark brown streaks. A light rufous facial disk frames its yellowish bill and dark brown eyes. It has a black-streaked forehead on a buff-rufous head and lacks ear tufts. The tarsi and feet are unfeathered.

HABITAT: Usually occurs in pristine forests bordering on lakes, rivers greater than 10 m across and pools in swamps. It is sometimes found away from water and in temporarily flooded forests.

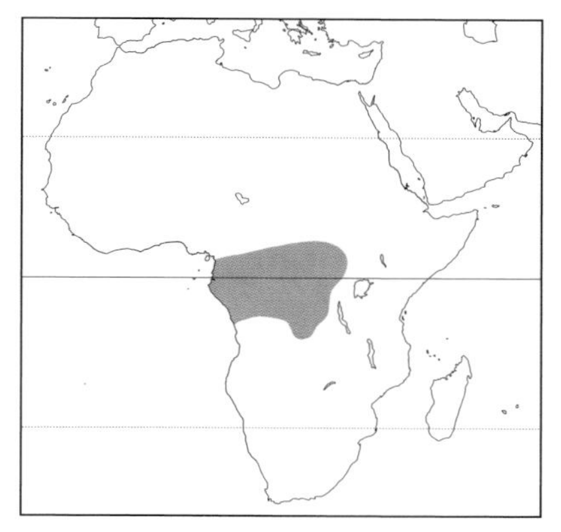

NATURAL HISTORY: It hunts at night from low perches (1 to 2 m above water) from which it takes mainly crabs and crustaceans, but also small fish, frogs, small birds and mammals. Its 1 to 2 eggs are laid in the old stick nests of larger birds such as the spot-breasted ibis (*Lampribis rara*). Like the rufous fishing owl, it is thought that only 1 of the slow-maturing young survives. There is no available information on incubation or fledging. Its drawn-out croaking hoot is followed by 4 to 8 shorter and faster notes sounding like: "drook krook ook-ook ook-ook." It may be heard calling throughout the year.

CONSERVATION STATUS: Not Globally Threatened (IUCN). Its small population, secretive nature and remote haunts have all limited knowledge of this owl's biology. Some state that it is less dependent on fish and water and is more common than previously thought. However, more work is needed, especially along large rivers in the Congo Basin, to better estimate its status.

110 Spotted Wood Owl, *Strix seloputo*

DESCRIPTION: A medium-sized owl (40 to 47 cm) that lacks ear tufts. Its large dark brown eyes and greenish-black bill are set in a rusty facial disk. It has rich brown upperparts, extensively marked with large black-edged white spots. The underparts are white to rufous with black barring. The legs and toes are feathered.

HABITAT: Occurs in a variety of human-altered and natural landscapes, including plantations, parks in towns and villages, orchards, fragmented secondary forests, swamps and coastal mangroves. Ranges in elevation from sea level to 1000 m.

NATURAL HISTORY: Its call starts with a rolling staccato "hoo-hoo-hoo" ending with a long extended "hoo"; also utters a barn owl–like screech. It is a resident and nocturnal species that is especially vocal at dawn and dusk. While hunting from low perches (2 to 3 m above the ground) it takes mainly rodents, but also small birds and large invertebrates, especially beetles. Its 2 to 3 eggs are laid in tree cavities or open-ended large branches.

CONSERVATION STATUS: Not Globally Threatened (IUCN). Perhaps it occurs at normally low densities, but because of its retiring nature its numbers may be underestimated. Reported as uncommon to common in various parts of its range. More field research on its biology and population estimates is needed to better assess its status.

111 Mottled Wood Owl, *Strix ocellata*

DESCRIPTION: A medium-sized owl (40 to 48 cm) with gray upperparts that are beautifully vermiculated and mottled with black, white and reddish-brown. The white and rufous-orange facial disk with concentric black rings stands out in contrast to the darker head and blackish bill. The throat is chestnut with prominent white collars on either side. Its underparts have narrow blackish bars on a white to golden-orange buff background.

HABITAT: It occupies lowland areas, including wooded plains, mango, tamarind and banyan tree groves. Apparently it does not avoid villages and cultivated areas, providing there are densely leaved trees present.

NATURAL HISTORY: Thought to be sedentary and mainly nocturnal. It eats rodents and birds as large as domestic pigeons, but also invertebrates (crabs, scorpions and beetles). During the breeding season it utters a wavering "chuhua-aa" call. Its 2 to 3 eggs are laid in natural tree cavities, and perhaps also in old raptor stick nests. There is no other natural history information available for this species.

CONSERVATION STATUS: Not Globally Threatened (IUCN). Its status is uncertain in many parts of its range; it is reported as both common and uncommon in India. Clarification of its status is needed. Sadly, it may already be extirpated from Pakistan.

112 Brown Wood Owl, *Strix leptogrammica*

DESCRIPTION: A large owl (40 to 55 cm) with as many as 14 races, some of which may prove to be species, that vary in size, color and plumage markings. Its dark brown eyes are each surrounded by a broad black circle, a stark contrast to its white to rufous-brown facial disk and white "eyebrows." Its head is dark brown, and the upperparts are light chestnut. Pale to white barring highlights the wings and tail. The greenish-horn bill stands out against the dark brown chin. The throat and underparts are white to buff with thin, diffuse brown barring. The breast may be brown to rufous.

HABITAT: It frequents dense and undisturbed tropical forests along seacoasts, typically in lowland areas to 500 m. Also occurs in deep temperate, dry tropical, subtropical and tropical rainforests in subtropical to temperate mountain zones, occasionally up to 4000 m (Nepal). While

it is generally thought to avoid inhabited areas, it has been reported from densely wooded gardens in Sri Lanka.

NATURAL HISTORY: It appears to be sedentary and strictly nocturnal, roosting in dense foliage high above the ground. The diet consists of a variety of prey: small mammals (including a fruit bat), birds, reptiles (lizards) and large insects. Fish are reportedly taken as well. More birds, some as big as pheasants, are predated relatively more often within the northern parts of its range. There is no information available on its hunting habits. Its 2 eggs, sometimes 1, are laid in large cavities in big trees, on cliff ledges or on the ground at the base of a tree or rock. Its call is a short series of 3 to 4 short hoots: "hoo-ho-hooh," starting quiet and terse and ending loud and drawn out. It also utters a: "oot-oot-tu-whoo."

CONSERVATION STATUS: Not Globally Threatened (IUCN). This shy owl appears to avoid contact with people and human-altered habitat. It is reportedly uncommon in most of its range (Indian Subcontinent), rare and local in Bangladesh and Java, and locally common in Sri Lanka. Forest destruction is a major threat in many parts of its range. Some of the several protected areas that support populations include Baluran National Park (Java), Way Kambas National Park (Sumatra), Chitwan National Park (Nepal) and Nam Nao National Park (Thailand).

113 Tawny Owl, *Strix aluco*

DESCRIPTION: A medium-sized owl (36 to 45 cm) with a robust body and a large, round head. Its gray, brown and red morphs intergrade and vary mainly in background color. The facial disk is generally pale with darker concentric rings, and the eyes are blackish-brown. Its bill ranges from pale yellow to cream. The top of the head has a dark central area bordered by pale bands. It is overall generally streaked dusky and with a crossbar or herringbone pattern.

HABITAT: This Old World species lives mainly in temperate climates. Its broad, round wings are adapted to flying in woodland and forest habitats. It can be found in mature forests, including boreal coniferous, deciduous and mixed forests; montane forests with oak, beech, pine and fir; as well as the cloud forests of Burma. Found over a wide range of elevations up to 4200 m. It readily uses nest boxes and seems to thrive in human-altered landscapes.

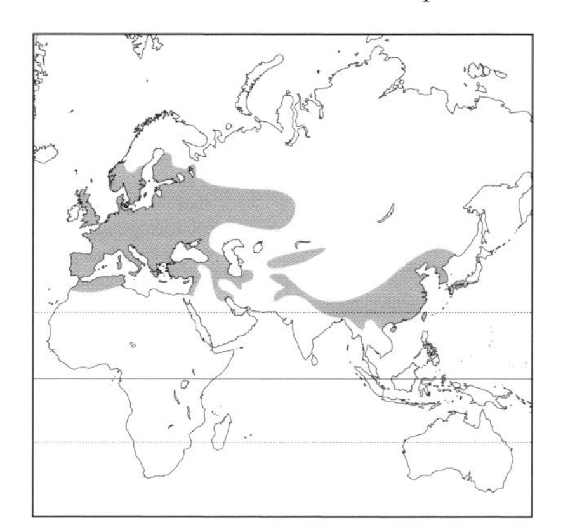

NATURAL HISTORY: It is mainly nocturnal, but it has been seen hunting during the day, especially during the breeding season in northern latitudes. Its calls are as varied as those of the barred owl and Ural owl. The best known call is denoted as "hooo-huhuhohoo," where the final note is extended and resonates. It readily uses nest boxes and other artificial structures for breeding, and it seems to have adjusted to life in human-altered landscapes, including cities (e.g., London) with forested areas. In some areas where it relies almost exclusively on nest boxes, pine marten have become significant owl predators, having learned to routinely check boxes for nesting owls. The tawny owl is very aggressive when protecting its nest and young against human and other intruders, resulting in the temporary closure of some public parks until the fledged young are capable of flight. Its 3 to 5 (range 1 to 8) eggs are laid in natural tree cavities, in nest boxes, on abandoned stick nests and in the hollows atop old tree stumps. It is a resident species, and juvenile dispersal is typically less than 100 km. Populations appear to be rather stable, relating to the great variety of prey consumed. It takes mainly small woodland rodents, also invertebrates, and various birds, amphibians and mammals, including squirrels, young rabbits and hares and weasels. It will prey upon other owls and diurnal birds of prey, e.g., the little owl, boreal owl, long-eared owl, sparrowhawk and Eurasian kestrel.

CONSERVATION STATUS: Not Globally Threatened (IUCN). The conversion of southern boreal forests to farmland has allowed this adaptive species to expand its range northward, with the simultaneous retreat of the more aggressive Ural owl. The extent of mortality from collisions with vehicles, overhead wires and contamination with organic pesticides, mercury compounds or heavy metals is unknown and may be significant.

114 Hume's Owl, *Strix butleri*

DESCRIPTION: A medium-sized owl (30 to 35 cm) named after Allan O. Hume, a

British administrator in India, who first described it in 1878 based on a specimen from western Pakistan. This is an unusual desert-adapted member of the wood owl genus (*Strix*), most of which are forest-dwelling owls. Its fine silky feathers are most similar to those of the barn owl. It has honey-yellow eyes and pale buff plumage. Although the feet are relatively small and unfeathered, the bottom surface of the toes have a thinly feathered covering, an unusual feature whose function is unknown. The slender legs are only thinly feathered.

HABITAT: Occupies rocky desert and steppe habitats, especially areas with temporary water, springs, palm groves in oases, gorges and old ruins.

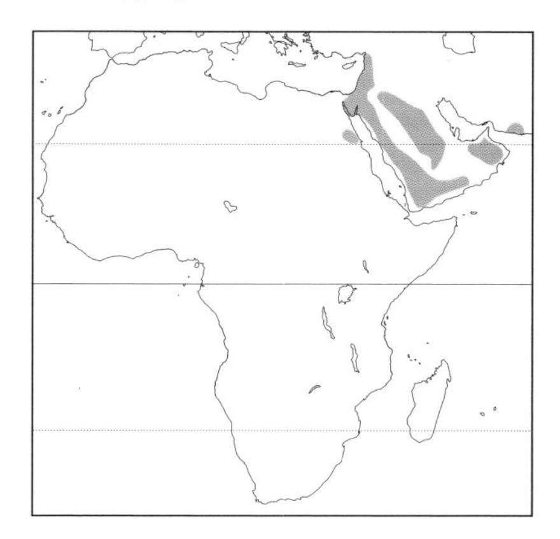

NATURAL HISTORY: It is not known what limits the Hume's owl to arid regions, but there is some suggestion that interactions with more aggressive owls (e.g., tawny owl in Israel) may prevent it from expanding its range. On the other hand, it appears to exclude the little owl in some areas of its range. The evolutionary history of this species must have been interesting and dramatic, but to date it remains largely speculative. Its main call is like that of the tawny owl, consisting of 1 long hoot followed by 2 double notes: "whoo whoohoo." Hunting at night, its diet consists mainly of small rodents, but invertebrates may be equally important. Reptiles and birds are eaten less often, but it likely takes any vertebrate prey it can catch. Its 5 eggs are laid in cavities and caves in the sides of steep valleys in arid mountains.

CONSERVATION STATUS: Not Globally Threatened (IUCN). Its habitat is generally considered secure, but an increase in roads and traffic has resulted in local high mortality from collisions with vehicles.

115 Spotted Owl, *Strix occidentalis*

DESCRIPTION: A medium-sized owl (46.6 to 48 cm) that is chestnut to buff brown with irregular white spots. The facial disk is round with indistinct darker brown concentric rings around each of its very dark brown eyes. Its bill is yellowish-green. There are 3 recognized subspecies. Its thick plumage and densely feathered tarsi and toes are more characteristic of northern rather than temperate climate adapted species.

HABITAT: Most of these owls are associated with humid mature or old-growth conifer or oak forests, but 1 subspecies also occurs in heavily logged pine-oak forest. The multiple-storied structure of older forests permits the selection of cool roost sites to avoid heat stress in response to seasonal variation in temperature. This southernmost population can apparently tolerate warmer and drier conditions by seeking refuge in the shade of deep, well-vegetated and rocky canyon walls.

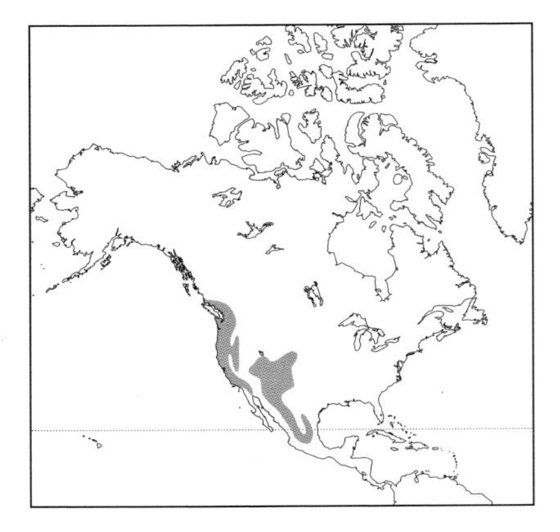

NATURAL HISTORY: It is mainly nonmigratory, although it undergoes seasonal altitudinal movements up to 65 km (elevationally up to 1500 m) in some parts of its range, moving to lower elevations in winter. Its territorial call is a 3 to 4 deep, drawn-out hoots, denoted as "who who-who whoooo." The contact call, given to locate mates, consists of a series of frightening rising whistles ending with a siren-like echo; the alarm call is similar to dog-like barks. It appears to be strictly nocturnal is and hard to locate in thick, dark cover while roosting up against tree trunks. Roosting owls appear to be lethargic and reluctant to fly, resulting in cases where they have been stroked and easily captured, even clubbed to death. They appear to be heat-stressed at temperatures above 27°C. Like other mainly sedentary owls, it appears to eat anything it can catch (invertebrates, amphibians, birds and mammals) in its small year-round home territory. Hence its list of prey taken is long, and comparable to that of the barred owl and tawny owl. Its 2 eggs (range 1 to 4) are laid in existing structures such as natural tree cavities, atop old raptor or squirrel nests and on mistletoe brooms. It takes small to medium-sized mammals, mainly rodents, with northern flying squirrels dominating its diet in the northern parts of its range, and the dusky-footed wood rat in the south, coinciding with the prey's distribution limits. Less frequently taken prey include invertebrates, amphibians and birds. This species, along with the white-headed woodpecker and the flammulated owl, seems to have survived the last Ice Age in southwestern North American forest refuges. Its dense plumage, while adaptive during colder times within its restricted glacial range, ironically may limit its ability to disperse into warmer habitats.

CONSERVATION STATUS: Lower Risk—Near Threatened (IUCN). It is listed as Threatened under the U.S. *Endangered Species Act* and as Endangered in Canada. Despite exceptional effort on spotted owl research and conservation, regionwide declines apparently continue. The estimated global population is 15,000. Mexican populations may be stable because low-impact forestry activities used there modify but do not destroy habitat. Other populations are declining because of continued habitat changes (clear-cutting and selective logging), which allow opportunistic predators (the great horned owl) or fierce competitors (the barred owl) to invade. In some areas, where the larger barred owl has recently (since the 1960s) expanded its range into barred owl habitat, they have cross-bred at least 5 times to form fertile hybrids. Therefore, the use of nest boxes to mitigate habitat loss for this species needs study to determine its effectiveness as a conservation measure.

116 Barred Owl, *Strix varia*

DESCRIPTION: This is a large owl (40 to 60 cm), mostly brown and white with a round head without ear tufts, brownish-black eyes and a yellow bill. There are 4 or 5 brownish concentric rings around the eyes. Its facial disk is large and round and varies from grayish-white to pale buff gray. The feathers of the wings and tail are barred and spotted. The throat and upper chest have many horizontal bars, and there are wide vertical streaks on the lower breast and abdomen.

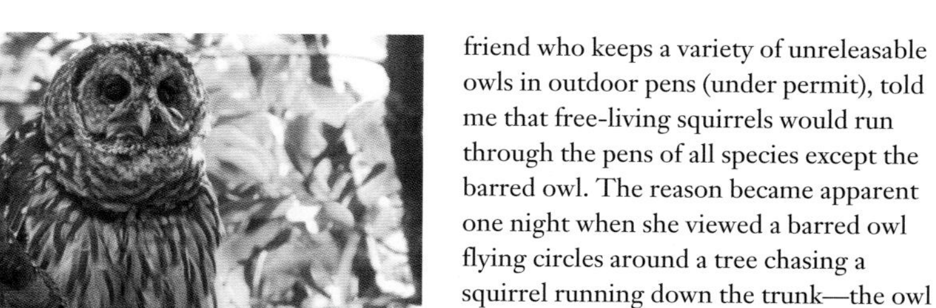

HABITAT: The barred owl is found in coniferous or mixed-wood forests. In more southern parts of its range it inhabits dark deciduous woods. Some authors consider it an old growth forest-dependent species, but others have shown that successful nesting can occur in younger and heavily fragmented forests. The availability of suitable nest sites seems to be a major factor limiting the distribution of this locally common owl. It nests mainly in large tree cavities but will also use the open tops of large snags. It is less successful at raising young when it uses old stick nests built by hawks or old squirrel nests, as these relatively flimsy nests sometimes fall apart before the eggs hatch or before the young can survive out of the nest.

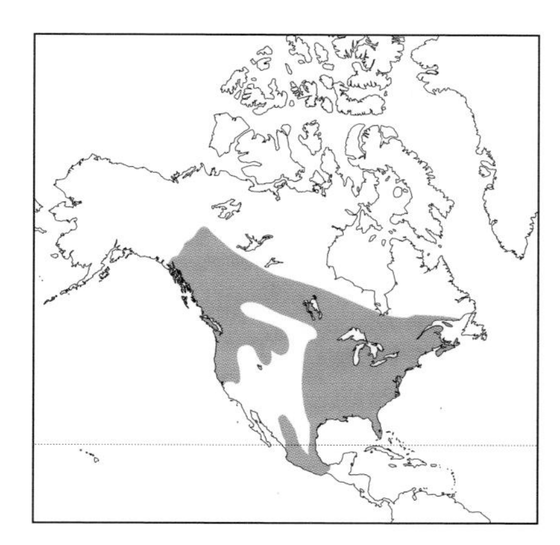

NATURAL HISTORY: This owl is mainly nocturnal, but it has been seen hunting in the daytime, especially in winter months in northern parts of its range. It is a very agile and fast flyer, and is capable of catching a great variety of prey including rodents, hares, birds, turtles, fish, frogs, snakes, hares and rabbits. Katherine McKeever, a friend who keeps a variety of unreleasable owls in outdoor pens (under permit), told me that free-living squirrels would run through the pens of all species except the barred owl. The reason became apparent one night when she viewed a barred owl flying circles around a tree chasing a squirrel running down the trunk—the owl caught the squirrel! Eating a variety of prey allows the barred owl to switch prey species when one type becomes rare. That explains why this hefty-looking owl is typically a permanent resident, often using the same nest site year after year. It may have evolved an elaborate and rich repertoire of hooting calls to protect its territory from other barred owls. One typical call is often cited as "Who cooks for you? Who cooks for you-all?" Imitating this call near a breeding pair will often bring 1 or both adults close in a frenzied flight and hooting display. Exercise caution, however, as this owl does not like intruders. In a case of mistaken identity, one owl researcher was struck savagely in the face after playing a recording of the owl's territorial call too close to a nest site! (See Chapter 3.)

CONSERVATION STATUS: Not Globally Threatened (IUCN). As a species that requires at least some older trees for nesting, local populations have suffered in the eastern and southeastern parts of its North American range. Conversely, it has expanded its range into western North America, starting in the 1960s. Researchers have suggested a variety of habitat-related explanations for this, including the influence of human-altered landscapes. There is concern that in western Canada and the northwestern United States this aggressive species may be displacing the closely related endangered spotted owl.

117 Fulvous Owl, *Strix fulvescens*

DESCRIPTION: A large owl (38 to 48 cm) with a round head that lacks ear tufts. It is rusty-brown with whitish spots above and pale fulvous-brown below. The upper breast and neck have brown barring, but the rest of the underparts are broadly streaked. Its tarsi and the base of the toes are feathered. Its bright yellow bill and white "eyebrows" are set in a pale gray-brown facial disk that is a darker brown near the black-brown eyes. The rim of the facial disk is dark brown.

HABITAT: It occurs in humid cloud forest, tropical and temperate pine-oak forests in montane regions at elevations from 1200 to 3000 m.

NATURAL HISTORY: Its call, a series of rhythmic low and short barking hoots: "who-wuhu-woot-woot," is similar to both that of the spotted owl and the great horned owl. It also utters single hoots and a nasal "gwao" call reminiscent of parrots. Its diet includes small mammals and birds, frogs, lizards, insects and other invertebrates. It is nonmigratory and nocturnal, roosting by day in leafy shadows or in tree cavities. Its 2 to 5 (usually 2 to 3) eggs are laid in tree cavities, but in general its breeding biology is poorly known. It is smaller than and sounds distinct from the closely related barred owl, which occurs as close as only 100 km away in Mexico.

CONSERVATION STATUS: Not Globally Threatened (IUCN). Its status is uncertain, but loss of forest habitats within its restricted range threatens local populations. The cumulative impact of this loss is unknown. More research is required to compare it to other *Strix* owls and to set reasonable conservation objectives within its range.

118 Rusty-barred Owl, *Strix hylophila*

DESCRIPTION: A medium-sized owl (35 to 37 cm) with a light rusty-brown facial disk with wide concentric brown rings. It has small white "eyebrows," dark brown eyes and a yellow-tipped gray bill. The underparts are barred dark brown on orange-buff, changing to white toward the belly. The upperparts have brown, orange-brown and white barring. It is 1 of 2 species of *Strix* found only in South America.

HABITAT: The rusty-barred owl is found both in and adjacent to primary and secondary montane, lowland tropical evergreen and temperate forests with dense undergrowth. It tolerates settlements to some extent and occurs in lowland areas and at elevations up to 1000 m. Roosts by day in dense foliage and in tree cavities.

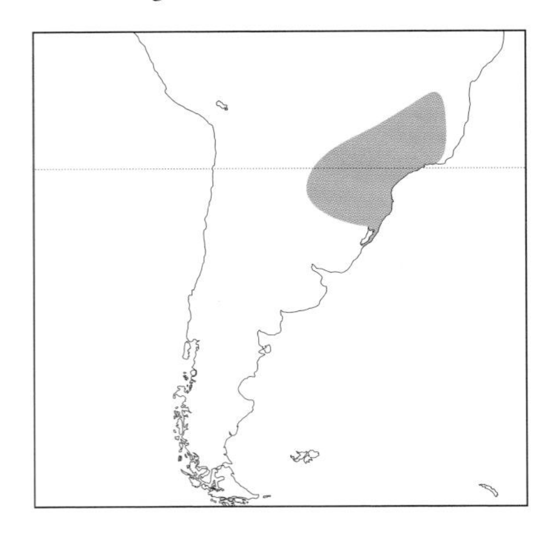

NATURAL HISTORY: It hunts in the forest canopy at night, taking a variety of prey including small mammals, birds, reptiles and invertebrates. It is assumed to be nonmigratory. Its 2 to 3 eggs are laid in tree cavities and are incubated for about 29 days. The young, which are independent when 4 months old, can reproduce the following breeding season. Its call is a grunting "grugruu-grugruugrugru" resembling that of the chaco owl, but it also utters a rolling: "rrrrro" and a long, drawn out: "i-u-a." It responds aggressively to playback of its calls, and also to those of other owl species.

CONSERVATION STATUS: Not Globally Threatened (IUCN). Increased monitoring and assessment of its population is needed because of the degree of habitat loss throughout its range. Logging and burning of habitat is a major threat. Depending on large forested areas, it is known to occur in several protected areas, including Iguazu National Park (Argentina), and Aparados da Serra National Park and Rio Doce State Park in Brazil.

119 Rufous-legged Owl, *Strix rufipes*

DESCRIPTION: A medium-sized owl (33 to 38 cm) with a pale orange-brown facial disk with faint concentric lines. It has a white throat, pale yellow bill and dark brown eyes. The upperparts are dark rufous-brown or sepia with white and orange-buff spots and distinct light-colored bars. The underparts are heavily barred white, buff and black. As its name suggests, the legs and toes are orange-brown to cinnamon-buff. Its distinct vocalizations support its treatment as a separate species from the similar chaco owl.

HABITAT: It is found in moist, dense, older montane forests with lichens and mosses to elevations of 2000 m and in lowland areas with open woods. Older secondary growth is sometimes occupied as well.

NATURAL HISTORY: Thought to be a resident and nocturnal bird, but its breeding biology and general habits are poorly known. It roosts by day in dense vegetation or in tree cavities. Mammals, especially small tree-living rodents, dominate its diet, but it also takes birds, reptiles and invertebrates, sometimes on the wing. Its 2 to 3 eggs are laid in tree cavities or old raptor nests. Its call starts with low-pitched notes, followed by a fast, clear, guttural series of 10 to 12 nasal notes: "brrr brrr brrr brrr, u AU u u u..." Young owls are known to disperse widely, and it has been reported as a possible breeder on the Falkland Islands.

CONSERVATION STATUS: Not Globally Threatened (IUCN). An elusive bird, its status is uncertain. Although it was formerly described in 1828, there has been little research on its biology. Breeding densities as high as 0.22 pairs/km have been documented in old-growth forest. Forest destruction and pesticides are considered possible threats to this species. It is known to occur in several protected areas, including Nahuelbuta and Cerro La Campana National Parks (Chile) and Tierra del Fuego National Park in Argentina.

120 Chaco Owl, *Strix chacoensis*

DESCRIPTION: A medium-sized owl (35 to 40 cm) with a dark-rimmed, white facial disk with dense, dusky rings. Its eyes are dark brown, and its bill is pale yellow. The tail and flight feathers have noticeable rufous barring. Its upper body is a deep dark brown, spotted with white, and heavily barred. The white underparts have dense, dark-brown bars. Creamy-buff feathers cover the legs and toes.

HABITAT: It may prefer locations near water and reportedly avoids open habitats. It uses dry chaco woodlands (tropical and subtropical grasslands, savannahs, shrublands and dry broadleaf forests) with thorn bushes and large cacti.

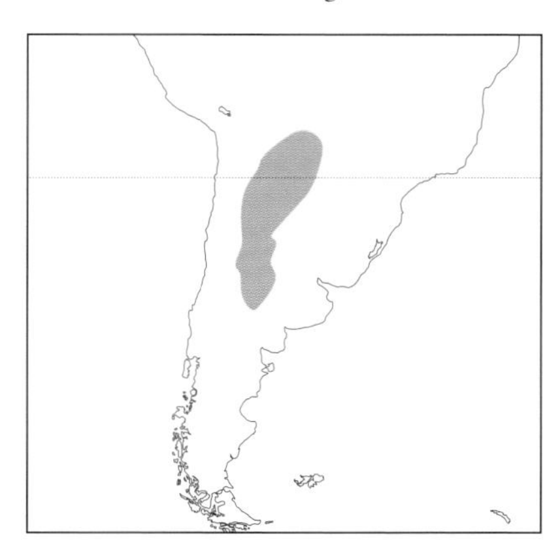

NATURAL HISTORY: This nonmigratory owl hunts from dusk to dawn, taking small mammals, birds, reptiles and invertebrates, including spiders, scorpions and centipedes. There is little information available on its breeding biology. A captive owl laid 3 eggs, and the young left the nest at about 5 weeks after hatching. In the wild it lays its 2 to 3 eggs in tree cavities, and perhaps also on the ground under a bush or fallen log. Its frog-like call is denoted as: "crucrucru craw-craw." Clearly there is much yet to learn about this striking owl.

CONSERVATION STATUS: Not Globally Threatened (IUCN). A lack of population data precludes a sound estimate of its status. Development associated with human population growth, accompanied by intense cattle and goat grazing, is altering much of the habitat of the chaco ecoregion, especially in the south. By the early 1990s only 10% of the original forested land remained. The word "chaco" means "hunting land" in the Quechua language. New paved roads are being built to allow better access to formerly inaccessible hunting areas. While this has encouraged new agricultural development in pristine wilderness, it is not known if habitat loss is a threat to this owl. It is noted as locally common in some areas such as the Chancani Provincial Park, Cordoba.

121 Ural Owl, *Strix uralensis*

DESCRIPTION: A large owl (50 to 62 cm) with a rounded head without ear tufts and a relatively long tail and elongated body compared to other *Strix* owls. Its upperparts are brownish-gray to gray with dark and white streaks. The pale gray-brown to white underparts have dark streaks. The back and scapular feathers are white-edged. A plain grayish-white facial disk frames dark-brown eyes and a yellow bill.

HABITAT: It lives in boreal coniferous forests and mixed and deciduous woods (e.g., beech) near open areas including bogs, fields and clearings. It does not appear to avoid human-occupied areas. In winter it can be found in other open habitats including near towns and parks. It occurs at elevations from 450 to 1600 m in central Europe and from lowlands to 1600 m in Japan.

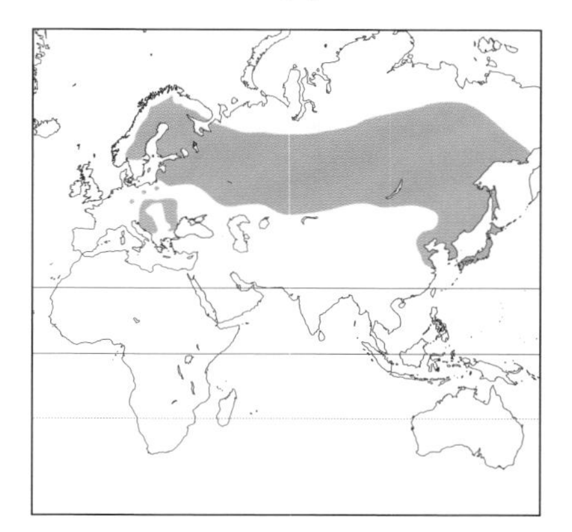

NATURAL HISTORY: The Ural owl is a resident species even during severe winters, with most young remaining within 30 km of their natal nest site, and a few moving up to 200 km. Adults typically form stable pair bonds with 1 mate, and a relatively few females may relocate to new breeding sites up to 150 km. Its 2 to 5, rarely 1 to 6, eggs are laid in nest boxes, old nests built by other species, tree stumps, tree cavities or buildings; on the ground; or on cliff faces. The young are dependent for up to 5 months. It takes a variety of prey, mainly at dusk and before dawn, but also during the day. Its diet consists primarily of small mammals (e.g., voles and young rabbits), but also small birds and frogs. Occasionally it subdues prey as large as the black grouse. Its ability to switch prey relative to their availability allows this owl to remain in its large territory (450 ha) all year. It frequently kills other owls (e.g., the tawny owl) within its home range. In Sweden it is sometimes called the "attacking owl" because it aggressively swoops at anyone who climbs to its nest. Its seldom-heard call is: "Hu-ooo, Hoooo hu-HOO-hoo" uttered at 15 to 50 second intervals.

CONSERVATION STATUS: Not Globally Threatened (IUCN). It appears to be widespread throughout most of its range and is locally rare to abundant (up to 10 pairs/100 square km). Breeding productivity varies in relation to cyclic vole population fluctuations. Nest sites (old trees and stumps) in Fenno-Scandia lost to logging resulted in local population declines, which have been effectively mitigated by the placement of thousands of nest boxes; affected populations either recovered or even increased.

122 Sichuan Wood Owl, *Strix davidi*

DESCRIPTION: A large owl (58 to 59 cm) with a white-streaked dark head and back that contrasts with a pale facial disk with dark concentric lines and a clearly marked rim. The eyes are dark brown, and its bill is yellow. The scapulars are distinctly whitish, the upper tail coverts indistinctly marked and plain in contrast to the barred tail. Underparts are grayish-white with dark crossbars and streaking.

HABITAT: The Sichuan wood owl is found in dense mature spruce, fir and pine forests interspersed with alpine meadows at elevations between 2700 and 4200 m, perhaps even to 5000 m. It also occurs in old mixed coniferous forests adjacent to open areas with limited ground cover.

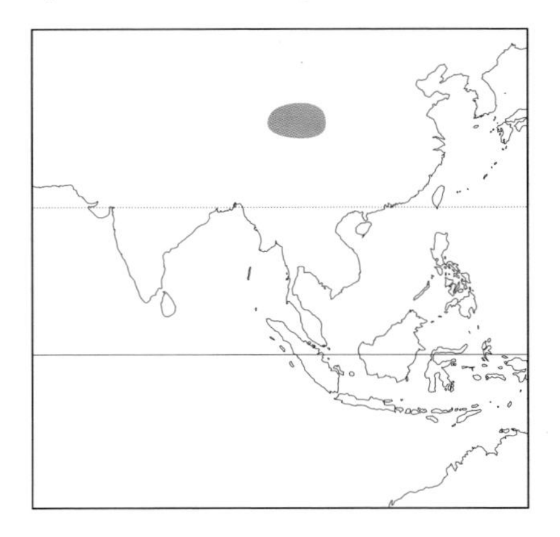

NATURAL HISTORY: A resident of the mountain woods, its call, similar to that of the Ural owl, is reported as a drawn-out, wavering, barking: "khau khau." More information on its genetic relationship to the Ural owl might help explain its apparent restricted distribution (mountains of western China and western Sichuan). Some speculate that it may be an isolated, strongly marked race of the Ural owl. No information is available on its breeding biology, diet or habits.

CONSERVATION STATUS: Vulnerable. (IUCN). Considered rare throughout its restricted range, it is perhaps more aptly designated Endangered due to ongoing extensive logging. It occurs in the Jiuzhaigou Reserve, Sichuan; more research is imperative to ensure its survival.

123 Great Gray Owl, *Strix nebulosa*

DESCRIPTION: One of the largest owls in the world (61 to 84 cm) and the only *Strix* owl that breeds in both the Old and New World. Its feathers are mostly gray with patterned combinations of whites, grays and browns. Its upperparts are irregularly marked dark and light, and its underparts are boldly streaked over fine barring. Its large circular facial disk has fine barring that forms 6 or more nearly concentric rings of gray or brown on a white background. Its bright yellow eyes are separated by 2 white crescents (lores). Its bill is yellow with a black patch below flanked by distinctive white patches. Its long legs are fully feathered and its long tail is wedge-shaped and mottled gray and brown. The European subspecies has broader and whiter frontal stripes and lores. Flight is slow and heron-like, but surprisingly maneuverable through thick forests.

HABITAT: The great gray owl primarily occurs in dense boreal forests, but it also finds suitable coniferous habitat south into the northern Rocky and Sierra Mountains and along some central Asiatic mountain chains. It hunts in or near bogs, forest edge, montane meadows and other forest openings.

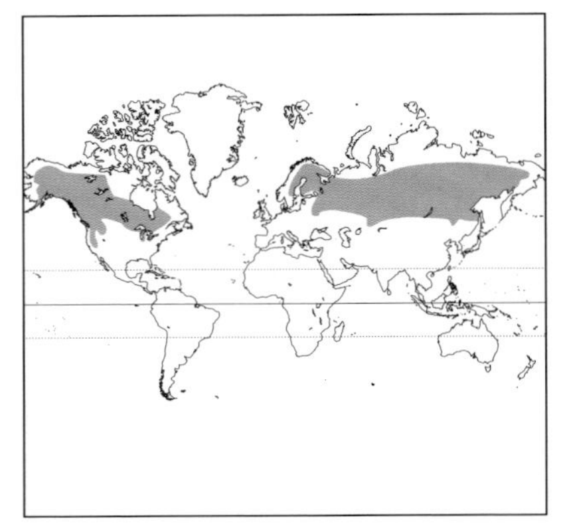

NATURAL HISTORY: This rodent specialist can locate and capture prey (primarily *Microtus* voles) under snow as deep as 45 cm, plunging through snow crusts thick enough to support the weight of an 80 kg person. The great gray owl can also capture pocket gophers tunneling under the soil surface. Its breeding density is limited by prey and nest site availability. Its voice is a deep, booming series of "whoo's," that accelerates and then falls in pitch. It nests in the abandoned stick nests of other birds, also in rotted tops of broken trees and artificial platforms; rarely it nests on the ground. The female lays 2 to 5 eggs and is fed by the male during incubation and brooding. One captive male, however, was persistent in his efforts to assist his mate with incubation. The female consumes the feces and pellets of the young until they are 2 to 3 weeks old, regurgitating this waste away from the nest, thus minimizing the risk of predation by bears and other carnivores. When *Microtus* voles are abundant, pairs will sometimes nest as close as 400 m and cooperative breeding behavior has been observed in captivity and in the wild. Young sometimes join the brood of neighboring pairs after leaving the nest. Social nesting behavior needs more study. In some winters, when its prey is scarce, a considerable number of these owls migrate south of the boreal forest in North America and also in parts of Europe and Asia. A single great gray owl wintering on a farm in Massachusetts for 2 months attracted no less than 3000 birdwatchers!

CONSERVATION STATUS: Not Globally Threatened (IUCN). Threats to local populations include forest habitat loss to peat extraction and agriculture, intensive timber management that reduces live and dead large-diameter trees used by nest-building birds, rodenticides and overgrazing of montane meadows. Alternatively, forest management can enhance or maintain optimal great gray owl habitat by creating new foraging habitat (cutover areas) while retaining snags as hunting perches. Leaving forested buffers around nest sites is an important management recommendation. Provision of man-made nest platforms can help offset local habitat loss.

124 African Wood Owl, *Strix woodfordii*

DESCRIPTION: A medium-sized (30 to 35 cm) owl with a large round head without ear tufts. Its large dark-brown eyes are set in a brown facial disk with white "eyebrows" and lores. It is an overall dark rufous-brown to chestnut, with underparts barred dusky on brownish-white while upperparts are white-spotted. It has a yellowish bill and feet.

HABITAT: It is found regularly in forests and dense woodland, and also in thick coastal bush, well-wooded riparian strips, plantations and wooded suburban gardens.

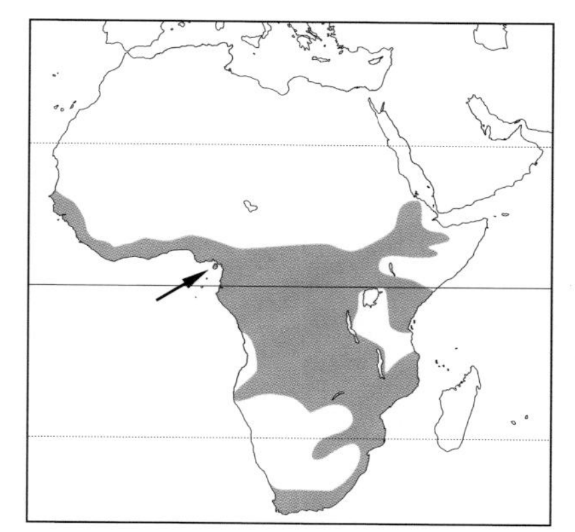

NATURAL HISTORY: The spotted wood owl has been described as the African ecological counterpart to the tawny owl of Europe. It is a secretive and nocturnal owl, which is more often heard than seen. The female utters a "eee-yow" call in contrast to the call of the male's low, gruff "hoo" or bubbling and undulating "hoo-hoo, hu, hu, hu, hu, hu." The spotted wood owl is reported to be a permanent resident, remaining faithful to a nest site even after being disturbed repeatedly. It eats insects, and is capable of catching flying termites and moths. It occasionally takes small birds, rodents, frogs and snakes. Its 2, rarely 3, eggs are laid in a tree cavity deep enough to conceal the incubating bird. It rarely nests on the ground or in a stick nest built by other birds such as the African goshawk. Little was known about its biology until a coffee farmer's wife, Jo Scott, found a nest from which she made extensive life history observations over the years, with some assistance from owl biologist Peter Steyn. They subsequently published their notes, providing new information on this owl's biology. This demonstrates the important role that amateur naturalists have in advancing our basic knowledge of these and other secretive owls.

CONSERVATION STATUS: Not Globally Threatened (IUCN). A common bird in suitable habitat.

125 Mottled Owl, *Strix virgata*

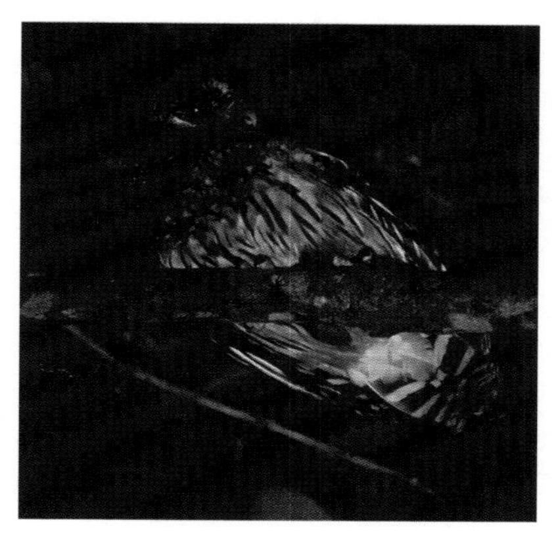

DESCRIPTION: A medium-sized owl (29 to 38 cm) with light and dark morphs. Populations in its northwestern South American range are dark, and the unspotted upperparts are almost black; the lightest, buff-brown morphs occur in northwestern Mexico. At the southernmost portion of its South American range it is light buff above and more heavily spotted with white. It has feathered legs and bare toes.

HABITAT: It occupies dense humid and tropical dry lowland forests and montane semi-arid to cloud forests up to 2500 m. It has been recorded in human-altered areas including coffee and other plantations, dense bamboo stands, second-growth forests and even cities.

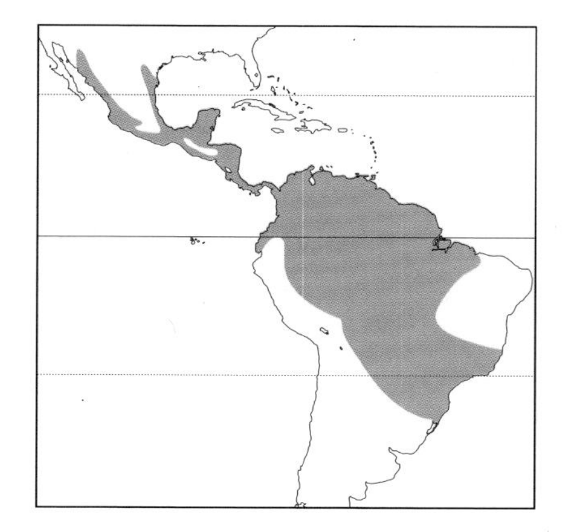

NATURAL HISTORY: Thought to be nonmigratory and nocturnal, but there are few detailed studies on its movements or circadian rhythm. Day roosts are in tree holes and shady, thick foliage. Its call is similar to that of the spotted owl; a series of deep guttural, frog-like hoots denoted as "bru bru" and "whoo-oo." It also can produce a surprisingly deep barking sound for an owl of its size (235 to 305 g). It is reportedly strongly territorial. Its 1 to 2 eggs are laid in live tree cavities, palm stubs and the nests of other bird species. Based on limited data from stomach content analysis, its diet appears to be diverse, including nocturnal and arboreal small mammals, invertebrates, reptiles and amphibians. It is occasionally seen at night near artificial lights, perhaps opportunistically drawn to prey attracted to the light.

CONSERVATION STATUS: Not Globally Threatened (IUCN). Although infrequently seen, it occurs in and near small towns in Mexico and Central America. It appears to tolerate some forest fragmentation, but the further loss of habitat may threaten some populations. It is unclear if local populations can sustain an apparently high frequency of vehicle collision mortality. Luckily, it occurs in many protected areas throughout its range.

126 Black-and-white Owl, *Strix nigrolineata*

DESCRIPTION: A medium-sized owl (33 to 45 cm) distinguished from the closely related black-banded owl by its plumage and unique vocalizations. It has generally dark sooty-brown upperparts, including a black facial disk with a poorly defined white-flecked border. The dark brown eyes might be hard to locate if not for its yellow bill and white-speckled "eyebrows." The white underparts have fine, dark-brown bars.

HABITAT: Reportedly often overlooked, although it uses habitat near villages and settlements. Elsewhere, it uses a great variety of natural humid, deciduous forests, forest openings and forest edges. It can also be found in mangroves and woodlands near water. It occurs at elevations from sea level to 2400 m in Colombia.

NATURAL HISTORY: A nocturnal resident, it often hunts near artificial lights at night and can take insects and bats on the wing. The wide variety of small mammals and insects eaten suggests it likely preys on birds as well. Its 1 to 2 eggs are laid in tree cavities, among epiphytes and orchids, and perhaps in old stick nests built by other birds. Its calls include a series of 3 to 5 barking notes: "who-who-WHOW-who." It roosts by day high above the ground in dense vegetation.

CONSERVATION STATUS: Not Globally Threatened (IUCN). Its status across its range varies locally from rare to fairly common, but it is always widespread. Forest clearance and heavy pesticide use are recognized threats to this owl. It is a regularly noted bird in at least one protected area, Henri Pittier National Park in Venezuela.

127 Black-banded Owl, *Strix huhula*

DESCRIPTION: A medium-sized owl (31 to 36 cm) with a black facial disk with

concentric white lines. It is blackish-brown overall, with irregular white barring. The eyes are orange-yellow, and the bill is light yellow.

HABITAT: It appears to use the upper canopy of tall tropical and subtropical rainforests but also resides near forest clearings, plantations close to forests and even cloud forests. Occurs at elevations from sea level to 1100 m in northern Argentina.

NATURAL HISTORY: This nocturnal resident forages in the forest canopy mainly for insects but also takes small vertebrates. It roosts in dense foliage high above the ground. Its vocal and other behavior needs more study, but one call is a described as a series of 3 or more low notes: "whohohoho whuo who." It likely nests in tree cavities or on rotten stumps. Additional research is needed to determine its ecological relationship to the similar black-and-white owl.

CONSERVATION STATUS: Not Globally Threatened (IUCN). Its status is more accurately described as undetermined and warrants further study. It appears to be scarce throughout its range, but it is likely frequently overlooked. It was reported as common in northern Argentina in the 1950s, but appears to have declined since. It is threatened locally by extensive forest destruction (e.g., northeastern Argentina). Several protected areas support populations, including Amacayacu National Park (Colombia), Manu National Park (Peru) and Imataca Forest Reserve in Venezuela.

128 Rufous-banded Owl, *Strix albitarsis*

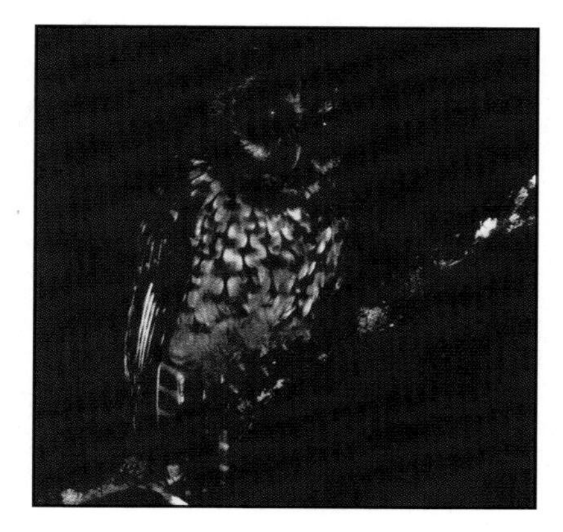

DESCRIPTION: A medium-sized owl (30 to 36 cm) with blackish-brown upperparts and a head with buff-rufous spots and barring. The facial disk is rufous but blackish around the yellow-orange eyes, and with buff-white lores and "eyebrows." Its bill is pale yellow. The black chin contrasts with the white throat. The dark brown chest has tawny to white spots and bars that together make a vague chest band. Its remaining underparts have rectangular white spots.

HABITAT: It is found in montane and cloud forest, and also in adjacent sparsely treed open areas. It occurs at higher elevations in the Andes, from 1700 to 3700 m. Known to occur near occupied dwellings.

NATURAL HISTORY: This nocturnal and crepuscular owl is thought to be nonmigratory. It roosts by day in dense vegetation and likely nests in tree cavities. There is no available information on its diet (probably mainly insects and small mammals), breeding biology (only young owls observed on 2 occasions) and hunting behavior (forages in forest canopy). Its call is a rapid sequence of terse notes: "hu, hu-hu-hu, HOOa," repeated at 8 to 11 second intervals.

CONSERVATION STATUS: Not Globally Threatened (IUCN). Its status is essentially undetermined, and more research is needed on its biology to determine which forest harvesting practices negatively affect populations. It is reported as uncommon in Ecuador. Managed areas with known populations include the El Pahuma Orchid Reserve and the Pasochoa Forest Reserve, Ecuador. Also documented in Venezuela, Colombia, Peru and Bolivia.

129 Maned Owl, *Jubula lettii*

DESCRIPTION: A medium-sized owl (30 to 44 cm) with bushy, long ear tufts that give it a maned look. The head is reddish-brown with whitish "eyebrows" and forehead. The rich yellow eyes and pale yellow bill are set in a light rufous facial disk with a dark brown rim and fine dusky-brown vermiculations. Blackish shaft streaks mark the buff-ochraceous underparts. The upperparts are chestnut-brown to rufous with variable marks and barring.

HABITAT: The maned owl has never been reported outside gallery forest or lowland rainforest, occurring especially in those with abundant creepers. It is notably associated with rivers and lakes.

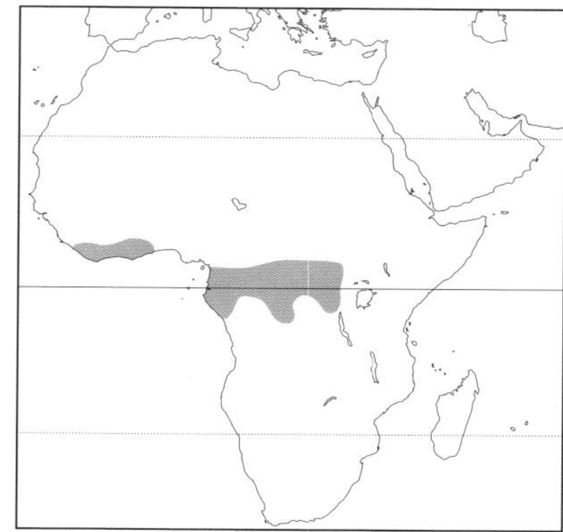

NATURAL HISTORY: This nocturnal resident frequently roosts in creepers, emerging at dusk to hunt from open perches. Its call is anecdotally reported as 2 well-spaced (10 seconds) mellow hooting notes (the first "who" higher pitched than the second). This call may have been confused with the call of the vermiculated fishing owl, also known to occur in that particular study area. The maned owl's 3 to 4 eggs are likely laid in tree cavities and old stick nests. It hunts mainly insects but also small vertebrates. There is little available information on its general biology, especially its breeding biology.

CONSERVATION STATUS: Not Globally Threatened (IUCN). It has been documented as locally common, but its distribution is poorly known. More research is needed to determine its status and its dependency on old forests in light of increased loss of forest habitat. It is reported as rare in Ghana, Liberia and Cameroon.

130 Crested Owl, *Lophostrix cristata*

DESCRIPTION: A medium-sized owl (36 to 43) with distinct light and dark forms. The pale morph has dark brown eyes set in a tawny to chestnut facial disk. The long, mainly white erectile ear tufts seem to

extend from the white "eyebrows." It is plain buff to gray-brown above, and whitish below. The dark morph is an overall dark brown, but paler below. Its eyes are yellowish-orange in contrast to its black-rimmed rufous facial disk.

HABITAT: It ranges from sea level to cloud forests at elevations of 1950 m. It occurs in humid, heavy forests and second-growth gallery woodlands.

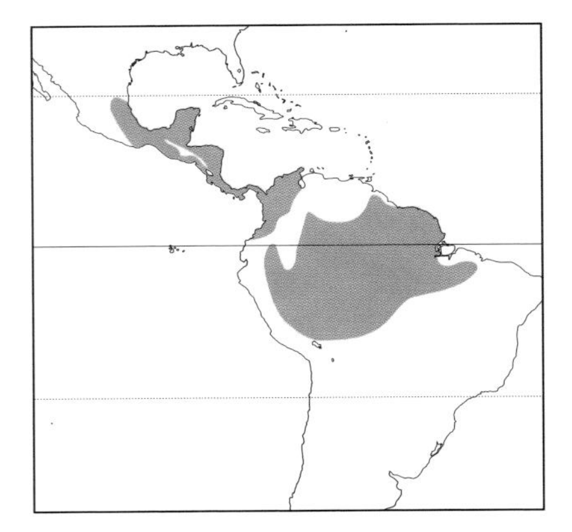

NATURAL HISTORY: This poorly known nocturnal resident nests in tree cavities, and at least once nested in the loft of a house. Larger insects (e.g., caterpillars, beetles and grasshoppers) feature prominently in its diet, and it likely takes small vertebrates as well. Its call is a repetitive low-pitched: "k-k-kk-kk-kkkrrrrrroa." Considerable debate has occurred on its ecological and genetic relationship to *Otus*, *Bubo*, *Pulsatrix* and *Jubula* owls. It roosts by day in dense foliage, frequently close to the ground.

CONSERVATION STATUS: Not Globally Threatened (IUCN). There appears to be insufficient information on its status, and therefore it should be a priority for research, survey and monitoring efforts. Such effort may show that it can survive in residual patchy forest remnants, but it may be at risk from intense deforestation and habitat loss.

131 Spectacled Owl, *Pulsatrix perspicillata*

DESCRIPTION: A large owl (41 to 52 cm) with a dark brown facial disk with contrasting white lores and "eyebrows" that impart a spectacled appearance. The head and neck are a darker brown than the rest of the upperparts. A broad, rich brown, breast-band overlays the lighter buff-brown underparts. The yellow eyes are notably brighter than the creamy-yellow bill. Young spectacled owls can take up to 5 years to acquire the species' definitive adult plumage.

HABITAT: Occurs more frequently in dense tropical rainforest with mature, large trees, including forest edges. Also found in dry forest, treed savannah habitat, plantations and open areas with scattered trees, and at elevations up to 1500 m in subtropical montane forest.

NATURAL HISTORY: This nonmigratory nocturnal owl frequently perches along forest streams. Its 2 eggs, rarely 3, are laid in large tree cavities, but typically only 1 chick survives to leave the nest at 5 to 6 weeks old. Incubation is reportedly 5 weeks, considerably longer than for most other owl species. The young owlet remains with its

parents for up to 1 year. The call of the spectacled owl is a low-pitched, descending series of notes: "PUP-pup-pup-pup-po," sometimes uttered by both the male and the female. It takes mainly small mammals, but also opossums, rabbits, skunks, rodents, agoutis and bats. Less frequently it eats birds, reptiles, amphibians, insects and crabs.

CONSERVATION STATUS: Not Globally Threatened (IUCN). It appears to have a patchy and variable distribution across its range, with local populations ranging from rare to common. It occurs in numerous protected areas, and is more tolerant of forest fragmentation than the crested owl, needing relatively fewer trees for nesting and roosting. Local extirpations result from intense and extensive destruction of forest habitat. Its basic biology is still largely a mystery; hence, as for many owls, detailed population studies are needed.

132 Tawny-browed Owl, *Pulsatrix koeniswaldiana*

DESCRIPTION: A large owl (44 cm) with a dark brown upper belly-band over a buff belly and light-ochre vermiculated underparts. The upperparts are an even darker chocolate-brown. Its facial features include white rictal bristles around the pale-yellow bill, small whitish chin patch, creamy "eyebrows," chestnut-brown eyes, and an ochre-bordered, brown facial disk. It is smaller than the otherwise similar spectacled owl.

HABITAT: The tawny-browed owl is found in pristine, humid tropical forest, open woodland and heavily impacted secondary forest, from sea level to 1500 m elevation. The presence of older, larger trees appears to be an important habitat feature.

NATURAL HISTORY: This nocturnal owl is suspected to be nonmigratory. Its call is a descending series of low-pitch notes, which sound like: "brrr brrr brrr brrr" or "ut ut ut ut ut." It captures small mammals, birds and large insects that it detects from hunting perches in the forest canopy. Its 2 eggs are laid in tree cavities, and, from what little is known, its breeding biology (e.g., incubation, fledging and dependency period of young) seems similar to the spectacled owl.

CONSERVATION STATUS: Not Globally Threatened (IUCN). This range-restricted species is confined to the Atlantic Forest Lowlands Endemic Bird Area. It is known to occur in several protected areas, and populations seem to persist in Argentina in areas subject to forest harvesting. Other reports indicate that it is dependent on relatively intact montane forest, and is rare overall to locally endangered. More intensive and directed surveys are needed to better estimate population size and trends.

133 Band-bellied Owl, *Pulsatrix melanota*

DESCRIPTION: A large owl (48 cm) with a dark brown head and white neck collar. The facial disk is also dark brown and has notable white lores and "eyebrows." The eyes are a dark reddish-brown, and the bill is a pale horn. The upper breast has a broad rufous-brown band that is barred and mottled whitish-buff. The belly is creamy with brown bars.

HABITAT: Based on very limited knowledge, it is thought to occur in tropical rainforest, and even open woodland, from lowlands to 1600 m elevation.

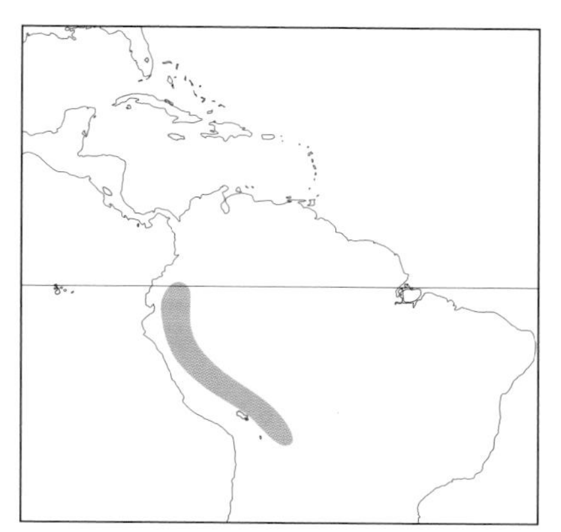

NATURAL HISTORY: There is no available information on its breeding biology, diet, movement patterns or behavior. Its call has been described as a low-pitched, terse purring sound followed by 4 to 5 "pop" notes. All aspects of this species' natural history warrant study.

CONSERVATION STATUS: Not Globally Threatened (IUCN). Its true status is uncertain owing to a lack of information. Few confirmed records exist; it has perhaps been overlooked due to its reportedly shy nature, nocturnal habits and use of rarely explored forest habitat. There is insufficient information to determine if it is threatened by the ongoing loss of habitat within its range.

134 Northern Hawk Owl, *Surnia ulula*

DESCRIPTION: A medium-sized owl (36 to 41 cm), it is brownish-black above and white with fine dense gray-black to cinnamon barring below. Its head, back and wings are heavily spotted and streaked with white. Fine white bars are noticeable across the relatively long and tapered tail. The round head has no prominent ear tufts. The whitish facial disk is rimmed in black. The northern hawk owl is atypical in morphology and behavior compared to other northern hemisphere owls, resembling in many ways the accipiter hawks—hence the name "hawk owl." When flying, it either glides low over the ground at high speed or flaps its pointed wings in deep, powerful, falcon-like strokes. Although most owls show appreciable reversed sexual size dimorphism (female larger than the male) this is not strongly pronounced in the northern hawk owl.

HABITAT: The only species in its genus, the northern hawk owl breeds in the circumpolar boreal forest zone from Alaska eastward through Newfoundland, and from Scandinavia through Siberia. It nests in dead tree stubs or woodpecker holes, especially in open coniferous or mixed coniferous-deciduous forests, burned-over areas or muskeg.

NATURAL HISTORY: One of the least-studied birds of North America, the northern hawk owl has a bold nature and seems to lack fear of humans, delighting birders, who often travel great distances to observe this strongly diurnal owl.

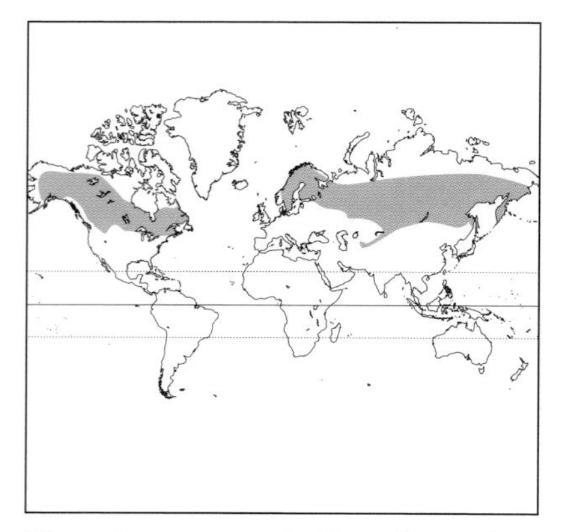

The male utters a musical 3- to 4-second trill, "kuhurrrrrrrrrrr," at a rate of 15 notes per second. Although this owl winters throughout its breeding range, it periodically "invades" southern parts of its range. The magnitude and extent of these winter irruptions are thought to correlate with high reproductive success followed by severe winter conditions and decreased prey availability, but more research is needed to document these relationships. Early studies suggest that this owl primarily eats small rodents, but recent evidence indicates that grouse, ptarmigan and hares comprise a greater proportion of its diet than previously thought. This species usually hunts from perches atop prominent trees. Despite its lack of skeletal ear asymmetry, it can locate and capture concealed prey up to 30 cm beneath snow cover. The local appearance of tens or hundreds of hawk owls during winter south of their breeding range is an exciting event noted by researchers, birders and laypersons alike. One recent recovery of a banded northern hawk owl documented a movement (from Alberta to Alaska) of 3168 km!

CONSERVATION STATUS: Not Globally Threatened (IUCN). While much remains to be learned about the natural history of this species on its breeding range, there is no evidence that this species is at risk. Low population density and the remoteness of its breeding range have challenged those wanting to study or observe it. More information on habitat selection, home range, breeding dispersal and winter diet is needed.

135 Eurasian Pygmy Owl, *Glaucidium passerinum*

DESCRIPTION: A small dark-rufous to gray-brown owl (15 to 19 cm) without ear tufts. Its relatively small yellow eyes are set in an undefined pale gray-brown facial disk with concentric rings of fine dusky spots. The gray-brown tail has 5 narrow, whitish bars. The head has fine white spots. The upperparts have white spots, and the underparts are off-white with brown streaks. The legs and toes are feathered.

HABITAT: Occurs mainly in the interior of tall coniferous and mixed forests that have clearings and open areas. Also uses cool moist ravines in more southern, temperate zones. It is found typically at elevations ranging from 250 to 1000 m, but has been reported at 2150 m in the Alps.

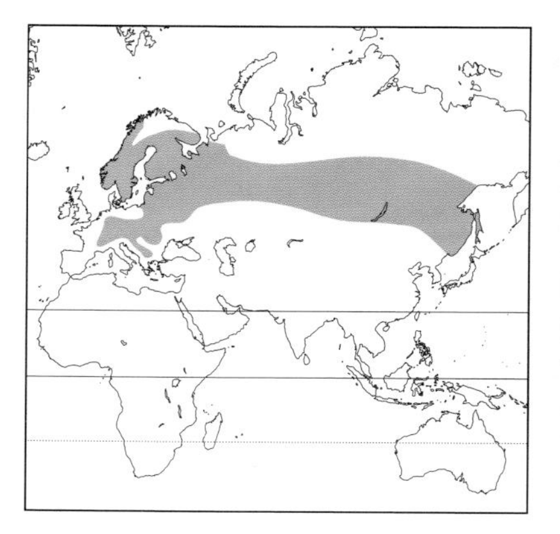

NATURAL HISTORY: Mainly resident, except in northern Europe where irruptions are likely triggered by cold weather and scarce rodent populations. Sometimes relocates to residential areas or mixed woods in winter. It is mainly crepuscular, but also frequently hunts during daylight. Roosts against the trunks of well-foliated trees. Small mammals are its main prey at the beginning of the breeding season, but the proportion of small birds taken, especially recently fledged young, increases over the summer. The male's call is a series of 5 to 10 clear, flute-like "deu" notes uttered at 2-second intervals. Its 4 to 7 eggs are laid in natural tree cavities, holes made by woodpeckers and nest boxes. The breeding female will show aggression to the male, apparently to stimulate him to hunt. The young are driven out of the home range when 7 to 8 weeks old.

CONSERVATION STATUS: Not Globally Threatened (IUCN). Densities as high as 4.2 pairs/km are reported in the central mountains of Europe. Deforestation, and associated ecological changes (e.g., invasion by the tawny owl at higher elevations), has resulted in its extirpation from large parts of its range in Germany. Populations in Finland, however, appear to be stable despite the extensive logging of old forest habitat. It has been successfully captive-bred and reintroduced in Germany's Black Forest.

136 Collared Owlet, *Glaucidium brodiei*

DESCRIPTION: This tiny round-headed owl (15 to 17 cm) has been described as a miniature Asian barred owlet. Overall, it is barred gray-brown and has white "eyebrows" and throat patch, and a spotted (creamy-buff) crown with black patches bordered with buff or white, creating a staring "dorsal face." Three morphs occur: rufous, chestnut and gray-brown. Although it is the smallest of the south Asiatic owls, it is a ferocious hunter.

HABITAT: Occupies open submontane forest (deciduous and mixed oak, fir, deodar and other conifers) to dense montane forest (within the forest or on the edge). From 1370 to 3000 m, but from 400 to 1800 m in the Malay Peninsula, and 800 to 1650 m in Sumatra. In China, at 740 m near cultivated areas.

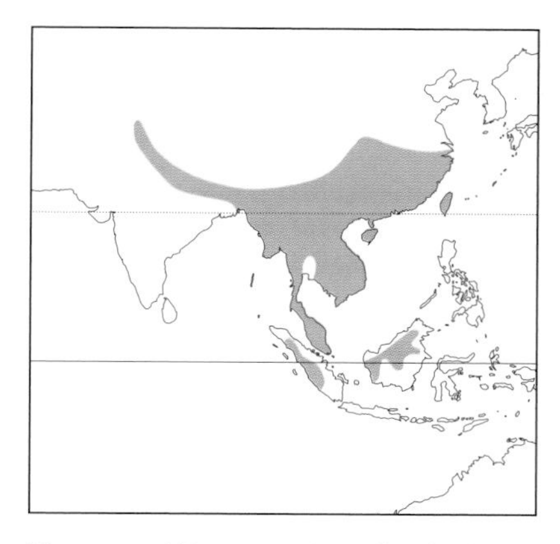

NATURAL HISTORY: A resident bird, but it may undergo seasonal altitudinal movements. Considered the most diurnal bird of its genus, it is active during all hours of the day. Its main call is a series of 3 to 4 bell- or whistle-like notes ("tew-tewtew-tewt"). Usually 4 eggs (3 to 6) are laid in a tree cavity (2 to 10 m above the ground) or the nest of a woodpecker or barbet. It may kill the nest builder and take over the nest site. Prey includes small rodents, lizards, skinks, cicadas, grasshoppers, beetles and other large insects, and small birds. A fierce hunter, it will take adult and nestling barbets, woodpeckers, thrushes, minivets and magpie robins; there is one record of predation on the pugnaceous ashy drongo (in Borneo). Said to frequently prey upon nestlings.

CONSERVATION STATUS: Not Globally Threatened (IUCN). The species is reported to be common to fairly common throughout most of its range. It is found within several protected areas, including the Hsitou Forest Recreation Area (Taiwan), Khao Yai and Kaeng Krachan National Parks (Thailand), Kerinci-Seblat National Park (Sumatra) and Mount Kinabalu National Park (Borneo). It is vulnerable to habitat destruction.

137 Pearl-spotted Owlet, *Glaucidium perlatum*

DESCRIPTION: A small fearless owl (17 to 20 cm) with a stocky build, no ear tufts, yellow eyes and relatively big and powerful feet. The 2 large black spots bordered and connected with white on the back of its head form a "false face." The overall color of the upperparts is varying shades of brown to cinnamon-brown, sprinkled with many round whitish spots. This gives the bird a "pearled" appearance. It often jerks its relatively long, brown tail when excited or nervous.

HABITAT: It occurs in open savannah with short grass and scattered trees and acacias (thorny shrubs) to tall woodland with closed canopy. Also occupies dry, semi-open woodland (e.g., mopane) and semi-open forest along rivers next to savannah. It avoids tall grass, tropical rainforest, montane and treeless desert habitats.

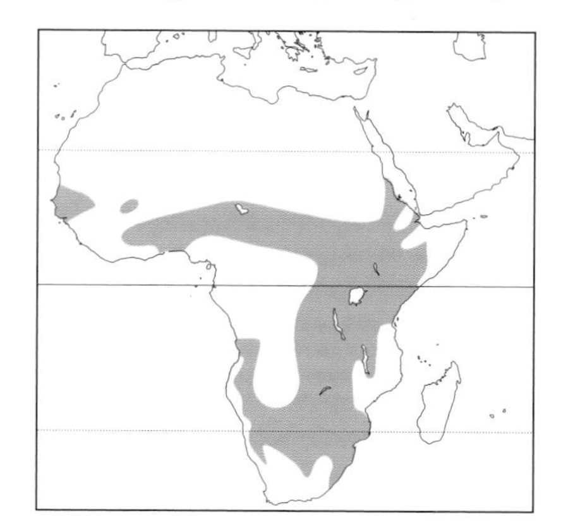

NATURAL HISTORY: It is thought to be nonmigratory. Although known as a diurnal African owl, it is active mainly active at night. Its 3 eggs (2 to 4) are laid in natural cavities, nest holes of woodpeckers and barbets, and nest boxes (1 to 10 m above the ground). Females adopt a prone position over eggs or young if disturbed. Its territory size is estimated at 60 ha. Nests have been recorded only 200 to 500 m apart, resulting in overlap of hunting ranges. It bathes regularly in shallow pools or rain. It eats primarily large arthropods (such as orthopterans, millipedes, spiders and beetles). Also eats small mammals and birds, amphibians, lizards, occasionally small snakes, and can take bats in flight. It may take prey larger than itself, such as doves. It also robs prey from other species (e.g., wood hoopoe), takes nestlings and eats carrion. Its call is a flute-like series of "few" notes followed by a pause and then a descending "peeooh peeooh...."

CONSERVATION STATUS: Not Globally Threatened (IUCN). Uncertain status, but reported as locally common to rare within its range. Uncommon in the upper western edge of its range, only one report from Liberia, uncommon in Ghana, and only a few records from the Lake Victoria Basin. In some regions it is possibly endangered as a result of bush fires and pesticide use.

138 Northern Pygmy Owl, *Glaucidium californicum*

DESCRIPTION: A small owl (16 to 18 cm) that has gray and rufous color morphs. It is stocky with a round head without ear tufts, short wings and a relatively long tail. The indistinct facial disk is brownish-white. The face-like pattern on the back of its head has black "false eyes" and a black "bill." Its real bill is whitish-yellow, and its eyes are yellow. The upperparts are gray-brown, and there are brownish-black streaks on the whitish underparts. It has fine white spots on the head, nape, shoulders, sides of the

chest, flanks and wings.

HABITAT: Generally prefers low to intermediate forest canopy cover, e.g., edges along meadows and lakes, in coniferous and deciduous forests. Ranges from foothill oak savannahs to mixed montane conifer forests up to 2200 m. In California, it occurs in mature, residual and secondary-growth pine, cedar, fir and oak forests.

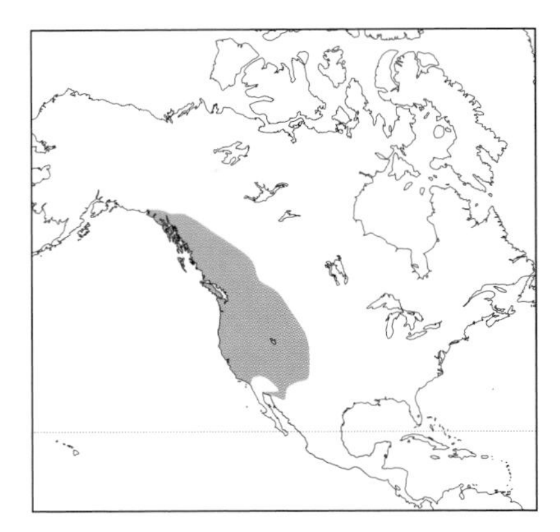

NATURAL HISTORY: Its call is a series of single "toot" notes given at regular, 0.5-second intervals. It catches prey as big as quail and squirrels—remarkable for this small (62 to 73 g) owl. Females reportedly take more small mammals and males more birds. It also eats large insects, including beetles and moths. It is thought to hunt mainly by sight during the day, but is also active at dawn and dusk. Its bold nature allows people to approach it closely. Its 2 to 7 eggs are laid in tree cavities, occasionally near or in the same tree as other cavity-nesting species including the northern saw-whet owl and the pileated woodpecker. Unlike other owls, its eggs are thought to hatch synchronously, with the young leaving the nest at the same time when about 23 days old. There is little information available on its migratory or possible seasonal vertical (altitudinal) movements, although it is reported to be partly resident, migratory and nomadic.

CONSERVATION STATUS: Not Globally Threatened (IUCN). There are no documented threats to this species. Variable population trends across its range, and large areas of unoccupied yet apparently suitable habitat, suggest that long-term research on its biology and population is needed.

139 Mountain Pygmy Owl, *Glaucidium gnoma*

DESCRIPTION: A tiny pygmy owl (15 to 17 cm) similar in appearance to the northern pygmy owl, but smaller, with a shorter tail, pointed (not rounded) wing tips and a different voice. Its underparts are also more diffusely streaked, and it has a long whitish area from its throat to the centre of its breast. Its overall color is dark brown to foxy-rufous.

HABITAT: The species is usually found from 1500 to 3500 m in mountainous regions. It occupies coniferous forests, especially pine-oak, pine and humid pine-evergreen forest; also forest edges. In Arizona, oak forests on southern slopes are the typical habitat of this owl.

NATURAL HISTORY: It is thought to be resident. It is most frequently seen at dusk

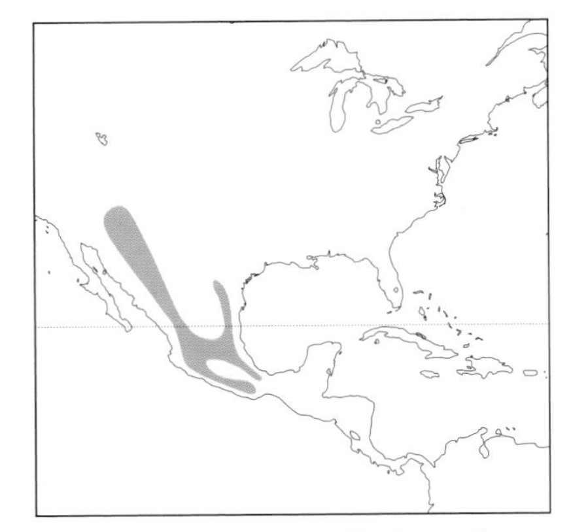

or dawn but also occasionally during the day. An extended series of terse, single or double staccato "gew" notes is its main call. It responds readily to its recorded calls, which could be used to help study its breeding biology. Its 2 to 4 eggs are laid in tree cavities, most frequently old woodpecker holes. Its diet is mainly insects (orthopterans, beetles, grasshoppers and crickets), but also includes small mammals, reptiles and birds.

CONSERVATION STATUS: Not Globally Threatened (IUCN). Its status is uncertain, although it is generally considered locally common. The most serious threat it faces is forest destruction.

140 Guatemalan Pygmy Owl, *Glaucidium cobanense*

DESCRIPTION: A small, bright reddish-brown pygmy owl (16 to 18 cm), with a long tail and yellow eyes and feet. It was formerly considered a race of the mountain pygmy owl, but information

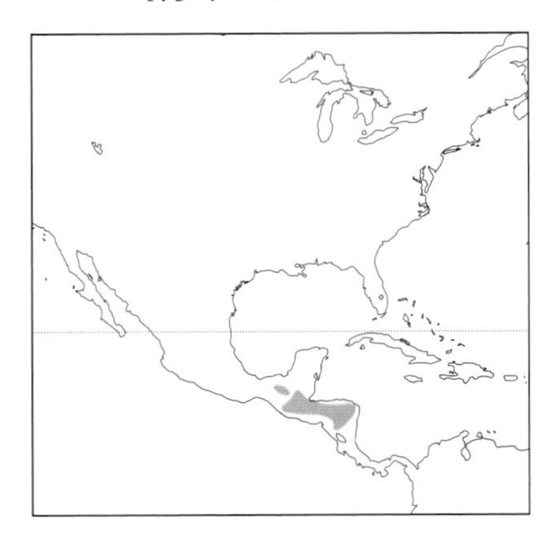

on its vocalizations is needed to clarify its status as a species. The wings are both spotted and barred, and there is little mottling on the head. Underparts have diffuse, broad rufous streaks.

HABITAT: It occurs in highland forest habitat.

NATURAL HISTORY: Extremely little information is available for this species. It is thought to be resident in its restricted range. Its 3 to 4 eggs are laid in old woodpecker holes, and it preys upon insects, rodents and occasionally birds. A description of its vocalizations, yet to be recorded, will assist biologists in evaluating if this owl should be treated as a separate species from the mountain pygmy owl.

CONSERVATION STATUS: Not Globally Threatened (IUCN). Its status is essentially unknown. It occurs in the North Central American Highlands Endemic Bird Area. Recently recorded in Mexico. Forest destruction may threaten this species.

141 Baja Pygmy Owl, *Glaucidium hoskinsii*

DESCRIPTION: A tiny owl (15 to 16 cm) similar to the northern pygmy owl, but smaller and with a shorter tail. Once considered a race of the mountain pygmy owl, it has been given separate status based on vocalization differences. Overall plumage is sandy gray-brown in color, with females being more rufous. It has rounded wing tips, brownish facial disk, thin white "eyebrows," and buff spotted crown, nape and underparts. It has "false eyes" on the nape.

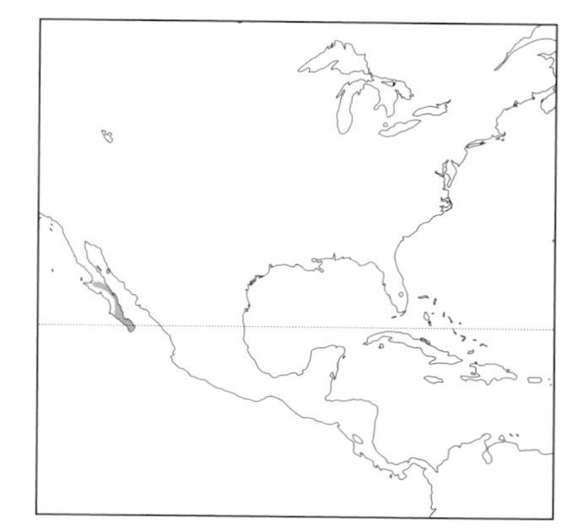

HABITAT: Found from 1500 to 2100 m in pine and pine-oak forests; down to 500 m into deciduous forest in winter.

NATURAL HISTORY: Basically a resident species, but in winter it may migrate to lower altitudes, and rarely to northern Baja California. The limited information available on this species indicates that it nests in tree cavities, probably old woodpecker holes. Its diet includes insects, reptiles, and small mammals and birds. The male utters a relatively slow series of single or double "whew" notes.

CONSERVATION STATUS: Not Globally Threatened (IUCN). It is reported as being rare to fairly common. Due to its restricted range, it likely is negatively affected by habitat loss. It occurs in the Baja California Endemic Bird Area. The ecology and population status of this species is poorly understood.

142 Costa Rican Pygmy Owl, *Glaucidium costaricanum*

DESCRIPTION: Formerly considered a race of the Andean pygmy owl, but it is now recognized as a full species based on its unique vocalizations. This tiny owl (15 cm) occurs in brown and rufous morphs. Its head is brown, and the crown is densely spotted with white. The "eyebrows" and lores are white, and it has "false eyes" on its nape. The overall brownish plumage is boldly patterned and has much spotting, barring and streaking.

HABITAT: It occupies pine-oak upland forest, where it prefers the canopy and forest edges. Semi-open habitat next to forests is also used, as well as pastures containing some trees. It is found at 900 m on the Caribbean slope and 1200 m on the Pacific slope.

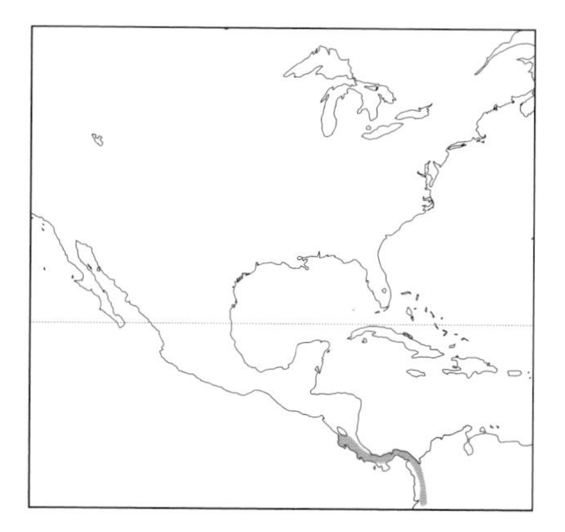

NATURAL HISTORY: Assumed to be resident in its restricted range. It nests in tree cavities, especially old woodpecker holes (one nest with 3 eggs recorded at 1 to 2 m above the ground). It is primarily insectivorous, but it also takes small vertebrates such as birds and lizards. It usually hunts from a perch and pounces on its prey. Its call is an extended series of whistle-like "poop-poop" double notes, which accelerates when it is excited.

CONSERVATION STATUS: Not Globally Threatened (IUCN). Reported as rare (e.g., Panama) to locally fairly common (e.g., Cerro de la Muerte) in Costa Rica. Also occurs in the Costa Rica and the Panama Highlands Endemic Bird Areas.

143 Cloud Forest Pygmy Owl, *Glaucidium nubicola*

DESCRIPTION: This small pygmy owl (16 cm), which was discovered only in 1999, was likely overlooked due to its general similarity with and geographical proximity to the Andean pygmy owl. Nevertheless, it has distinctly different vocalizations. Its plumage is similar to other Neotropical pygmy owls, but there are fewer bars and spots on its sides, and its tail is shorter. The "false eyes" at the back of its head—which most pygmy owls have—are rimmed in white. The back, mantle, scapulars, upperwing coverts and rump are all dark brown, with bold white-spotted scapulars and upperwing coverts.

HABITAT: Found on the steep slopes (1400 to 2000 m) of primary cloud forest.

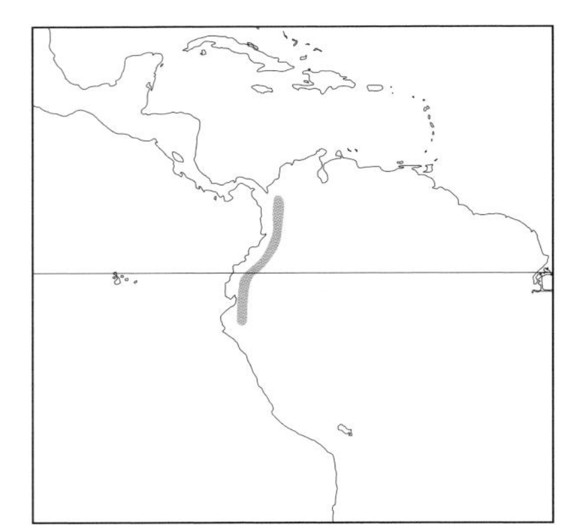

NATURAL HISTORY: Assumed to be resident. Almost no other information is available for this species. The stomach contents of dead owls revealed that it takes insects and small lizards, but it probably preys on small birds as well. Its vocal behavior needs study. One call is described as an extended series of double notes.

CONSERVATION STATUS: Not Globally Threatened (IUCN). No information on population levels. Forest destruction (especially in western Ecuador) and road construction (western Colombia) are major threats. There are some protected areas within the range of this species. It should be considered a high priority for further study.

144 Andean Pygmy Owl, *Glaucidium jardinii*

DESCRIPTION: A tiny pygmy owl (14.5 to 16 cm) that occurs in a dark-brown morph with white markings and a dark-chestnut morph with buff markings. The rather long wings have pointed tips. It has white short "eyebrows," a white moustache and a wide white throat patch.

HABITAT: It is found at elevations from 900 to 4000 m in montane (semi-open) forest, elfin forest, open cloud forest and wooded ravines. It apparently prefers to occupy the middle and high forest canopy levels but also uses forest edges near pastures.

NATURAL HISTORY: A resident species, it is active during the day and at night. It may be found by day when small birds mob it to drive it away. Usually, 3 eggs are laid in a cavity in branches, dead stumps and old woodpecker holes in trees. Its diet includes insects, small mammals and other vertebrates. It also eats small birds (up to its own size) more frequently than other pygmy owls. The male's call is a series of 4 to 5 sudden, rising "pueehtututu" notes that are followed by a slower staccato of "tew" notes.

CONSERVATION STATUS: Not Globally Threatened (IUCN). Its status is uncertain, but it is widespread and thought to be locally quite common. Forest destruction has likely negatively impacted it. It is present in some protected areas, including Las Cajas National Recreation Area and Podocarpus National Park (Ecuador).

145 Yungas Pygmy Owl, *Glaucidium bolivianum*

DESCRIPTION: The Yungas pygmy owl is a small owl (16 cm) most similar in appearance to the Andean pygmy owl, but with a longer tail, more rounded wing-tips, more densely spotted crown and more barring on the sides of the chest. Three morphs are currently known: brown, gray and rufous.

HABITAT: It occurs in montane and cloud forest with epiphytes and creepers (from 1000 to 3000 m). "Yungas," or thick cloud forest with dense undergrowth (e.g., climbing bamboo) is where this species is most common (1000 to 2500 m, locally at 900 m). It prefers the middle and upper canopy levels.

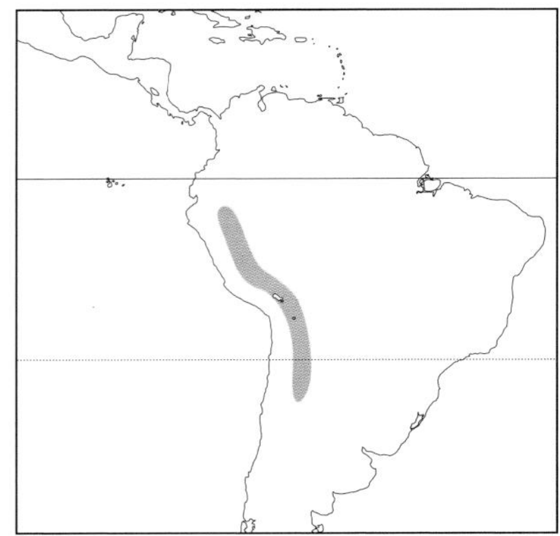

NATURAL HISTORY: It is predominantly a resident species, but some altitudinal movements may occur. Both diurnal and nocturnal, it is increasingly active at dusk, though it is less diurnal than other pygmy owls. Its song starts with 2 to 3 thrush-like "wueeurrr" notes followed by a long series of "whew" notes. Very little information is available on its breeding biology. High, abandoned woodpecker holes are used for nesting. On forested slopes in northern Argentina, territories were between 0.5 and 1 square km. Preys on insects and other arthropods, small birds and possibly other small vertebrates (reptiles).

CONSERVATION STATUS: Not Globally Threatened (IUCN). Endangered in northern Argentina, and seems to have declined locally due to forest destruction and habitat degradation (logging, cattle grazing and burning). Present in several protected areas: Cotapata National Park, La Paz (Bolivia) and Calilegua National Park, Jujuy (Argentina).

146 Colima Pygmy Owl, *Glaucidium palmarum*

DESCRIPTION: A tiny pygmy owl (13 to 15 cm) with yellow eyes; there is a narrow cinnamon band below its "false eyes." Both the crown and nape are densely spotted whitish to pale buff. The plumage above is sandy-gray brown to olive-brown, the underparts are whitish with buff to cinnamon streaks and the sides of the upper breast have brown mottling.

HABITAT: The Colima pygmy owl occupies dry tropical woodlands, foothills (thorn woods), palm groves, coffee plantations and semi-deciduous forests (oak or pine-oak). Locally found in ravines and swamps. This species is most common from sea level to 1500 m.

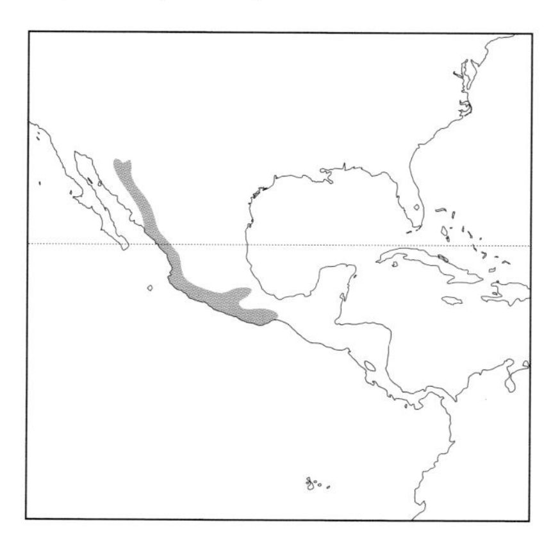

NATURAL HISTORY: This is a resident species that is partly diurnal. There is little information available on its behavior, but it is likely similar to other pygmy owls. Males utter a series of up to 24 or more "whew" notes. Its 2 to 4 eggs are laid in tree cavities or woodpecker holes. Its diet includes insects, reptiles and birds.

CONSERVATION STATUS: Not Globally Threatened (IUCN). It is considered to be common in its range.

147 Tamaulipas Pygmy Owl, *Glaucidium sanchezi*

DESCRIPTION: This small pygmy owl (12 to 18 cm) is unique among owls in that the sexes are not the same color . The crown and nape of the male is gray-brown, with the forecrown delicately spotted pale buff or white. The back is grayish olive-brown, and the wings are spotted pale cinnamon to paler buff. The female's crown, nape and upperparts are washed cinnamon and noticeably redder than the male's.

HABITAT: Found in humid evergreen forests (especially cloud forest) and semi-deciduous forests at elevations of 900 to 2100 m.

NATURAL HISTORY: It is a resident and partly diurnal species whose natural history is poorly known. Its 2 to 4 eggs are laid in tree cavities or old woodpecker holes. Its main call is a series of 2 to 3 high-pitched "phew" notes. The diet includes insects and reptiles. More information is needed on all aspects of its biology.

CONSERVATION STATUS: Not Globally Threatened (IUCN). Although some consider it to be common in its restricted range, its status is unknown. It may be locally endangered due to logging. It is found in the Southern Sierra Madre Oriental Endemic Bird Area.

148 Central American Pygmy Owl, *Glaucidium griseiceps*

DESCRIPTION: This tiny pygmy owl (13 to 18 cm) has a light gray-brown facial disk with whitish flecks and short whitish "eyebrows." The upperparts are a rich brown, contrasting with the brownish-gray of the crown and nape, and covered with small whitish spots. There is a whitish area from the throat to the center of the breast. Its underparts are whitish with rufous-brown streaks.

HABITAT: It occupies humid tropical evergreen forest and bushland, secondary growth, partly open regions and old cacao plantations; from sea level to 1300 m.

NATURAL HISTORY: This resident and partly diurnal species is poorly known. The male's call is a variable series of 2 to 4 "pew" notes followed by another series of up to 18 or more regular "pew" notes. Its 2 to 4 eggs are laid in tree cavities, old woodpecker holes and possibly old termite nests on trees. Its diet includes large insects, lizards, spiders, small mammals and birds (e.g., tanagers and honeycreepers).

CONSERVATION STATUS: Not Globally Threatened (IUCN). It is reportedly common locally (e.g., near Valle Nacional, Oaxaca, and Mayan ruins of Palenque and Monampak, Chiapas). It is also rather common in the Caribbean lowlands of Costa Rica, but locally uncommon (rare but widespread) in Panama. Logging may pose a threat to the species.

149 Subtropical Pygmy Owl, *Glaucidium parkeri*

DESCRIPTION: This small pygmy owl (14 cm) was first described in 1995. Both the crown and the sides of the face are gray-brown and have distinct white spots. The overall plumage is dark brown, with bold white spots on the scapulars. The underparts are white, and the sides of the chest are medium brown. Its lower chest, flanks and belly are broadly streaked olive-rufous.

HABITAT: It occurs in the subcanopy and edges of subtropical montane and cloud forests from 1450 to 1975 m.

NATURAL HISTORY: A resident and partly diurnal species with little information available on its habits. It probably nests in old woodpecker holes and eats primarily insects, small lizards and birds. The male utters up to 6 "hew" notes in a slow staccato-like call.

CONSERVATION STATUS: Not Globally Threatened (IUCN). It is thought to be uncommon and is threatened locally by deforestation. Its relationship with other neotropical pygmy owls requires study.

150 Amazonian Pygmy Owl, *Glaucidium hardyi*

DESCRIPTION: The Amazonian pygmy owl is a small owl (14 cm) with a relatively small head and eyes, a short tail and long, rounded wings. The plumage is dark grayish rufous-brown, and the brown mantle is unmarked, making the crown and nape (which are densely spotted with white) look considerably grayer. Its "false eye" spots are black. The off-white underparts and sides of the chest are streaked with dusk cinnamon-brown. A rusty morph has an unbanded tail. This species is elusive and difficult to observe.

HABITAT: It occurs in the tropical rainforests of the Amazon, in the foothills of the Andes (sea level to 850 m), and along forest edges. It prefers the uppermost canopy of forests (higher than the ferruginous pygmy owl and the tawny-bellied screech owl).

NATURAL HISTORY: It is a resident and partly diurnal species. Its breeding behavior is unstudied, but it probably nests in old woodpecker holes high in trees. Its diet consists primarily of insects, and it likely takes small arboreal vertebrates as well. The male utters a descending trill of flute-like "bew" notes.

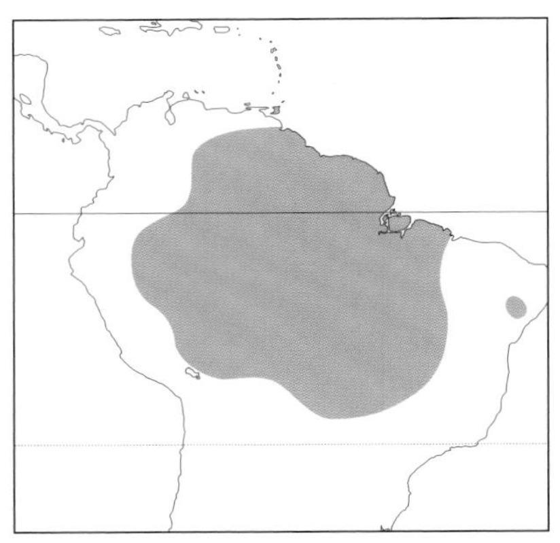

CONSERVATION STATUS: Not Globally Threatened (IUCN). It is thought to be uncommon, but it is easily overlooked due to its preference for high canopy habitat. Locals sometimes keep it as a pet, and it is vulnerable to habitat loss (deforestation).

151 Least Pygmy Owl, *Glaucidium minutissimum*

DESCRIPTION: This small owl (14 to 15 cm) has uniform dark-brown upperparts, a warm brown to cinnamon-brown crown with delicate whitish to pale buff spots and an unmarked mantle. The throat has a white patch. The off-white underparts have rufous streaks, particularly on the lower breast and flanks, and the tail is relatively short.

HABITAT: It occupies tropical and subtropical evergreen rainforest and is found in the forest canopy, along edges and in open bush canopy. Secondary growth is apparently avoided. This species ranges in elevation from sea level to 1100 m, from

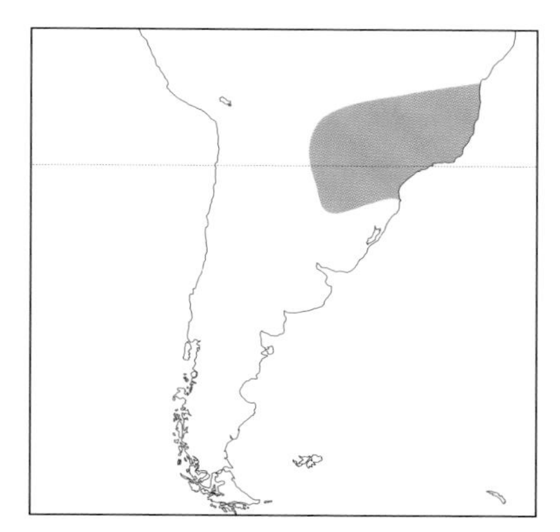

500 to 800 m in the foothills in southeast Brazil.

NATURAL HISTORY: It is a resident species that is both nocturnal and diurnal, with dawn and dusk the periods of most activity. Little information is available on breeding; it presumably nests in tree cavities and old woodpecker holes. Eats mainly insects, but also small birds and mammals. Hunts from a perch and pounces on prey. The male broadcasts a series of high-pitched "hew-hew" notes.

CONSERVATION STATUS: Not Globally Threatened (IUCN). Uncertain status; reported as possibly rare, but locally common in the region between Rio de Janeiro and Santa Catarina. Seriously threatened by habitat loss in primary forest. Present in the Estancia Itabo Private National Park, Paraguay.

152 Ferruginous Pygmy Owl, *Glaucidium brasilianum*

DESCRIPTION: A small owl (15 to 19 cm) with a white-flecked brown and rufous facial disk and prominent white "eyebrows" over lemon-yellow eyes. The bill is yellowish, with a hint of either gray or green. It has both a gray-brown and a rufous morph. The head and back of the neck are spotted whitish-buff, and there is much variation in streaking over its body. The rufous morph lacks the 7 or 8 white tail bands.

HABITAT: It is found in tropical lowlands and foothills with riparian trees, saquaro, brush, and mesquite or palm thickets. Uses primary and secondary forests, even human-altered habitats, e.g., plantations, suburban parks and gardens. It has been found at elevations up to 1900 m (Central America) and 2250 m (Venezuela).

NATURAL HISTORY: It is a resident species that hunts mainly during the day, especially at dawn and dusk. It sometimes dashes from its perch into thick foliage to surprise and capture prey, which includes mainly insects but also amphibians, reptiles, birds and mammals. It will take prey heavier than itself (e.g., rats and meadowlarks). Its 2 to 5 eggs (usually 3) are laid in cavities in sand banks, termite mounds and trees, and in nest boxes. Young are known to occasionally injure each other while fighting over food delivered by their parents. Its call is a series of 10 to 60 whistled "poip" or "whoip" notes uttered at a rate of 3 per second; given either at night or by day.

CONSERVATION STATUS: Not Globally Threatened (IUCN). It has declined significantly in the United States due to loss of habitat. Considerable conflict has recently arisen in Arizona over the private development of some of the 295,430 ha of habitat deemed critical to the long-term maintenance of this population. It is listed as Endangered under the United States *Endangered Species Act*. While it is at risk in Arizona and Texas, this species is thought to be common and secure in various protected areas over much of its range, including Mexico.

153 Tucuman Pygmy Owl, *Glaucidium tucumanum*

DESCRIPTION: A small pygmy owl (16 to 17.5 cm) that occurs in 3 morphs, brown, red and gray, with the latter the most common. The gray morph has dark gray-brown or slaty-gray upperparts, unmarked mantle, streaked forehead and round spots on the hind crown and sides of the head. The underparts are off-white and strongly streaked dusky. The red morph is a light rufescent-brown with indistinct streaks and spots on the crown. The brown morph has similar markings as the gray morph but with warm brown plumage.

HABITAT: It occupies dry forest with open areas, thorny scrub with trees and giant cacti, from 500 to 1800 m in areas with suitable habitat. Locally also found in parks or gardens near towns.

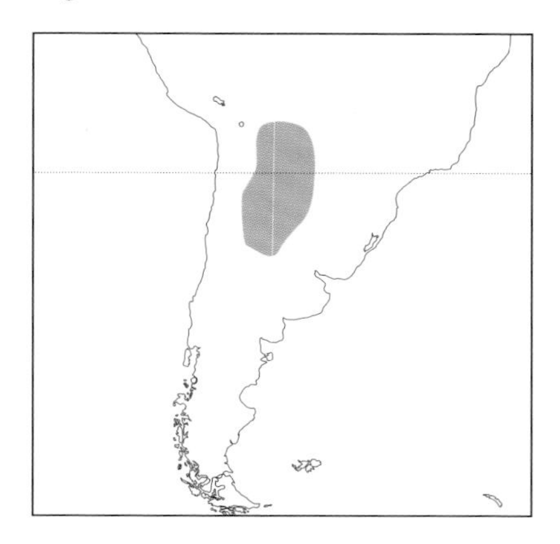

NATURAL HISTORY: It is presumed to be resident and is active from dawn to dusk. The male utters a series of 10 to 30 or more ascending staccato "toik" notes. Its 3 to 5 eggs are laid most often in old woodpecker holes in giant cacti or trees; also in natural cavities and possibly burrows made in banks by small mammals or birds. The same site may be used for several years. Its diet includes small birds (especially during the dry season), insects and reptiles. This species prefers to forage at lower heights, close to ground level. It hunts from a perch but will also grab birds in flight.

CONSERVATION STATUS: Not Globally Threatened (IUCN). Its status is essentially uncertain, but it is reported as common near Tartagal. Habitat loss (including burning) is its main threat, and birds are reportedly trapped for traditional customs in which live or dead owls, and parts thereof, are believed to bring fortune in love.

154 Peruvian Pygmy Owl, *Glaucidium peruanum*

DESCRIPTION: A small pygmy owl (15 to 17 cm) with gray, brown and red morphs. Gray morph birds are found at high elevations and have a crown with small and large whitish dots, while those at lower elevations have shaft streaks or drop-shaped marks of whitish or pale buff on the crown. The "false face" on the nape is rimmed below by a thin ochraceous band, and the underparts are distinctly streaked. The brown morph is similar but has a darker brown background-color; the red morph has a streaked crown and rufescent barring on the tail.

HABITAT: The Peruvian pygmy owl can be found in a variety of habitats, from semi-arid forest to riparian woodland, mesquite, thickets, open areas with scattered trees or groves, eucalyptus plantations, elfin forests, agricultural areas containing trees, thorny scrub with cacti and interspersed with trees, and urban parks. It is found at elevations from sea level to 3000 m.

NATURAL HISTORY: A resident, partly diurnal bird that can be observed in bright daylight in exposed positions. It often nests in old woodpecker holes and sometimes in the mud nests of other birds in branches or on telephone masts. Males are territorial and aggressive. Its diet includes mainly insects and small birds, also small vertebrates (such as mammals). Hunts from a perch at about the mid-canopy level. The male utters a series of ascending staccato "toit" notes.

CONSERVATION STATUS: Not Globally Threatened (IUCN). Overall rare, but it is also considered to be locally common in some areas. May be threatened by pesticide use in agricultural areas.

155 Austral Pygmy Owl, *Glaucidium nanum*

DESCRIPTION: A small pygmy owl (17 to 22 cm) occurring in 2 morphs. The gray-brown morph has many white spots on the scapulars and wing coverts, white streaking on the forehead and crown, whitish

"eyebrows" and lores, and densely streaked underparts. The rufous morph has overall rufous plumage, but it is otherwise similar.

HABITAT: The species can be found in open forests interspersed with shrubs and trees, desert, oases, temperate forests, beech forests in Tierra del Fuego and southern Patagonia, thorn scrub, ravines, gardens or parks within cities, and agricultural land with trees and groves.

NATURAL HISTORY: It is generally resident, with southernmost birds (mostly immature) migrating north in late autumn to central and northern Argentina. Both nocturnal and diurnal, it can sometimes be observed in open perches in bright daylight. The male broadcasts a series of 20 to 30 terse, staccato "ku" notes. Insects make up about half their diet, with small mammals, birds and reptiles making up the remainder. It is capable of securing prey up to 160 g in weight. Occasionally it takes spiders and scorpions. It often nests in holes in tree trunks (1 to 2 m above the ground), especially those made by the Chilean flicker

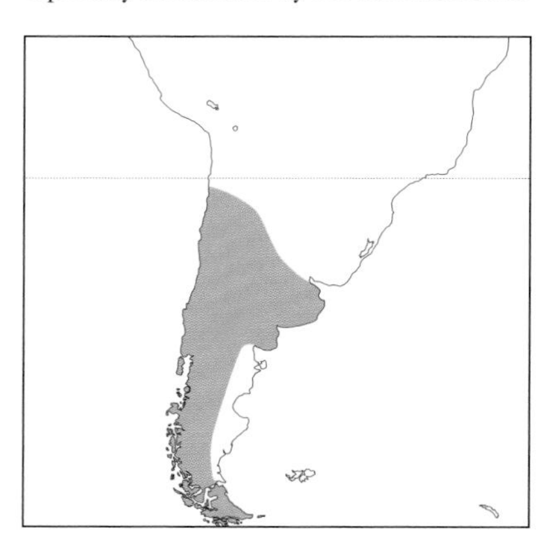

(*Colaptes pitius*), burrows in river or road banks (excavated by mammals or birds), holes in buildings, ground cavities or tree crotches. Its clutch size is 3 to 5 eggs.

CONSERVATION STATUS: Not Globally Threatened (IUCN). Reported as locally common. It is stated to be the most numerous owl in Chile. Its population appears to be increasing in central Chile. Threats include pesticide use and human persecution. The species seems tolerant to habitat changes due to human activity. It is abundant in Tierra del Fuego National Park, a protected area.

156 Cuban Pygmy Owl, *Glaucidium siju*

DESCRIPTION: A small pygmy owl (16 to 17 cm) that is restricted to Cuba and the Isle of Pines, where it is the only pygmy owl. Like most of the pygmy owls, it occurs in 2 morphs, grayish-brown and reddish. Its head and wing tips are rounded, the crown has light spots and the mantle and back are barred. The upper breast is densely barred, and the off-white lower breast and flanks have brown or buff spots. The voice is described as a piercing series of notes, accelerating and increasing in pitch; they also have a single, spaced-out hoot. It is the only American pygmy owl with a less swollen cere around the nostrils.

HABITAT: It occupies woodlands with open areas (including edges), coastal, montane and deciduous forest, secondary growth, agricultural land, parks and plantations from sea level to 1500 m.

NATURAL HISTORY: It is resident within its restricted range. The species is partly diurnal. Its 3 to 4 eggs are laid in natural tree cavities or old woodpecker holes. It eats insects, frogs, lizards, small mammals and birds, which it hunts from a perch. The male's call is a regularly spaced series of "tew" notes uttered about 4 seconds apart.

CONSERVATION STATUS: Not Globally Threatened (IUCN). Its status is reported as uncertain, but it is thought to be fairly common to common. It is negatively impacted by habitat loss.

157 Red-chested Owlet, *Glaucidium tephronotum*

DESCRIPTION: The red-chested owlet is a small owl (17 to 18 cm) with a round head and a relatively long tail with 3 spots. Its light gray facial disk is flecked with white, and the white "eyebrows" are short. Overall plumage is dark brown to grayish-brown, and the forehead, crown and primaries are unmarked. A "false face" on the nape is indistinct. The underparts are off-white, with the flanks and breast-sides heavily streaked rusty-orange and with black spots from the sides of the neck to the belly.

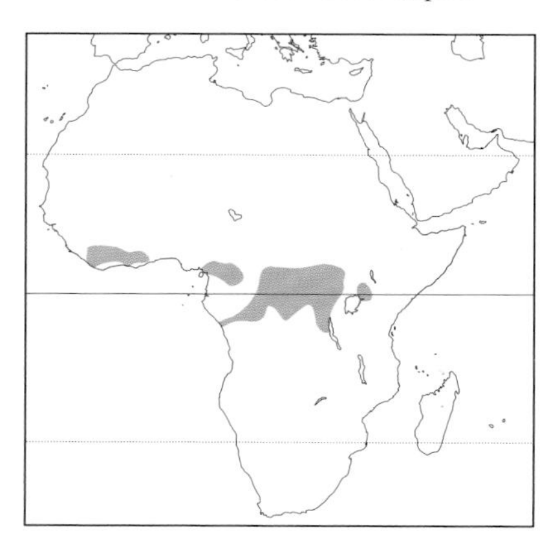

HABITAT: It occupies primary rainforest, forest-scrub mosaic, edges and openings in lowland areas, and occurs up to 2150 m in elevation.

NATURAL HISTORY: A resident bird that is not strictly nocturnal, it is sometimes seen active during the day. It roosts in tree holes. Its 2 to 4 eggs are laid in tree cavities, hollows or old woodpecker and barbet holes. Its diet consists primarily of insects, especially beetles, grasshoppers, mantises, cockroaches and moths. It has been observed chasing moths and flying insects in the forest canopy. It also eats small mammals and birds. This owl's vocalizations need further study. The male is reported to broadcast up to 20 whistling notes in a series that is repeated over and over.

CONSERVATION STATUS: Not Globally Threatened (IUCN). Its status is uncertain, but it is considered to be rare and generally difficult to observe. It is known to occur in at least a couple of protected areas: Mount Elgon National Park (Kenya) and the Impenetrable (Bwindi) Forest National Park (Uganda). Although it is possibly threatened by habitat destruction, it may be able to tolerate some changes to the habitat.

158 Sjöstedt's Owlet, *Glaucidium sjostedti*

DESCRIPTION: A relatively large pygmy owl (20 to 25 cm) without a dorsal "false face." The dusky-brown head and neck are heavily barred whitish. The chestnut-rufous upper back and blackish wings are lightly barred. Remaining upperparts are chestnut colored. The underparts are a pale cinnamon-buff with heavy brown barring.

HABITAT: It occupies the interior (e.g., not the edges) of lowland primary forest and is found at higher altitudes on Mt. Cameroon.

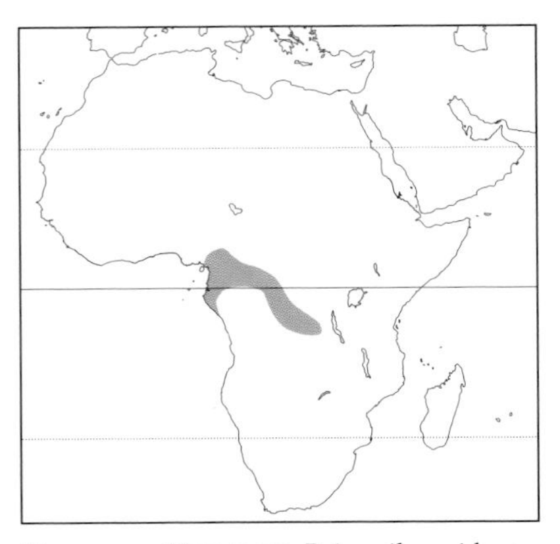

Natural History: Primarily resident and crepuscular, but may also be active during the night or day. It roosts within dense foliage. Unlike most pygmy owls, it does not flick its tail from side to side. At least 2 eggs are laid in tree cavities or hollows (1.5 m above the ground). The territory size has been estimated at 10 to 12 ha. Its diet consists mainly of insects (grasshoppers, dung beetles); also small mammals (rodents), spiders, crabs, snakes and birds (particularly nestlings). Hunts from about 2 m above the ground. The male repeats a series of 2 to 4 "kroo" notes every second.

Conservation Status: Not Globally Threatened (IUCN). Its status is uncertain. It is reported as generally uncommon throughout most of its range but is listed as common in Gabon. More information is needed to accurately assess its status. Most serious threat is habitat destruction.

159 Asian Barred Owlet, *Glaucidium cuculoides*

Description: The Asian barred owlet (22 to 25 cm) is similar in appearance to, but bigger than, the jungle owlet. Its plumage is dark brown to olive-brown and with close whitish barring on both the upper and underparts. There is a distinct white throat patch, but it lacks a dorsal "false face." The whitish abdomen is streaked dark brown.

Habitat: It occurs in montane and submontane forests (pine, oak and rhododendron), scrub, the foothills of subtropical and tropical jungle forests, and occasionally close to human settlements (including near rice fields). It is found in lowlands to 2100 m, and to 2700 m in the Himalayas of north Pakistan. It is known to nest in fruit gardens and coconut plantations.

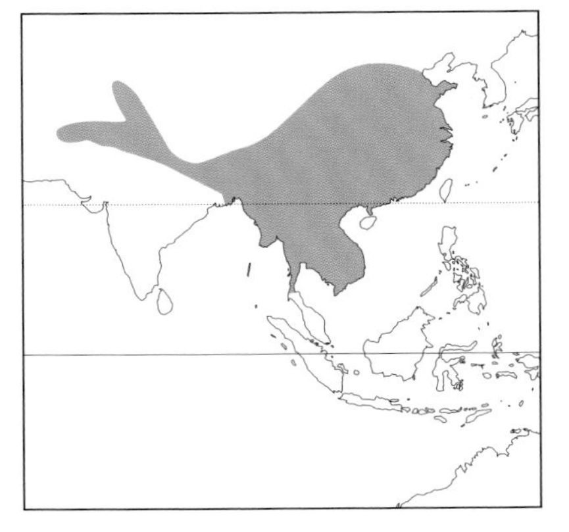

Natural History: Primarily a resident species, it exhibits altitudinal migrations on the south Himalayas, and also possibly in western China mountains, as birds move to lowlands of Sichuan and west Hubei in winter. In north-central China, migrations may be more extensive due to severe winters. It is for the most part diurnal and is active at all times throughout the day. The main call is a series of squawks, but males also utter a 4- to 7-second series of "wo" notes. Its 4 eggs (3 to 5) are laid in tree cavities, but barbets and woodpeckers may be killed and their cavities taken over. Prey is usually caught from a perch, but it may also catch birds or insects in flight. Its diet is mainly insects, but it also takes lizards, small birds, rodents and frogs.

Conservation Status: Not Globally Threatened (IUCN). It is reported as common throughout its range, and it is tolerant of human presence.

160 Javan Owlet, *Glaucidium castanopterum*

Description: This rather large owlet (23 to 27 cm) is the only one on Java and Bali with the following characteristics. The back is a uniform rufous-chestnut, with scapulars edged in white forming a line on either side of the mantle. The upper breast is brown with ochraceous barring, while the remainder is white with rufous-chestnut streaks. The head, neck and tail all have ochraceous barring.

Habitat: It favors hilly landscapes around 500 m elevation but is also found in dense primary and secondary rainforest (usually at lower elevations) and sometimes submontane or montane forest, gardens and villages. Occurs at lowland elevations to 900 m, but up to 2000 m in some regions.

Natural History: A resident, crepuscular species, it is also active day and night. Its 2 eggs (rarely 4) are laid in tree cavities or old woodpecker or barbet holes. It eats mainly insects (e.g., grasshoppers, beetles, mantids, cockroaches and walking sticks), but also scorpions, spiders, centipedes, myriopods, lizards, small birds and rodents. Its call is a staccato series of "sempok" notes.

Conservation Status: Not Globally Threatened (IUCN). It is reported as rare in some locations but is considered to be common locally in forested hillsides or virgin lowland habitat. Its main threat is habitat loss.

161 Jungle Owlet, *Glaucidium radiatum*

Description: The jungle owlet is a small owlet (20 to 24 cm) with a round head and a squat body. Its plumage is dark brown, with the wings more rufous, and uniformly barred pale rufous. The facial disk is also barred with brown and cream, and it has white "eyebrows," chin, and moustache. The abdomen and middle breast are white, with the remaining underparts dark and barred with dark olive-brown and white.

HABITAT: Occupies dense and open deciduous forest, secondary growth, scrub, and agricultural and teak-bamboo areas in foothill regions. It is not found in wet forests. It occurs in lowland habitat and up to 915 m elevation in Nepal; to 2000 m elsewhere.

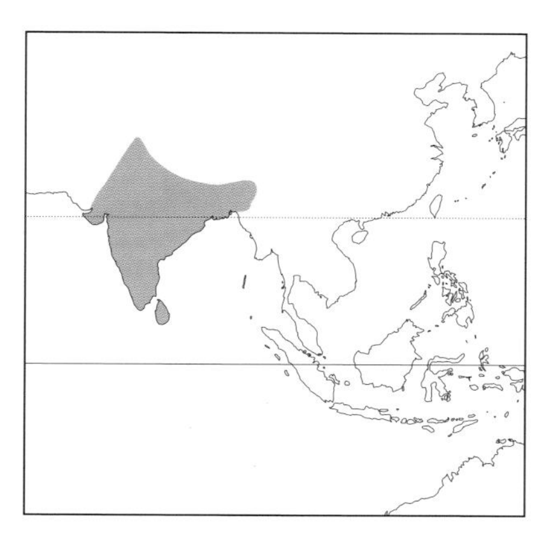

NATURAL HISTORY: A resident and mostly crepuscular species, it is also active at night and on cloudy days. Its call is a loud sequence of 2 to 3 "kao" notes followed by a series of "kao-kuk" notes. It roosts in dense foliage or in tree cavities. Its 3 eggs (range 2 to 4) are laid in tree cavities or old woodpecker holes 3 to 8 m above the ground in open forests. Insects (e.g., grasshoppers, locusts, cicadas and beetles) are its main prey, but it also eats molluscs, lizards, small birds and rodents.

CONSERVATION STATUS: Not Globally Threatened (IUCN). It is reportedly common in India, Nepal and Sri Lanka. It is known to occur in several protected areas, including Corbett and Periyar National Parks (India), Chitwan National Park (Nepal) and Yala National Park (Sri Lanka). Its main threat is habitat loss.

162 Chestnut-backed Owlet, *Glaucidium castanonotum*

DESCRIPTION: This small owl (19 cm) is confined to Sri Lanka. It has bright chestnut upperparts with blackish barring. The white breast and flanks have dark barring, and the white belly has olive-brown streaks.

HABITAT: It occupies humid and dense forests at elevations up to 1950 m.

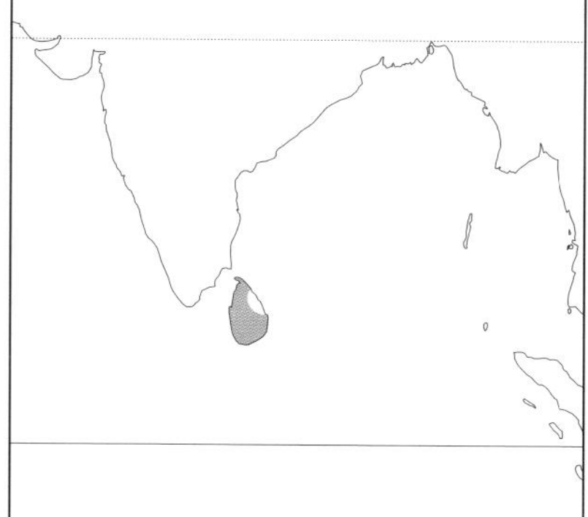

NATURAL HISTORY: It is resident in its restricted range. Although mainly diurnal, it appears to be shy and is difficult to observe. The male's call is a rapid extended series of purr-like "krew" notes. Like the jungle owlet, it often perches on tall treetops on steep hillsides. Its 2 eggs are laid in tree cavities, foliage of coconut palms or old woodpecker or barbet holes. It eats mainly insects, especially beetles, but also takes lizards, mice and small birds. Larger vertebrates are captured more frequently when adults are feeding their young. No other information is available on its breeding biology.

CONSERVATION STATUS: Lower Risk—Near Threatened (IUCN). Its population has declined apparently due to extensive deforestation. It occurs in the Sri Lanka Endemic Bird Area.

163 African Barred Owlet, *Glaucidium capense*

DESCRIPTION: This small owlet (20 to 22 cm) is one of the least-studied owls in South Africa. Its widespread distribution, diverse habitats and many distinct forms (four subspecies that may prove to be full species) warrants further study. The grayish-brown crown and nape have whitish barring, while the cinnamon-brown mantle and back are thinly barred buff. It lacks the "false eyes" on the nape. The underparts are off-white with large dusky spots, and the buff upperparts have dusky-brown barring.

HABITAT: It occupies a wide variety of habitats throughout its range, below 1200 m. Common habitat features include patchy, dense bush without much ground cover, particularly those scattered with tall trees; woodland and riparian trees in open

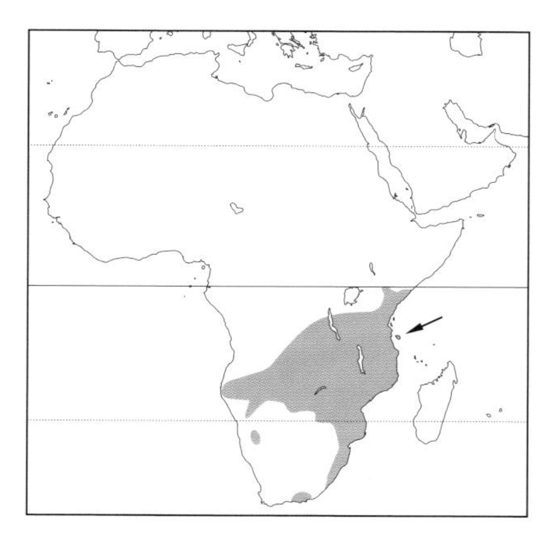

country; secondary growth; and forest edges.

NATURAL HISTORY: It is resident, mainly diurnal, but also active at night. It vocalizes mainly at dawn and dusk. Roosts low among dense foliage or in tree cavities. When emerging from roost sites it is usually inconspicuous. The male utters a descending series of "kweeo" or "kew" notes. There is little information available on breeding, and few nests have been located. Its 2 to 3 eggs are laid in natural tree cavities, 3 to 6 m above the ground. Prey includes insects (e.g., beetles, caterpillars, grasshoppers), scorpions, frogs, skinks, small birds and mammals. Its small, relatively weak feet are thought to limit its prey to small species.

CONSERVATION STATUS: Not Globally Threatened (IUCN). Its status is largely unknown, and all aspects of its biology need study. Main threats are habitat destruction and pesticide use. Areas in which it is reportedly uncommon include the eastern Cape, northern Tanzania and southern Somalia. It is locally common at Moremi Wildlife Reserve (Botswana) and in the riparian forests of the Okavango River in Caprivi Strip (Nambia).

164 Chestnut Owlet, *Glaucidium castaneum*

DESCRIPTION: A small owlet (20 to 22 cm) with a pale facial disk, white "eyebrows" and a dark brown head that is heavily spotted. It does not have "false eyes" on the nape unlike other owlets . The chestnut mantle and back are unmarked. The white upper chest has brown bars, and the belly is covered with many dark spots. Parts of the wing feathers (scapulars and lesser wing coverts) are edged in white.

HABITAT: It occurs in humid lowland, tropical and montane forests at elevations from 1000 to 1700 m. These include primary forest, older secondary forest and logged forests. The male utters a rolling series of "kyurr" notes that speed up during the call.

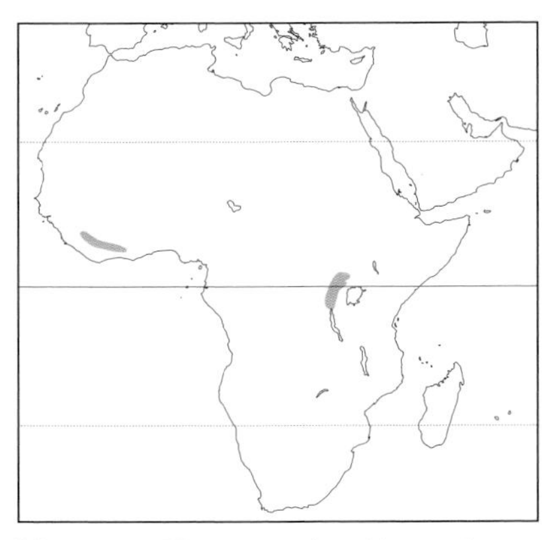

NATURAL HISTORY: A resident and partly diurnal species with no information available on its breeding biology. Its diet includes small vertebrates, birds and insects. It frequently forages in the undergrowth.

CONSERVATION STATUS: Not Globally Threatened (IUCN). Its status is uncertain. In Liberia it is reportedly uncommon overall, but locally common, whereas on the Ivory Coast it is thought to be widespread and common.

165 Albertine Owlet, *Glaucidium albertinum*

DESCRIPTION: This small owlet (20 cm) is very poorly known. The upperparts are warm maroon-brown. The brown flight feathers have pale brown barring, the crown is spotted cream and there are no "false eyes" on the nape. Both the hindneck and the edges of the inner wing feathers are spotted with white (more bar-like on the hindneck).

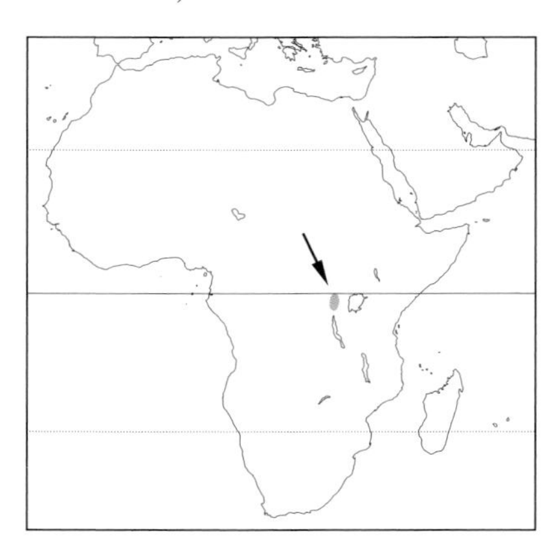

HABITAT: It occurs in montane forests with dense undergrowth, from 1100 to 1700 m.

NATURAL HISTORY: This species is assumed to be resident in its small restricted range. It is partly diurnal and eats insects and possibly small vertebrates. It is known from only 5 specimens. No other information is available. It probably nests in tree cavities. Thorough inventories will require knowledge of its vocalizations.

CONSERVATION STATUS: Vulnerable (IUCN). Limited to the Albertine Rift, this endemic owlet is possibly endangered by habitat loss. The Albertine Rift is characterized by one of the highest human population densities in Africa, which has resulted in forest fragmentation: isolated montane forest blocks surrounded by cropland. The Albertine Rift Conservation Society has coordinated priority-setting workshops, and conservation plans have been prepared. The Albertine Rift montane forests are recognized as important for the conservation of biological diversity, mainly because of the high proportion of endemic plants and animals (e.g., designated as an Endemic Bird Area).

166 Long-whiskered Owlet, *Xenoglaux loweryi*

DESCRIPTION: A tiny owl (13 to 14 cm) named in part after the late Dr. George Lowery of Louisiana who facilitated the study of Peruvian birds that led to its discovery in 1963. Its rounded head has no ear tufts. The facial feathers, particularly those on either side of the orange-brown eyes, extend beyond the head to form a wispy fringe. Some of the well-developed bristles around the yellow-tipped, grayish bill extend up between the eyes and the pale, creamy-yellow "eyebrows." It is a brown owl with some white spots on the upperparts; the lower belly has many narrow, white vermiculations. The unfeathered legs end in bare pinkish toes.

HABITAT: It occurs in a restricted area of ridge-top cloud forest with dense undergrowth and epiphytes in the Andes. Also found in exposed and stunted (4 m high) forests. It ranges from 1890 m to 2200 m above sea level. Moist from condensing clouds, this wet habitat is seldom visited by biologists.

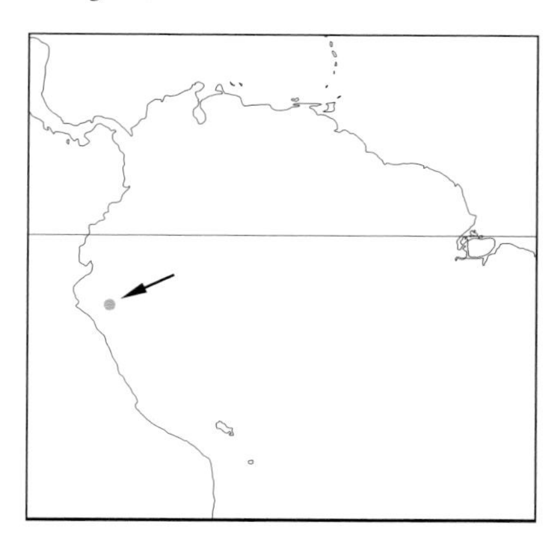

NATURAL HISTORY: It is thought to eat mainly insects, but this and other aspects of its biology need more study. There is no information available on its breeding and other behavior. The forest near the Abra Patricia pass, northern Peru, is where the first of only 3 specimens of this near-mythical owlet were found in 1963. However, few surveys have been conducted since then; in August 1976, 3 were live-captured in nets (1 female and 2 others thought to be a breeding pair), and another 2 in October 1978. It has not been observed since. Guerilla activities made the area too dangerous to visit until 1998 when tape recordings were made at night of the call (short, mellow whistles and a faster series of whistles followed by higher-pitched notes) of an unknown species thought by some to be this owlet. It is literally still a *Xenoglaux*, which means a strange or foreign owl.

CONSERVATION STATUS: Endangered (IUCN). It is possibly the rarest owl in the world. The habitat within its restricted range has been either extensively logged or badly degraded, but some pristine areas remain. Much habitat below 1000 m has been cleared for rice paddies and cash crops. Reinvigorated road construction and upgrades threaten to permit more logging in previously inaccessible humid temperate and subtropical forested areas. Road access is typically followed by further settlement and habitat fragmentation. More efforts are needed to establish a program to conserve this owlet's habitat and study its biology.

167 Elf Owl, *Micrathene whitneyi*

DESCRIPTION: This is reportedly the smallest owl (12 to 14 cm) in the world, weighing only 35 to 55 g. The lower edges of its cinnamon-buff facial disk are marked with white spots. The face is further distinguished with white "eyebrow" marks, lemon-yellow eyes and a greenish-yellow to olive-colored bill. Upperparts are grayish brown with buff dappling, and there are 2 rows of bold white spots on the upper wing. It has a grayish breast with scattered gray to cinnamon-brown vertical streaks and a grayish-white belly. Its tail is short, with incomplete buff bands, and its lower legs and feet are dull yellow to tan.

HABITAT: Subtropical thorn woodland, montane evergreen woodland and riparian forests support the densest and most stable populations, but most people associate it with saguaro cacti and desert vegetation. The elf owl has also adapted to partly urbanized habitats. It is found nesting in river bottoms to elevations of 2000 m.

NATURAL HISTORY: It lays 3 eggs (range 1 to 5) in old woodpecker cavities in trees, cacti, the flowering stalks of the yucca and agave, fence posts, bird boxes and utility poles. It often shares nest trees and nest habitat with up to 16 other kinds of cavity-nesting birds; these species cooperatively defend nests from common predators such as snakes, larger owls and ringtail cats. The elf owl eats mainly insects and rarely small lizards, snakes and young rodents. It hunts from low perches, but sometimes runs after prey on the ground. It can hang from flowers while waiting for insects attracted to blossoms. It removes the wings from large moths, and also the poisonous stingers from scorpions, thus avoiding eating large indigestible or dangerous parts. Its song is a fast series of distinctive high-pitched, dog-like "yips," especially during the breeding season. Northern populations are migratory, more southern populations being permanent residents. It would be interesting to know what interactions occur, if any, between wintering elf owls from northern areas and southern permanent residents.

CONSERVATION STATUS: Not Globally Threatened (IUCN). It is thought to be disappearing from the western edge of its range in California, where it is listed as Endangered, and it may be extirpated from northeastern Baja California, Mexico. In these and other areas, habitat destruction and urban growth are the main causes of declines. In contrast, it is either expanding into or recolonizing habitat northeastward in western Texas by nesting in woodpecker holes in wooden fence and utility posts, perhaps aided by global warming. It is the most abundant raptor in the upland deserts of Arizona and Sonora, Mexico.

168 Little Owl, *Athene noctua*

DESCRIPTION: A small owl (19 to 25 cm) with lemon-yellow eyes and a yellow to green-gray bill set in a rather indistinct facial disk. Its short, stocky body is dark gray-brown to sandy, with dense whitish-ochre spots above and well-spotted and streaked brown underparts.

HABITAT: The open country habitats that this owl uses, and its avoidance of even patchy forested areas, suggests that it is

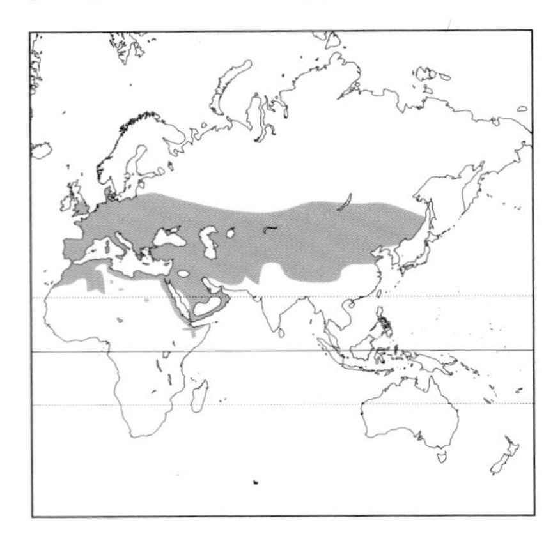

easy prey for hidden ambush predators. It is found in stony deserts, steppes, sparsely treed woodland, farmland, villages and other urban habitats. The little owl can be found at elevations from sea level to montane regions at 4600 m in the Himalayas.

NATURAL HISTORY: It is a sedentary, nonmigratory species whose young typically settle within 20 to 50 km of their natal home ranges. Its 3 to 6 eggs are laid in tree and other cavities, and remarkably it has shared such nest sites with other birds (e.g., common wren), including predators such as the common barn owl and common kestrel. The call is a somber question-like: "goeeek" or "goohk." This mainly insectivorous owl will also take other invertebrates and smaller vertebrates such as frogs, birds and mammals. Hunting from open perches, it bobs up and down energetically when judging distances to items of interest. If more prey is caught than the owl can use, it readily stores the surplus in cavities.

CONSERVATION STATUS: Not Globally Threatened (IUCN). It will readily use nest boxes, which may be employed to mitigate local habitat and nest-site loss resulting in range reductions and population declines in some parts of its range. Elsewhere it is widespread, abundant and demonstrably secure. Other known threats include vehicle traffic, pesticides and decreased prey availability associated with more efficient farming practices. Severe winters with thick snow cover can reduce prey availability and reduce or even temporarily eliminate northernmost populations.

169 Spotted Owlet, *Athene brama*

DESCRIPTION: A small owl (21 cm) with yellow eyes, a greenish-horn bill and brown-barred whitish underparts. Upperparts are gray to brown. There is much color variation among at least 5 recognized subspecies or races, but all have wings that are spotted and banded white and a tail with narrow white bars.

HABITAT: It is found in semi-deserts and other open-country habitats away from heavy forests. Occurs at elevations from sea level to 1400 m in the Himalayan foothills.

It also appears to thrive near villages with older groves and plantations.

NATURAL HISTORY: This nocturnal and crepuscular owl utters a "cheevak, cheevak, cheevak" sound, which it sometimes mixes with a harsh, rapid screeching: "chirurrr-chirurrr-chirurrr..." It is considered nonmigratory. It prefers to hunt from open perches, sometimes near artificial lights at

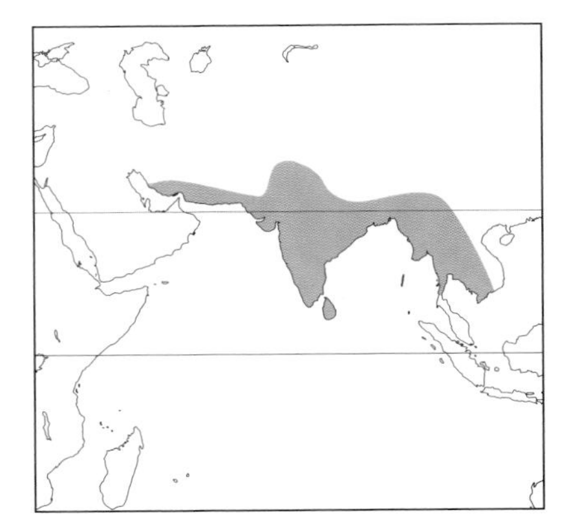

night where it captures beetles, moths and many other insects in flight. It also takes lizards, mice, worms and small birds. Its 3 to 5 eggs are laid in tree and other cavities, including those in human dwellings and earthen embankments. The relationship of this species' color variation relative to environmental conditions would be a fascinating study. It appears to replace the more northerly distributed little owl in Iran, India, Burma and Indochina.

CONSERVATION STATUS: Not Globally Threatened (IUCN). This widespread and common owl is remarkably poorly known. Its general natural history and behavior are likely similar to the little owl, but need to be documented.

170 Forest Owlet, *Athene blewitti*

Description: A small owl (20 to 23 cm) that is larger, but shorter-winged and more diurnal than the spotted owlet. It is known from only 7 specimens collected since its discovery in 1872. Observations of tail-flicking behavior have led some authors to propose that this species may be more closely related to pygmy owls. Prominent white "eyebrows" and whitish-brown facial disk frame its yellow eyes and bill. The spotting on the crown and back is reduced or absent. The throat, breast and belly are mostly white, but there is uniform dark-brown broad barring across the upper chest.

Habitat: It uses dense, humid deciduous forest and local mango groves near small rivers and streams. All 4 sites where it has been found were below 500 m elevation.

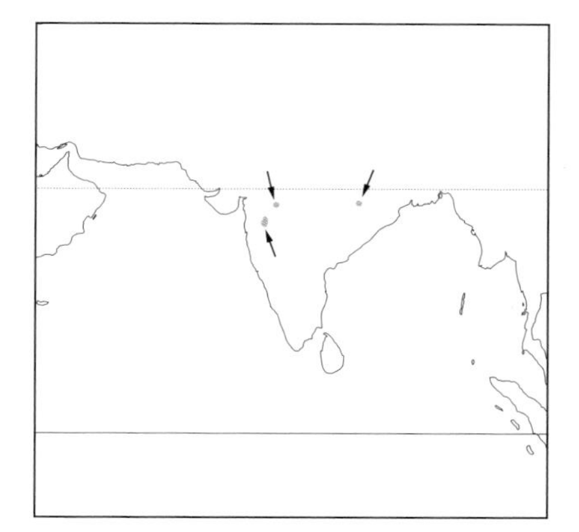

Natural History: There is little available information on its behavior or biology. An observation of one with a blood-smeared bill suggests that its relatively large feet and bill are used to take vertebrate prey, including lizards. It likely consumes invertebrates as well. It is mainly active during the day, when it can sometimes be seen exhibiting a strange bobbing display in which it lowers its head, covers its feet, then stretches tall, exposing its white belly and legs. Its call is a series of short, "uwww" or: "uh-wuwwww" notes uttered at 4- to-15-second periods during the day.

Conservation Status: Critically Endangered (IUCN). Apparently it was always rare, but its restricted distribution in India is a likely a shadow of its former range. In 1968, one was photographed, the first evidence of its continued existence since the last specimen was collected, reportedly near Bombay, in 1914. Scandalously, the 1914 specimen was later proven to be a restuffed study skin from the late 1800s that had been relabeled by an overly zealous collector. The most recent reports of extant occurrences stem from observations in 1997 and 1998. Clearly, more intensive research and management effort is needed to prevent this species from disappearing from our planet.

171 Burrowing Owl, *Athene cunicularia*

Description: The burrowing owl is a small brown owl (18 to 26 cm) with relatively long legs, a short tail and bright yellow eyes accentuated by white "eyebrow" markings. Its head lacks ear tufts and appears somewhat flat. Its body feathers are marked with brown and white spots above, and are barred and spotted below. Young owls have a buff-colored breast. The lower parts of its legs are sparsely feathered, and the toes are bristled. In general, female owls are larger than males, but burrowing owl females are slightly smaller than males.

Habitat: The burrowing owl is found both in arid grassland habitat of the Great Plains ecoregion in North America and in the more patchy grasslands of Central and South America. It can be found locally in

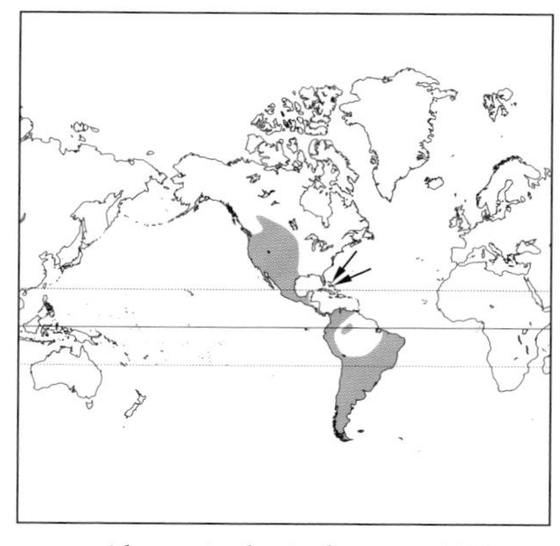

open arid areas in the Andes up to 4500 m. In the United States, it is found mainly in the west, but several thousand live in the higher and drier parts of south and central Florida.

Natural History: The burrowing owl was once believed to be strictly diurnal, active only during the day, but researchers studying radio-marked owls documented that they also hunted at night. This species feeds on rodents, beetles, grasshoppers and other insects, amphibians, lizards and occasionally on small birds. It produces a variety of calls, including shrieks, whistles and a laugh-like "cuhooh." The burrowing owl gets its name because it uses burrows of ground squirrels, foxes and badgers for roosting and nesting. This owl is often seen on an elevated area near its burrow entrance on the lookout for danger. When something approaches, it bobs its head up and down a few times, then retreats into the safety of its burrow; adults will usually fly off after briefly ducking into the burrow, whereas flightless young remain there. Some predators, like the American badger, are capable of digging out and killing roosting or nesting owls. The owl's habit of lining the burrow entrance with the dried dung of grazing animals was thought to deter predators, but research has not yet supported this claim; nests with dung are equally likely to be depredated as those without. It lays between 6 and 12 eggs. These take 30 days to hatch, the young owls remaining in or close to the nest for about 42 days before dispersing. Northern populations are mainly migratory, whereas it is nonmigratory in southern portions of its range. One owl banded in Manitoba,

Canada, was recovered alive 2 months later on an offshore oil rig in the Gulf of Mexico, near Louisiana. The wintering grounds of burrowing owls nesting in Canada are still largely unknown.

CONSERVATION STATUS: Not Globally Threatened (IUCN). Recognized as endangered in Canada, the burrowing owl is also listed as a "species of special concern" in several states because its numbers are declining. In Mexico, the burrowing owl was listed as a federally threatened species in 1994. Carbofuran, an insecticide used to control grasshoppers, was banned in Canada in 1995 because it lowered burrowing owl nesting success. Other threats to this owl include destruction of its grassland habitat for agricultural and commercial development. Some agricultural practices, such as cattle grazing, benefit burrowing owls by maintaining low grassland habitat. Burrowing owl conservation programs that recognize the efforts of landowners who voluntarily protect burrowing owls nesting sites have been successful at conserving habitat in the prairie regions of Canada. Specially protected areas, such as the Flamingo Gardens Everglades Wildlife Sanctuary in Florida, allow some populations of burrowing owls to flourish.

172 Boreal Owl, *Aegolius funereus*

DESCRIPTION: The boreal owl is a small owl with a large head and relatively long wings. Females (25 to 28 cm, 132 to 215 g) are considerably larger than males (21 to 25 cm, 93 to 139 g). Its grayish-white facial disk is edged with brownish-black, highlighted by white "eyebrows" and yellow eyes. It is cream-white with brown streaks below and has densely feathered legs, feet and toes. Upperparts are umber-brown with large blotchy white spots; the wings have up to 5 rows of white spots. The buff-white bill, small white spots atop its umber-brown head, and tail with 3 rows of white spots distinguish it from other similar-sized owls.

HABITAT: Found year-round in boreal and mixed-forests with spruce, aspen, poplar, birch, tamarack and fir. In more southern areas is found locally in subalpine forests up to elevations over 3000 m. Individuals change roost sites daily. In summer, to avoid hot temperatures, it roosts in cooler sites with higher canopy cover and greater tree density. Hunts mainly in mature spruce-fir forests, especially in winter when uncrusted snow makes prey capture easier relative to open, hard-packed snow-covered areas. Will hunt in clear-cuts and farm fields, where small mammal prey densities are higher, but typically only in early spring after snow melt and until grass cover becomes too thick.

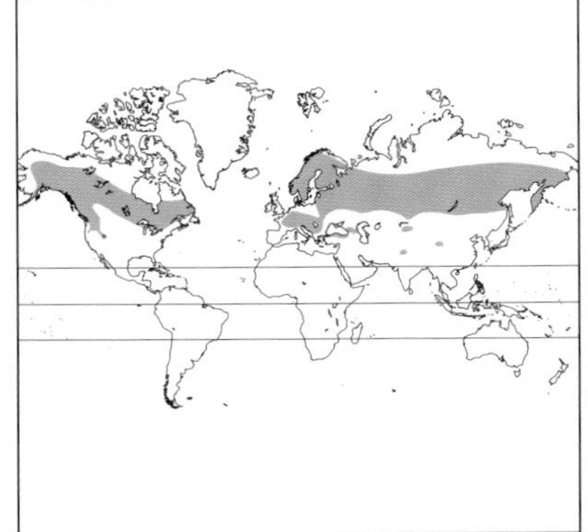

NATURAL HISTORY: Its primary call is a series of about 16 loud trills or "hooh"s repeated at intervals for up to 3 hours. Heavy snowfall and wind can inhibit calling. The male calls at dusk onward from various tree cavities in its territory to attract a mate. The female chooses one in which she will lay 1 to 10 eggs within 1 to 19 days. The male feeds the incubating and brooding female. When prey is abundant, some males may mate with and support a second female. The boreal owl can catch concealed prey effectively using its pronounced skeletal-ear asymmetry. Hunts mainly small mammals, but also small birds and insects. Larger prey, such as weasels and young snowshoe hare, are only rarely taken. Females and juveniles appear to disperse more widely than adult males during nomadic movements following prey population declines. Occasional winter irruptions of boreal owls south of the boreal forest typically involve a high number of starving individuals. These hungry owls often appear at bird feeders, where they await shrews and small rodents attracted to spilled seeds, but they will take small birds opportunistically.

CONSERVATION STATUS: Not Globally Threatened (IUCN). It is the most abundant forest owl in the northern hemisphere. More research is needed on its response to habitat change. Some populations in Europe depend on artificial nest boxes because timber harvest practices remove trees with natural cavities. Isolated montane populations in North America are of conservation concern, and many of these have been designated "sensitive" by the USDA Forest Service. Saving large snags (for nest cavities) in clear-cuts is one method used to manage these populations. Clear-cuts in montane spruce-fir forests will remain unsuitable for roosting and foraging for up to 100 years, and it can take almost 200 years for trees to grow big enough to provide suitable nest cavities.

173 Northern Saw-whet Owl, *Aegolius acadicus*

DESCRIPTION: A small owl (18 to 20 cm) with short legs, a short tail and rounded wings. Its large round head lacks ear tufts. The upperparts are streaked with brown, and there are white streaks on the head and neck. Its back, wings and tail have numerous white spots. The facial disk is white between and above yellow eyes, and pale brown, streaked with white and dark brown, on the sides. Its white underparts have broad white and brown stripes. Geographic variation is minimal, except for a distinctive buff-colored subspecies on the Queen Charlotte Islands in British Columbia.

HABITAT: Found in woodland areas, especially coniferous forests, and riparian woodlands. More common at middle elevations, but recorded up to 3200 m in subalpine forests. In winter it uses a greater variety of habitats, including rural and suburban human-altered landscapes with dense vegetation, for foraging and roosting.

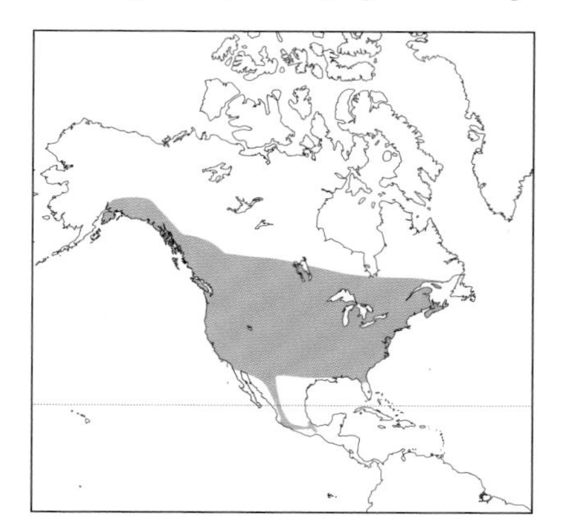

NATURAL HISTORY: The repetitive "tew" notes of its call is reminiscent of the rhythmic sound of a saw being sharpened (or "whetted") with a file, hence its common name. The northern saw-whet owl can locate prey by hearing alone, using its asymmetric ears. It preys at night, most frequently taking small mammals, especially deer mice. The female alone incubates 5 to 6 eggs for 27 to 29 days in a tree cavity, the male providing her with food then and while she broods the nestlings. Nesting females sometimes bathe in shallow streams just before dawn. When prey is plentiful, some males pair with and support a second female. Some individuals remain on the breeding range year-round, but significant numbers migrate south in winter and appear to concentrate around large lakes. There is still much to learn about its breeding biology, distribution and behavior.

CONSERVATION STATUS: Not Globally Threatened (IUCN). The northern saw-whet owl is one of the most common owls in forests across the northern United States and southern Canada. Population fluctuations of migrants largely reflect variation in annual breeding success. Erecting nest boxes can mitigate local habitat destruction, especially removal of trees with nest cavities.

174 Unspotted Saw-whet Owl, *Aegolius ridgwayi*

DESCRIPTION: A small owl (18 to 21 cm) with yellow eyes and a dusky bill. Its upperparts are uniform brown; the buff underparts have an undefined cinnamon-brown band across the breast. The short tail is brown, and it has broad, rounded wings. A white chin, lores and broad "eyebrows" boldly mark its dark-rimmed, brownish facial disk.

HABITAT: It occurs in high elevation (1400 to 3000 m above sea level) in moist pine-oak, oak and cloud forest. It uses edge habitats and treed gaps between forests.

NATURAL HISTORY: Presumably nonmigratory, its diet is the subject of great speculation, and it is thought to take small mammals, birds and invertebrates. There is no information on its breeding biology or

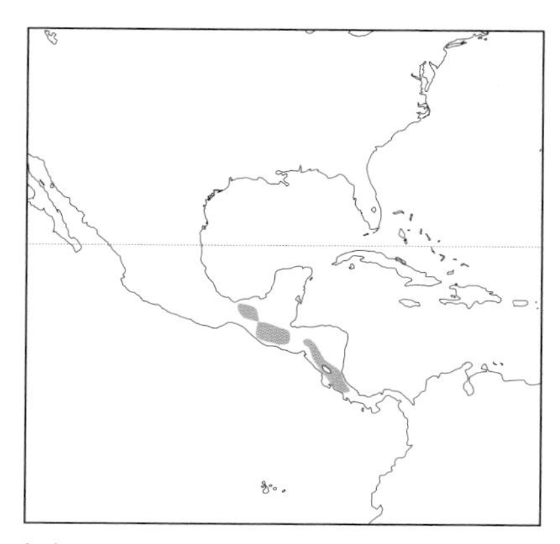

behavior, but it is reportedly nocturnal and solitary outside the breeding season. Its call may be a 4- to-5-note series of "toots" like that of the northern saw-whet owl.

CONSERVATION STATUS: Not Globally Threatened (IUCN). There is very little information available on its distribution, status and trends. Therefore, it should be more accurately assessed as data deficient. There is a pressing need to identify its habitat requirements in light of rapid human alterations of forested lands within its range. It is listed as Near-Threatened by BirdLife International.

175 Buff-fronted Owl, *Aegolius harrisii*

DESCRIPTION: A colorful, small owl (20 to 21 cm) with attractive greenish-yellow eyes. It has a relatively large, square dark head, a compact body and a blackish, white-tipped short tail. The black-rimmed buff facial disk features a broad and prominent brown V-shape comprising the brown lores converging toward the base of the grayish-yellow bill. The upperparts are chocolate-brown in contrast to the yellowish-buff underparts.

HABITAT: It occurs at elevations from sea level to 3800 m in the Andes. It has been reported in relatively open forests, forest edges, stunted montane forests, fruit tree orchards, palm plantations and treed gardens.

NATURAL HISTORY: There is very little information available on this resident

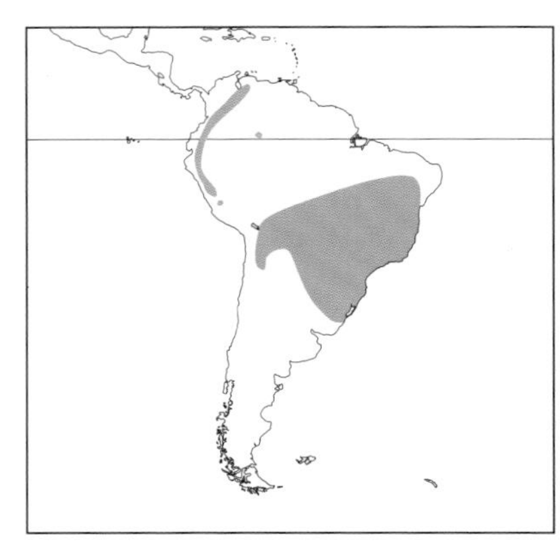

and nocturnal species. Its call is a high-pitched, 3- to-4-second whistling trill: "frurururururururu" that may be confused with some screech owls that occur in similar habitats. At least 2 documented nests (Brazil and Argentina) indicate that its 3 eggs are laid in tree cavities; one was 6 m above the ground. Prey remains in 1 nest included parts of insects and a small rodent.

CONSERVATION STATUS: Not Globally Threatened (IUCN). It is listed as Near-Threatened by BirdLife International, and is of conservation concern in Columbia because of suspected declines. Generally, it is considered rare throughout its range. However, in some areas it may be easily overlooked, and populations are likely underestimated. It is known to occur in several protected areas, including Podocarpus National Park in Ecuador, and El Ray and Iguazu National Parks in Argentina. Deforestation is a recognized threat to local populations.

176 Rufous Owl, *Ninox rufa*

DESCRIPTION: The plumage of this large hawk owl (40 to 52 cm) is finely barred all over. The back is dark rufous-brown, covered with thin light brown bars. The underparts are white to buff with much rufous-brown barring. The smallish head has an indistinct dark facial disk, dark sides and a pale throat and forehead. Variation in size and color is used to separate different races. Although aggressive during breeding, it is shy and seldom seen otherwise. Males are larger than females.

HABITAT: The species occurs in rainforest (upland and lowland), monsoon forest, thick savannah woodland, swamps, gallery forest and forest edges. Within its habitat it is usually found near watercourses. In New Guinea, it is found in lowlands and foothills up to 2000 m, 1200 m in Australia.

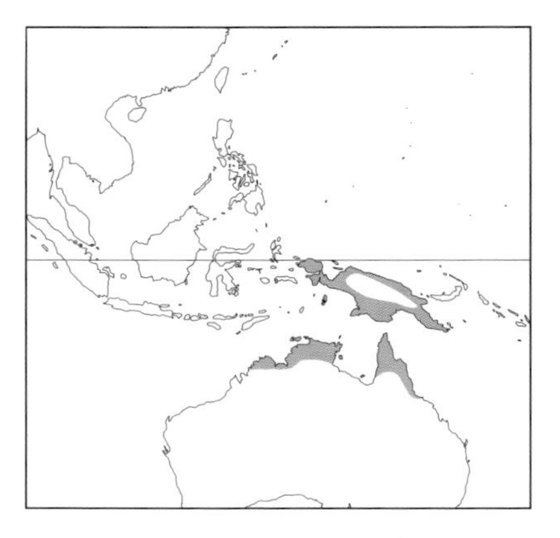

NATURAL HISTORY: Although considered to be resident, local movements are suspected during the wet season. A nocturnal bird, it roosts singly or in pairs in thick foliage of trees or in hollows, often with prey remains. Normally, 2 eggs are laid in large hollows in trees or branches (30 m above the ground), usually in eucalypt or paperbark trees. These birds are monogamous and territorial. The male's call is a low, double "wooh-hoo" note repeated every 2 to 6 seconds. An agile hunter, its diet includes mainly mid-sized mammals (e.g., sugar gliders and other marsupials), rodents and fruit bats; also birds (e.g., Australian brush-turkey) and large insects.

CONSERVATION STATUS: Not Globally Threatened (IUCN). Generally considered rare to very uncommon throughout its range. Invasive exotic weeds, fires in the dry season that destroy potential nest sites, and traditional hunting are further threats. Many parts of this species' range have been or continue to be destroyed by logging and agriculture. One subspecies, *queenslandica*, has been estimated at 1000 to 3000 pairs, and is listed as rare or "near threatened" in Australia and vulnerable in the state of Queensland.

177 Powerful Owl, *Ninox strenua*

DESCRIPTION: The powerful owl is the largest of the hawk owls (52 to 63 cm), with a relatively small, round head, long tail and very strong talons. The upperparts are dark brown to dark gray, and have off-white mottling and barring. The whitish underparts are distinctly marked with gray-brown V-shaped bars. The facial disk is dark gray-brown, and the crown is flecked. The call is a loud double hoot: "woo-hoo." It is the most "hawk-like" owl of the *Ninox* group, and males are larger than females (unlike most owl species).

HABITAT: The powerful owl prefers wet, lightly logged forest (sclerophyll forest), but it is also found in open dry forest, heavily forested gullies, ravines, coastal forest and woodland, pine plantations, woods adjacent to city parks, coastal scrub and cultivated regions. It occurs at elevations from coastal areas to 1500 m.

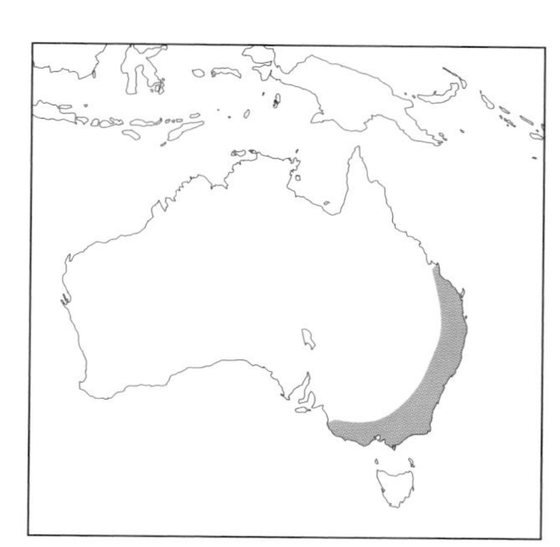

NATURAL HISTORY: The species is mainly resident in its southeastern Australian range, but young birds sometimes disperse inland. A nocturnal bird, it roosts alone, in pairs or in family groups. Its main call is a loud, sad-sounding double "whoo hooo" note. Roost sites are within foliage or more open sites; birds are easily approached. More than 1 site is used, and a site may be occupied for years. Birds will usually roost with prey remains in their feet. Normally, 2 eggs are laid in hollows of tall trees. It is monogamous, living permanently in pairs and on home ranges from 300 to 1450 ha. Males can be aggressive in defense of the nest. Its diet is mainly birds and mammals, especially phalangers; also flying foxes and insects.

CONSERVATION STATUS: Vulnerable (IUCN). It is thought to be uncommon throughout its range. Estimated populations include 500 pairs in Victoria and 1000 to 10,000 individuals in New South Wales. Threats include habitat loss, especially nest sites in old-growth forest. However, the species is tolerant of limited or selective logging.

178 Barking Owl, *Ninox connivens*

DESCRIPTION: This medium-sized hawk owl (38 to 44 cm) gets its name from its quick, dog-like call: "wuk-wuk" or "wuf-wuf." Another, less frequent, call is described as a long, strangled scream! Its distinctly reduced facial disk is sometimes barely visible. It has rather long wings, legs and tail. The upperparts and head are

smoky brown, the underparts whitish with rusty to dark-brown streaks. The scapulars and wing coverts are boldly spotted with white.

HABITAT: Inhabits riparian forest and vegetation (often near wetlands), temperate and subtropical forests, open habitat scattered with large trees, savannah, woodland, thick scrub, forest edge, sclerophyll forest, swamp and riverine woodland, and foothills adjacent to watercourses. Requires open areas for hunting. It occurs at elevations from lowlands to 1000 m.

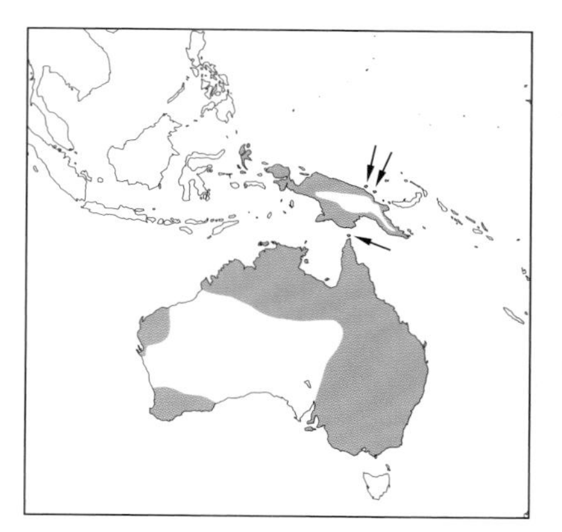

NATURAL HISTORY: It is presumed to be a resident species, but its movements are poorly known. The least nocturnal Australian owl, it is mainly crepuscular but is also active at night and during overcast days. The barking owl usually roosts in pairs or family groups of 3 to 4. Roost sites are within foliage but not always concealed. It will also clutch prey remains while roosting. A bold owl, it can be seen near rural homes. Its 2 to 3 eggs are laid in trees in large hollows, but this owl is also reported, although less frequently, to nest in the open fork of a tree or on the ground. Nests reported as 6 km apart. The male prepares the site by rotating and scraping within the hollow. Prey includes insects (e.g., beetles), possums, gliders, rabbits, young hares, mice, rats, bats, birds (e.g., Australian magpie, tawny frogmouth) and the occasional fish.

CONSERVATION STATUS: Not Globally Threatened (IUCN). This owl is widespread and is considered rather common in New Guinea, uncommon in Moluccas, declining in southeastern Australia and currently threatened by habitat loss. Populations in New South Wales and Victoria are considered vulnerable. It is uncommon overall in Australia, with the exception of areas in Queensland, and in Kimberleys (western Australia).

179 Sumba Boobook, *Ninox rudolfi*

DESCRIPTION: This medium-sized hawk owl (30 to 40 cm) is the only known *Ninox* species on Sumba Island in the Lesser Sundas. The brown crown and mantle are densely spotted with white. The brown scapulars and wing coverts are heavily barred, and the whitish underparts are broadly barred rufous (not streaked or mottled). These characteristics distinguish this owl from all southern boobook races.

HABITAT: Occupies open primary and secondary deciduous and evergreen forest;

also rainforest, remnant forest patches and forest edges. It is also found in farmland. It occurs in lowland habitat to elevations up to 930 m.

NATURAL HISTORY: The species is considered to be resident in its restricted range. There is very little information available on this owl, but its habits are likely similar to those of the southern boobook. Its main call is thought to be an extended series of regular, terse coughing notes. It roosts alone, in pairs or in groups (up to 4). Its breeding behavior is undescribed, and its diet is poorly known, although it is probably mostly insectivorous.

CONSERVATION STATUS: Lower Risk—Near Threatened (IUCN). It was formerly more common, but is currently uncommon or rare. Surveys in 1989 and 1992 resulted in only 5 locations where birds were found. Past and current habitat destruction have destroyed much of the natural forest, leaving only less than 10% of the island with forests. Only one protected area is present on the island.

180 Southern Boobook, *Ninox boobook*

DESCRIPTION: This small, stocky, brown owl (25 to 36 cm) is the only owl in Australia of its color pattern and size. The ill-defined facial disk is pale compared to the overall color of the owl, and there is a dark area behind the greenish-yellow eyes. The wings and back are spotted with white, and the cream-to-buff underparts are spotted and broadly streaked with dark

brown, pale or reddish-brown. Its plumage color varies according to habitat. Its common name stems from its well-described 2-note "boo-book" or "mo-poke" call.

HABITAT: The species occupies a wide variety of habitats from deserts to forests. It is present in farmland, tropical rainforest, mallee and mulga scrub, edges of treeless plains, woodland, cultivated fields, orchards, pine forest, parks, monsoon forest, suburbs and streets. It is usually found in lowlands but also at elevations up to 2300 m on Timor.

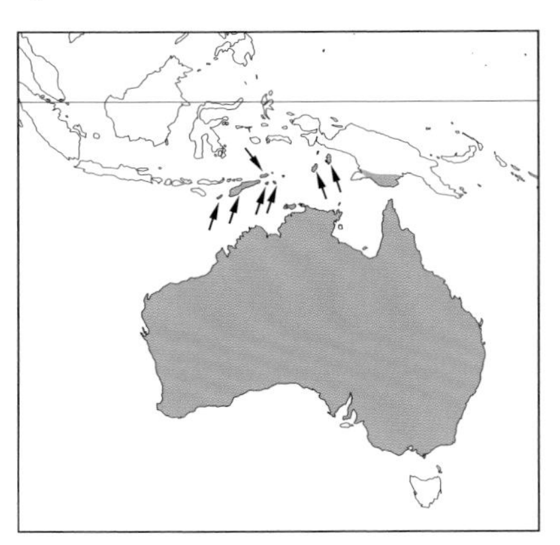

NATURAL HISTORY: It is primarily a resident, with some individuals dispersing in winter. It is possible that southern populations may be partially migratory. Nocturnal and partly crepuscular, it roosts during the day alone, in pairs or in family groups in dense foliage. It rarely roosts on the ground, in caves or tree cavities. It is monogamous and territorial. Usually, 3 eggs are laid in tree cavities (that males have previously cleaned, 1 to 20 m above the ground), a spout or chimney, or old corvid or babbler nests. Its diet includes insects, birds, bats, rodents, lizards and frogs. The southern boobook uses variable hunting techniques such as hawking, pouncing on prey on the ground and grabbing prey that are in the canopy or middle stories.

CONSERVATION STATUS: Not Globally Threatened (IUCN). It is reported as widespread and common throughout Australia in areas with suitable nesting and roosting trees. While it tolerates some habitat disturbance, activities causing the loss of suitable nesting trees negatively impact local populations. Other threats include secondary poisoning in rat-control programs and DDT-induced eggshell thinning.

181 Morepork, *Ninox novaeseelandiae**

DESCRIPTION: The name of this small owl (25 to 30 cm) is derived from one of its calls, a clear, repeated double hoot: "more-pork" or "boo-book." It is the only small owl without ear tufts endemic to New Zealand. Similar to other *Ninox* owls, the facial disk is indistinct. The wings are rounded (distinguishing it from the southern boobook owl, as well as being smaller and darker); the tail is long. The upperparts vary from deep rufous to chocolate brown and are flecked with white. The breast is marked with dark, irregular streaks.

HABITAT: The morepork is found in a wide range of habitats, including dense

forests, mallee and mulga scrub, arid landscapes, farmland and plantations.

Natural History: It is a mainly resident species, with young birds sometimes dispersing in winter; some birds from Tasmania overwinter in Australia. It is mainly a nocturnal hunter but is also active at dusk and sometimes during the day. The morepork roosts in cavities, caves, among rocks or within dense canopy. It mainly nests in tree cavities but also within hollow trees, epiphytes and tree forks; atop old sparrow nests; in nestboxes; and rarely in caves or holes in river banks. Pairs are monogamous and territorial. Usually, 2 eggs are laid, with males preparing the site. It is mostly insectivorous, but also takes spiders, lizards, small birds and mammals (including bats).

Conservation Status: Not Globally Threatened (IUCN). It is considered to be common in New Zealand. Populations have declined in Christchurch urban area since the 1930s, coinciding with an increase of little owls. Early in the 20th century, the subspecies *albaria* from Lord Howe Island disappeared, due to competition with other owls and from rats predating eggs and young. The subspecies *undulata* is endangered on the Norfolk Islands. The Norfolk Island Government, Australian National Parks and Wildlife Service, and the New Zealand Department of Conservation have been working together to increase *undulata* populations by installing nest boxes; this resulted in 2 chicks raised in 1989 and 2 in 1990 (hybrid offspring: *undulata* × *novaeseelandiae*). Threats include predatory rats, other hole-nesting competitors (e.g., Australian kestrel, starling and crimson rosella), feral cats, secondary poisoning and, especially, loss of old-growth forests for nesting.

* It is important to note that recent studies on the molecular variation of some *Ninox* owls suggests that Australia (including Tasmania), Norfolk Island and New Zealand share the 1 species of boobook owl, *Ninox novaeseelandiae*. Additional DNA evidence is needed to support or refute this suggestion.

182 Brown Hawk Owl, *Ninox scutulata*

Description: The brown hawk owl (27 to 33 cm) is widely distributed over southern and eastern Asia, Borneo, Japan and the Philippines, with 9 to 12 races recognized. It is said to be one of the most "hawk-like" of the owls, with long wings and tail. The upperparts are dark gray-brown. The indistinct facial disk is almost completely dark, and there is a white area between the eyes. The white underparts are marked with reddish-brown spots.

Habitat: It occupies a variety of habitats such as evergreen deciduous, and coniferous forests, bush containing tall trees, river edge thickets, cultivated areas with some trees or shelterbelts, mangroves, rainforests, plantations, parks, gardens and suburbs. It is found in lowlands and hills at elevations up to 1200 to 1500 m.

Natural History: Some northern races migrate south in winter, overlapping

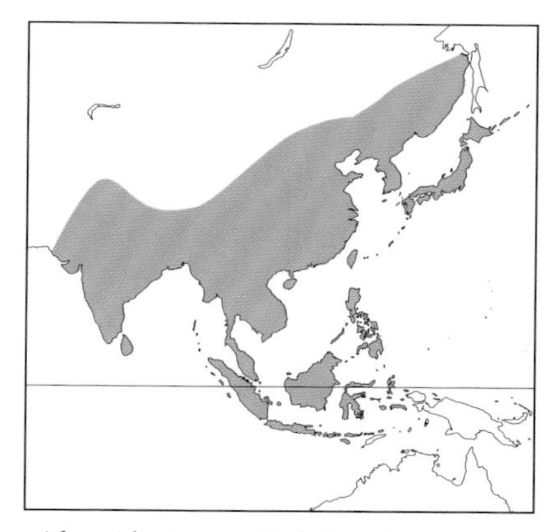

with resident races. Birds from Japan are known to overwinter in eastern China. Its main call is a soft, flute-like series of 6 to 20 "oo..kk" notes, but numerous other sounds (e.g., a cat-like "mew") have been described. The species is nocturnal and crepuscular, but it is occasionally active during cloudy days. It roosts alone or in pairs under dense canopy, often among creepers. Adults are monogamous and use tree holes or hollow trees for nesting (5 to 20 m high). Pairs have 2 to 5 eggs. In Japan it is reported to also nest on the ground among rocks and timber, and in nest boxes. The diet consists of flying insects (e.g., moths, beetles, cockroaches, dragonflies and cicadas), which are captured in the air. Less frequently, small birds and mammals (including bats), lizards, frogs and crabs are also taken.

Conservation Status: Not Globally Threatened (IUCN). Northern populations are reported as fairly common, and it is very common in Japan. It is rare in southeast Siberia and Java; uncommon in Moluccas, Lesser Sundas, Borneo and Sumatra; and common in northern Sulawesi, Talaud Island, Sangihe, northern India and southeast Asia. Southern races are threatened by loss of lowland primary rainforest.

183 Andaman Hawk Owl, *Ninox affinis*

Description: This medium-sized hawk owl (25 to 28 cm) closely resembles the brown hawk owl (*obscura* race), but it is smaller, browner and has bright rufous

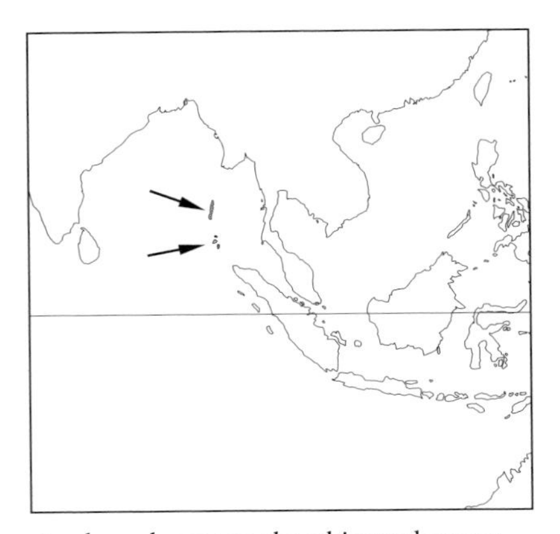

streaks and spots on the white underparts. The eyes are yellow with dark "eyebrows," and the head is gray.

HABITAT: It occupies lowland secondary forest, open forest, settlements, mangroves and rubber plantations.

NATURAL HISTORY: Little information is available on this poorly studied bird, but it is assumed to have habits similar to those of the brown hawk owl. The species is resident. Breeding behavior is unknown, and its diet is reported to include moths (captured in the air), grasshoppers and beetles. A repetitive "craw" is reportedly its main call.

CONSERVATION STATUS: Lower Risk—Near Threatened (IUCN). It occurs in the Andaman Islands and Nicobar Islands Endemic Bird Areas and is likely endangered. Its long-term survival is uncertain.

184 White-browed Hawk Owl, *Ninox superciliaris*

DESCRIPTION: A medium-sized hawk owl (23 to 30 cm) with an indistinct grayish facial disk, brown eyes and broad white "eyebrows." It has a brown, round head and long, pointed wings. The brown upperparts and crown are sparsely speckled with white. The white underparts are distinctly barred with brown. Both light and dark morphs occur and have similar plumage patterns.

HABITAT: The white-browed hawk owl occupies a variety of habitats: gallery forest,

open areas scattered with trees and subarid thorny shrubs, evergreen rainforest, deciduous dry forest, forest clearings, rocky ravines and sites near villages. Occurs at elevations from sea level to 800 m.

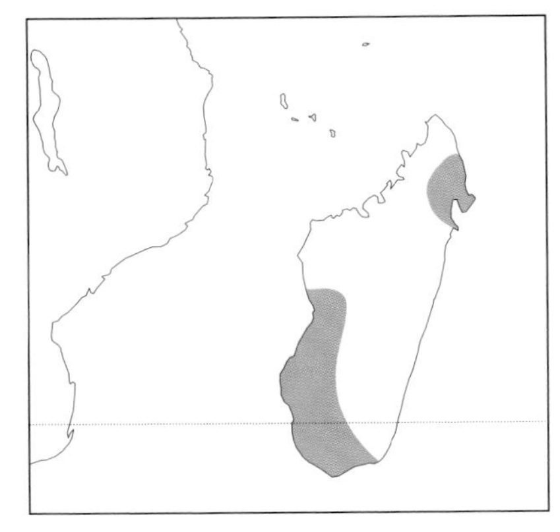

NATURAL HISTORY: Predominantly resident, it is assumed that the birds sighted in northern Madagascar are dispersing immatures. A nocturnal bird, it is very vocal at night and roosts by day in small caves. Its 3 to 5 eggs are laid in tree cavities or on the ground in a shallow depression. It is primarily insectivorous, but also eats lizards, small birds and mammals. Its call is reported as a series of howling "wuhuoh" notes, starting with a gruff "chruwuoh" note. During aggressive encounters it utters a loud series of up to 15 "kuatt" notes.

CONSERVATION STATUS: Not Globally Threatened (IUCN). Although locally common, information is lacking, and studies of its natural history are required. It was recently reported in Andranobe, Masoala Peninsula National Park, and is present and fairly common in Zombitse-Vohibasia National Park and the Berenty Private Reserve. Possible threats are human persecution, pesticide use and habitat loss.

185 Philippine Hawk Owl, *Ninox philippensis*

DESCRIPTION: This small hawk owl (15 to 20 cm) varies in appearance across its Philippine Islands range. Overall, the upperparts are a warm brown, including the indistinct facial disk. The scapulars and upper-wing coverts are spotted with white, and the flight feathers are densely spotted. The underparts are buff-white with rufous-brown streaks.

HABITAT: It occupies primary and secondary-growth forest, but also remnant patches of forest, gallery forest and edges. It is found at elevations from lowlands to higher altitudes.

NATURAL HISTORY: Its movements are unstudied, but it is probably a resident species. A nocturnal bird, it roosts by day in dark areas of the forest. It nests in tree cavities; no other information is available on its breeding biology. The diet of the Philippine hawk owl includes insects and rodents. Its call is a repeated monosyllabic "whoo."

CONSERVATION STATUS: Not Globally Threatened (IUCN). It is locally common despite extreme deforestation in the Philippines, showing its tolerance to changes in its habitat. It is present in several National Parks including Mount Canlaon (Negros), Quezon (Luzon), Mount Katanglad (Mindanao) and Rajah Sikatuna (Bohol). On Mindoro Island, more than 90% of the island has been deforested, and the *mindorensis* race remains common in the Siburan Forest (50 square km) only due to the penal colony there. Other races are seriously threatened by habitat loss on small islands, and this will probably result in their extinction, especially on Cebu and Tablas.

186 Ochre-bellied Hawk Owl, *Ninox ochracea*

DESCRIPTION: This medium-sized hawk owl (25 to 29 cm) is one of the more uniformly colored hawk owls. It lacks ear tufts, and its long tail has narrow whitish-buff bars. The upperparts are brown to dark chestnut, and the scapulars and wing coverts are spotted white. The "eyebrows" and throat are white. Its lower breast is tawny-ochre with indistinct dark markings.

HABITAT: It occupies rainforest, primary forest, riverine forest and tall secondary lowland forest. It occurs at elevations up to 1780 m.

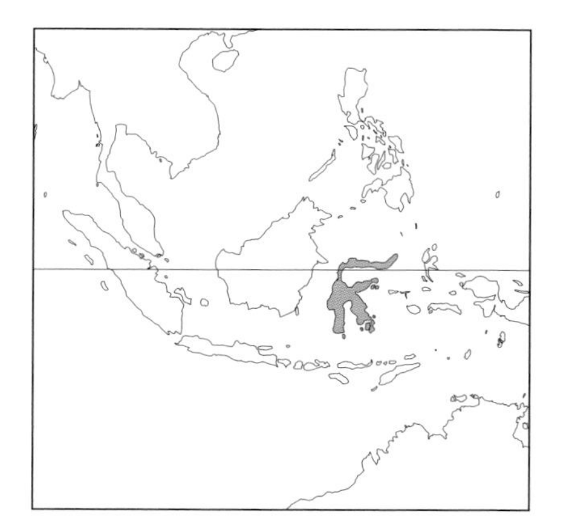

NATURAL HISTORY: The species is resident in its Sulawesi and Buntung Island range in Indonesia. It roosts, alone or in pairs, in thick vegetation. No information is available on its breeding behavior or food habits, but it probably eats insects. It has been observed hunting in the lower or middle canopy overlooking openings in the forest or roads. The male utters a series of gruff, double "krurr-krurr" notes.

CONSERVATION STATUS: Lower Risk—Near Threatened (IUCN). It occurs in the Sulawesi Endemic Bird Area. The status of this little known bird is uncertain, but it is considered to be widespread, though rare and uncommon across its range. Locally, however, it is reported as moderately common. Present in Dumoga-Bone National Park, Lore Lindu National Park and Tangkoko Nature Reserve. The species is elusive and needs more study to better assess its status.

187 Moluccan Hawk Owl, *Ninox squamipila*

DESCRIPTION: A medium-sized hawk owl (25 to 39 cm) with plumage and size varying among the different races. It is shaped like a typical hawk owl, but without the long tail. The upperparts are a dark reddish-brown with white bars on the scapulars and a few on the wing coverts. Both the tail and flight feathers have light rusty barring. The head and mantle are a dark brown. The underparts are whitish, and the breast is rufous; all with dark brown barring. The uncomplicated ear structure suggests that this species hunts mainly by sight.

HABITAT: Found in tropical lowland rainforest, primary and secondary forest, thickets, forest edge, groves, selectively logged forest, beach and coastal zones, up to mountain forests. Ranges in elevation from sea level to 1750 m.

NATURAL HISTORY: The species is resident in its restricted range, and roosts in thick vegetation (middle canopy) during the day. The male's call is a series of croaking, double "kwaor-kwaor" notes. Its breeding

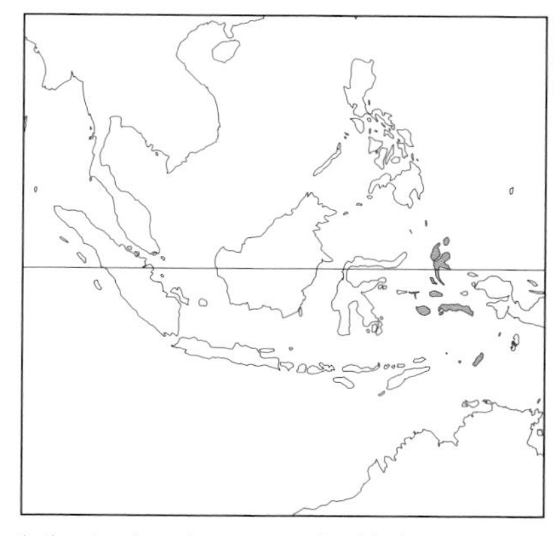

behavior is unknown, and it likely eats mainly insects. It has been observed hunting in lower to middle canopy, frequently in an open location. It is elusive and seldom observed, and more studies are needed to better understand the natural history of this owl.

CONSERVATION STATUS: Not Globally Threatened (IUCN). It occurs in the Buru, Seram, Banda Sea Islands and Northern Maluku Endemic Bird Areas. On Halmahera, Tanimbar and Seram Islands it is reported as moderately common. The nominate race is present in Manusela National Park, Seram.

188 Christmas Island Hawk Owl, *Ninox natalis*

DESCRIPTION: A medium-sized hawk owl (26 to 29 cm) that is restricted to Christmas Island in the Indian Ocean, where it is the only owl. It is no longer considered to be a subspecies of the Moluccan hawk owl due to its distinct calls and coloration. Recent DNA analysis has confirmed its status as a full species. The facial disk is tawny-brown to dark chestnut with whitish "eyebrows"; the lores and throat are also whitish. The upperparts are tawny-brown or cinnamon-rufous with delicate white spots on the nape and wing coverts. The buff-white underparts are heavily barred with rufous-brown. The species is said to be bold and easily approached, but other reports say it is secretive and shy.

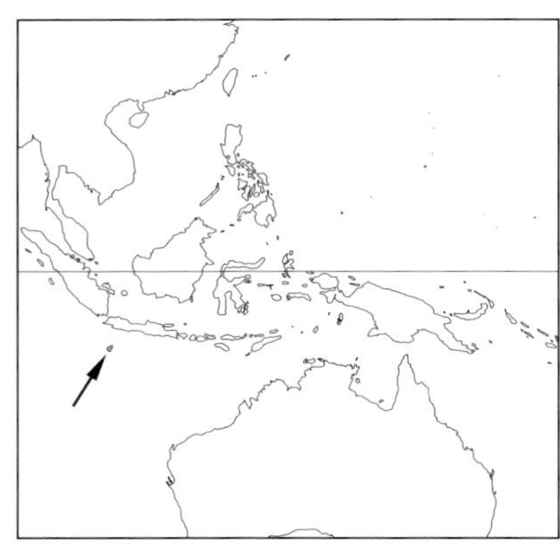

HABITAT: Found across the entire island, the Christmas hawk owl occupies rainforest, secondary forest, plateau and coastal terraces, and evergreen and deciduous forests. It is more abundant in remaining tracts of intact, natural forest.

NATURAL HISTORY: The Christmas Island hawk owl is a permanent resident in its restricted range. It is nocturnal and crepuscular, frequently roosting above busy sidewalks. This species is monogamous and territorial (territories estimated at 18 ha and defended all year), and only 3 nests have been recorded. Clutch size is unknown, but thought to be 2 eggs that are laid in tree hollows. Its diet is mainly large insects (e.g., Lepidoptera, Orthoptera and Coleoptera) which it has been observed catching in flight near lights. Other prey taken include spiders, geckos, birds and rats. Open areas, human settlements and roadsides are used for hunting. Its repetitive barking "ow-ow-ow" or "glu-goog" call has reportedly prompted some island dwellers to call this owl "the dog which no man feeds."

CONSERVATION STATUS: Critically Endangered (IUCN). Former clearing of forests is said to have caused populations to decrease by 25%. Populations were estimated at 10 to 100 pairs between 1965 and 1974, and 560 pairs in 1997. Habitat loss from mining and forest clearance, a decline in habitat quality, and the rapid spread of an introduced insect, the yellow crazy ant, have prompted Environment Australia to draft a recovery plan for this species. Establishing a captive owl colony was a safety precaution until an effective ant-control program could be established. Approximately 1200 breeding adults are estimated to now occur in a 140 square km area. Currently, much of Christmas Island is protected and populations do not face immediate threats. The Australian national list reports it as vulnerable.

189 Jungle Hawk Owl, *Ninox theomacha*

DESCRIPTION: The jungle hawk owl is a small to medium-sized owl (20 to 28 cm) with an unmarked gray-brown to blackish facial disk and lighter "eyebrows." The upperparts are a uniform sooty, chocolate brown, or with just a few white spots on the scapulars. The underparts are chestnut-brown and are unmarked (streaked in 1 race).

HABITAT: It occupies lowland forest, submontane and montane forests (lowlands to 2500 m). It uses forest, forest edge, tree groves in open landscapes and garden habitats.

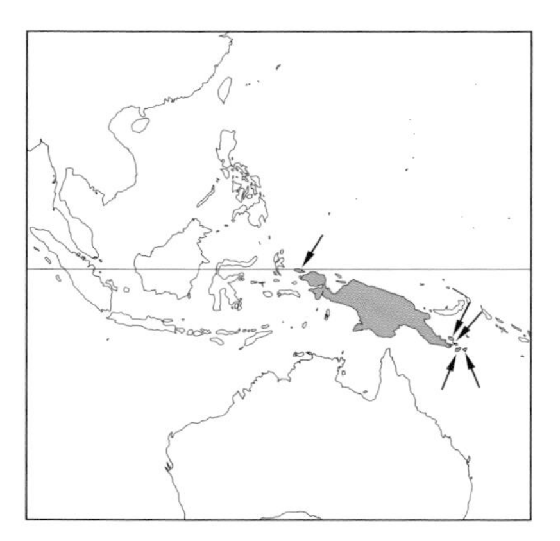

NATURAL HISTORY: The species is resident. During the day, it roosts under thick canopy. Birds are monogamous and territorial; normally 2 eggs are laid in tree hollows. It eats mainly large insects that it captured in flight. Its main call is a gruff series of "kreo-kreo" notes.

CONSERVATION STATUS: Not Globally Threatened (IUCN). It is reported as common throughout its range and is highly adaptable to habitat changes. More studies on this little-known owl are needed.

190 Manus Hawk Owl, *Ninox meeki*

DESCRIPTION: The Manus hawk owl (23 to 31 cm) has been described as one of the most handsome owls in the world. The brown facial disk is unmarked, and the rufous-brown crown and back occasionally are lightly mottled. Both flight feathers and tail are barred white, light rufous and brown. Below, it is whitish-buff, with distinct broad rusty-brown streaks. The species in endemic to Manus Island, in the Admiralty Islands.

HABITAT: It occupies forests, including degraded forest, and occurs within treed areas surrounding human settlements,

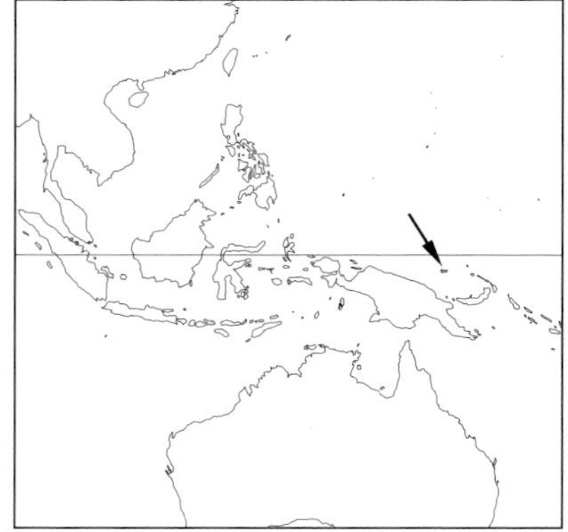

including cultivated riparian habitat.

NATURAL HISTORY: The species is a permanent resident on Manus Island. There is no information available on its breeding biology and habits, and it is reported to eat insects. More information is needed on all aspects of this species' natural history, including its vocalizations. One call is simply reported as a rough series of 10 or more notes.

CONSERVATION STATUS: Not Globally Threatened (IUCN). It occurs in the Admiralty Islands Endemic Bird Area. It is reported as common and likely not currently at risk.

191 Speckled Hawk Owl, *Ninox punctulata*

DESCRIPTION: This small, reddish-brown hawk owl (20 to 27 cm) has white spots on its head, back and wings. A broad white throat patch borders the blackish facial disk; the "eyebrows" are whitish. The reddish-brown upper breast has contrasting buff spots, the remaining whitish-buff underparts are marked with light barring or with reddish-brown spots.

HABITAT: It occupies in primary forest (often near streams), hill forest, open woodland, disturbed forests, open secondary forest, cultivated regions with scattered trees and forest edge, and occurs close to human settlements. It is found at elevations from lowlands to 1100 m, rarely to 2300 m.

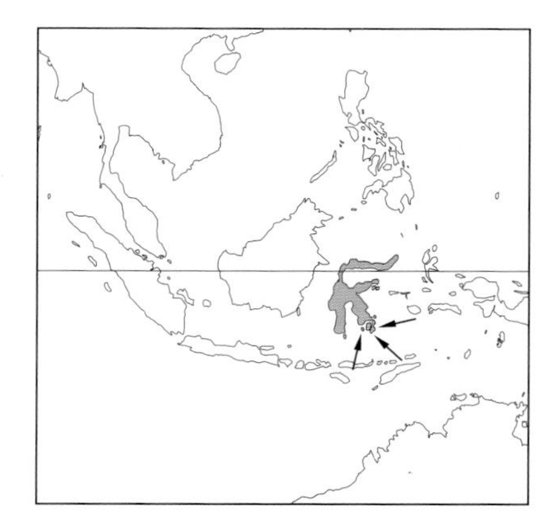

NATURAL HISTORY: A resident species, it is very vocal at night, and calls year-round. Virtually nothing is known about its breeding biology. Little information is available on its diet, but it presumably eats primarily insects; there is a record of it capturing a fruit bat. The male utters an ascending and accelerating series of "toy" or "koi" notes.

CONSERVATION STATUS: Not Globally Threatened (IUCN). It is reported as widespread and common to generally rare, and it tolerates disturbed habitats. It is found on the island on Sulawesi, Indonesia, in Lore Lindu and Dumoga-Bone National Parks.

192 Bismarck Hawk Owl, *Ninox variegata*

DESCRIPTION: The bismarck hawk owl is a small to medium-sized owl (23 to 30 cm). The upperparts are dark brown or rufous-brown, and the scapulars and wing coverts are spotted or barred white, as is the grayish-brown mantle. Light brown bars mark the flight feathers and tail. The whitish underparts are barred dark brown or orange-rufous.

HABITAT: It occupies lowland forest, forest edge, treed hills and low mountains at elevations up to 1000 m.

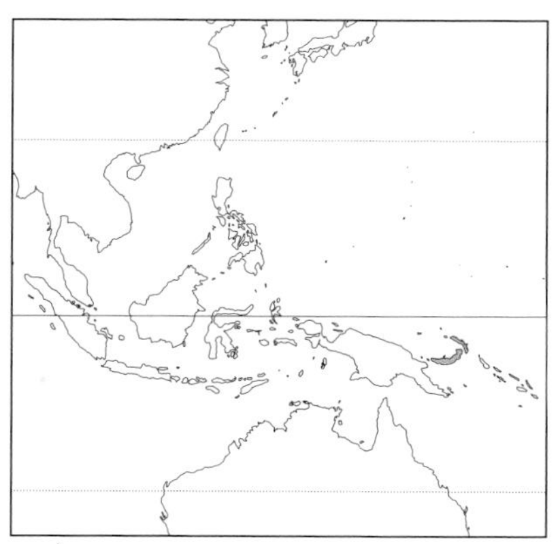

NATURAL HISTORY: The bismarck hawk owl is resident and nocturnal. Virtually no information is available on its breeding biology or food habits; presumably it eats insects. The call is described as a frog-like "kra-kra . . . kra-kra."

CONSERVATION STATUS: Not Globally Threatened (IUCN). It is reported as being widespread and fairly common throughout its restricted range. It occurs in the New Britain and New Ireland Endemic Bird Areas.

193 New Britain Hawk Owl, *Ninox odiosa*

DESCRIPTION: This small hawk owl (20 to 23 cm) has fine spots covering its plumage, and golden-yellow to orange eyes. The upperparts and wing coverts are chocolate brown, flecked with small buff spots. The facial disk is brown with contrasting bold white "eyebrows" and throat. The upper breast is marked with a wide brown band with buff-white bars, while the white lower breast and belly are densely spotted brown or rufous-brown.

HABITAT: The New Britain hawk owl is found in lowlands and hills at elevations to 1200 m, and also in plantations, cultivated regions and towns.

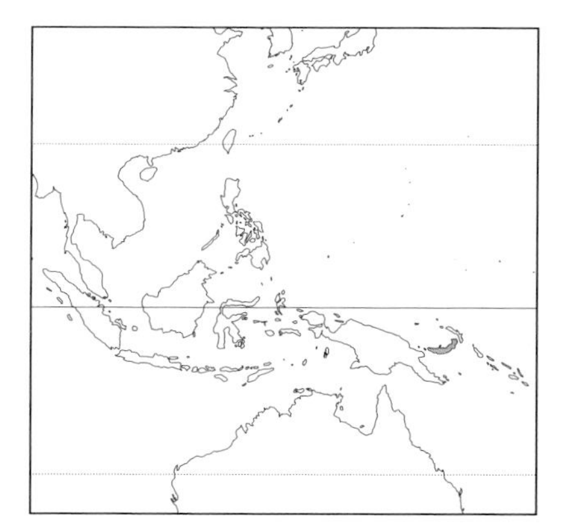

NATURAL HISTORY: The species is resident in its restricted range. A nocturnal bird, it roosts by day alone or in pairs in the upper middle forest canopy. No information is available on its breeding biology. It eats mainly insects, but also small mammals. Its call, reported as a fast series of monotone notes, needs further study.

CONSERVATION STATUS: Not Globally Threatened (IUCN). It occurs in the New Britain and New Ireland Endemic Bird Areas. It is reportedly widely distributed over its restricted range, and is apparently common, especially in lowland habitat.

194 Solomon Hawk Owl, *Ninox jacquinoti*

DESCRIPTION: A small to medium-sized hawk owl (23 to 31 cm) with rusty brown upperparts that may or may not be spotted or barred white. The gray-brown facial disk has narrow whitish "eyebrows" and a band of white at the throat. The buff-white breast is spotted or barred brown, while the belly is creamy and unmarked.

HABITAT: The Solomon hawk owl occupies lowlands and foothills at elevations up to 1500 m. Here it is often found in primary and tall secondary forest, and also in patchy forest and treed gardens.

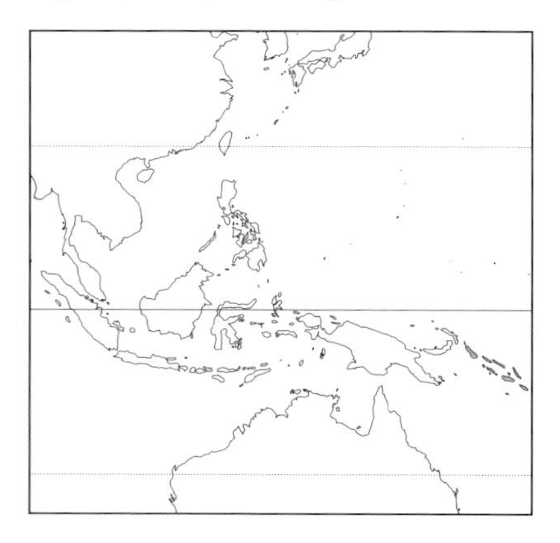

NATURAL HISTORY: The species is resident and nocturnal, roosting by day alone or in pairs in the canopy on main tree forks, or in tree cavities (15 m above the ground). Little information is available on its breeding biology, but it is reported to nest in tree hollows. It eats mainly insects, and presumably small vertebrates as well. One of its calls, a series of double "kwu" notes, has been attributed to the male. The function of this and other whistling hoots and sounds it produces need study.

CONSERVATION STATUS: Not Globally Threatened (IUCN). It occurs in the Solomon (Islands) Group Endemic Bird Area and is considered to be widespread and common in its restricted range. It is rarer on San Cristobal and Malaita Islands.

195 Papuan Hawk Owl, *Uroglaux dimorpha*

DESCRIPTION: A sleek, medium-sized owl (30 to 33 cm) with a relatively small head, long tail and short, rounded wings. It has a small and indistinct whitish facial disk with fine black streaks and white "eyebrows." The black and brown barred upperparts contrast with the light buff underparts that are streaked black and brown. It has large bright yellow eyes and a grayish-black bill.

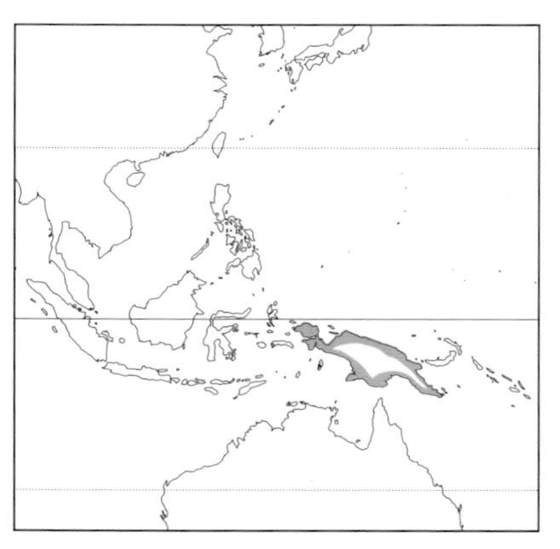

HABITAT: It is known to occur at elevations to 1500 m in lowland rainforest, including forest edges, and gallery forest in savannah habitat.

NATURAL HISTORY: It is likely a resident species in its limited range. The only information on its breeding biology is the observation of a young owl out of the nest. More research is needed on most aspects of its natural history. It reportedly feeds on insects, small rodents and birds nearly as large as itself, e.g., the wompoo fruit dove. Its call has been reported as a drawn-out whistle, "poweeeeho," that rises then falls in pitch.

CONSERVATION STATUS: Data Deficient (IUCN). There is insufficient information to assign this species a status rank. Known from only 9 sites (1980s and 1990s), including birds captured near Lae, the northeastern tip of the island of Malaita. Rarely seen, and thinly distributed across its range, it is assumed to be rare. It may be threatened by the logging of lowland forests.

196 Laughing Owl, *Sceloglaux albifacies*

DESCRIPTION: A yellowish-brown owl with dark brown streaks above and below. It was medium-sized (35 to 40 cm long) with long, broad and rounded wings. It had dark orange eyes. Its round head had no ear tufts. The facial disk consisted of gray feathers that changed outward to white, highlighted by brown feather shaft lines. Its lower legs were covered with yellow to reddish-buff feathers.

HABITAT: This species was found primarily in open country and areas with low rainfall, but several records and specimens in the 19th century were from forested areas. Nest sites appear to have been in rock crevices.

NATURAL HISTORY: The extinction of this species by 1914 precluded detailed study of its biology in the wild. Observations of captive owls suggested that it often fed on the ground. What little is

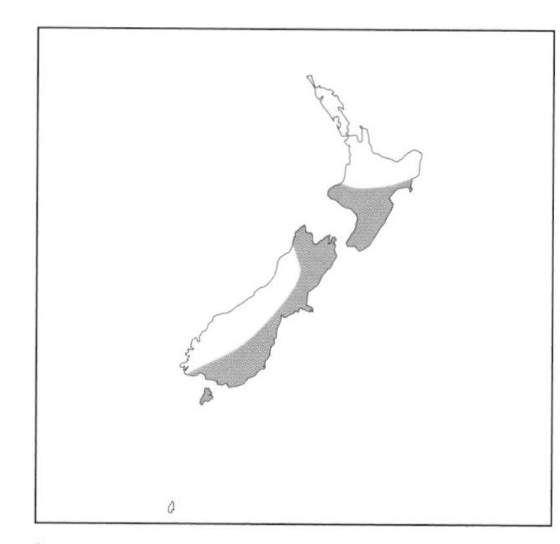

known of its diet was determined from pellets. It ate beetles, rats and mice. It was thought to be a resident, although there are few data to substantiate this. In captivity, the male fed the female while she incubated 2 round white eggs for 25 days. Most specimens came from cracks in rock faces, and were probably associated with nests. The laughing owl is named after one of several calls it gave in captivity, including dismal shrieks, whistles, mewing notes and chuckling noises. Its Maori name "Whekau" is descriptive of one such call.

Conservation Status: Extinct (IUCN). The only species in its genus, the laughing owl was endemic to New Zealand. Its demise coincided with that of many other endemic species following colonization of the islands by Europeans in 1840. It had become exceedingly rare a mere 40 years later due to land-use changes, predation by introduced cats and dogs and persecution by settlers. Unsubstantiated records since 1914 offer only the slightest hope that it may yet survive in some isolated location.

197 Jamaican Owl, *Pseudoscops grammicus*

Description: A medium-sized, tawny-brown owl (27 to 34 cm) with dark brown to black barring, flecks and streaks. Its amber to cinnamon facial disk is rimmed in white that is black-flecked toward the edge. It has hazel-brown eyes, long ear tufts and a grayish-yellow bill.

Habitat: Uses open habitat with scattered clumps of trees and patchy forests, mainly in lowlands but sometimes in montane areas. It will use treed gardens.

Natural History: More study is needed to determine its genetic relationship to other owls such as the long-eared owls and the screech owls. It is active at night, and individuals seem to use the same daytime roost repeatedly. Eats larger insects (mainly beetles) and spiders, amphibians (tree frogs), lizards, birds and rodents. Its 2 eggs are laid in tree cavities, but no other information is available about this resident owl's breeding biology. Its call may be a series of "to-whoo" notes, but it also utters a rough frog-like "k-kwoarrr" repeated every few seconds.

Conservation Status: Not Globally Threatened (IUCN). While it is reported as common and widespread, some populations are at risk from ongoing destruction of forest habitat.

198 Striped Owl, *Asio clamator*

Description: A medium-sized owl (30 to 38 cm) with large ear tufts and a brownish-white, black-rimmed facial disk. It has a black bill and brownish-cinnamon eyes. The wings are shorter and rounder than other long-eared owls. Upperparts are cinnamon-buff, with fine black vermiculations and heavy stripes; underparts are pale tawny to buff-white with dusky shaft streaks.

Habitat: A variety of habitats are used, including riparian woodland, tropical rainforest, marshes, savannahs, grassy open areas and even agricultural and suburban areas. It can be found at elevations from sea level to 1600 m.

Natural History: Its call is a 1-second-long single nasal hoot: "hooOOOoh." A nonmigratory owl, it eats mainly small mammals, especially rodents. Also takes birds (e.g., doves), as well as reptiles and insects. Often hunts in flight over open habitats. Communal roosts of up to 12 owls have been observed. Its 3 eggs (range 2 to 4) are laid in hollows on the ground, in grassy clumps, or sometimes in low (up to 3 m above the ground) tree cavities or on the bases of palm leaves. Apparently, only 1 chick typically fledges.

CONSERVATION STATUS: Not Globally Threatened (IUCN). The lack of information on its biology makes it hard to accurately assess its status. Although widespread, its distribution is patchy, and it seems to be uncommon where found. It may have expanded its range due to forest fragmentation by logging. Pesticides may be a threat in some areas.

199 Stygian Owl, *Asio stygius*

DESCRIPTION: A medium-sized, dusky owl (38 to 46 cm) with yellow eyes. Its white "eyebrows" are prominent against its blackish facial disk and black bill. The underparts are dingy buff with dark brown barring and heavy streaks. Upperparts have buff barring and spots on a blackish-brown background.

HABITAT: It ranges in elevation from sea level to 3100 m, occupying a variety of deciduous and evergreen forest habitat, including parks and open areas with patchy forests. In some areas it is found only on forested mountainsides due to habitat loss at lower elevations.

NATURAL HISTORY: The male's call note is a low-pitched "whuof" repeated at up to

10-second intervals. In response, the female reportedly utters a short cat-like "miah." Hunting at night, this mainly resident owl takes a variety of prey, including small mammals and birds, reptiles, amphibians, crustaceans and insects. Bats are taken in flight. Its 2 eggs are likely laid in the stick nests of other birds, and, apparently, occasionally on the ground.

CONSERVATION STATUS: Not Globally Threatened (IUCN). Its abundance varies considerably over its range, from common in Belize, western Mexico, to rare in Colombia and on Caribbean islands. Its distribution is incompletely documented, and appears to be patchy. Deforestation continues to threaten local populations, and continued human persecution (e.g., in the Dominican Republic) indicates that educational programs are needed to overcome superstitions harmful to the species. More information on its basic biology is needed to implement sound conservation efforts.

200 Long-eared Owl, *Asio otus*

DESCRIPTION: This medium-sized owl (35 to 40 cm) has a large round head, a well-defined buff to orange-colored facial disk and prominent ear tufts. Eye color varies from orange in Eurasia to golden-yellow in North America. The upperparts are a mix of black, brown, gray, white and buff. Its underparts are grayish-white and buff, with dark-brown barring and streaking. Females are generally darker than males.

HABITAT: It occurs in sparsely forested and open habitat across its range, including wooded riparian areas, shelterbelts, farmland, marshes and scrubby deserts. The presence of dense vegetation for nesting and open grasslands or shrub lands for hunting appear to be important habitat features.

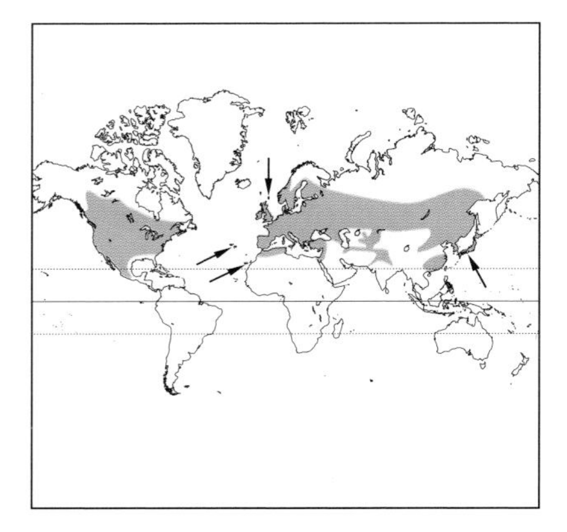

NATURAL HISTORY: The number of long-eared owls fluctuates with small mammal populations, especially meadow voles, and prey-induced nomadism is well documented both in Eurasia and North America. It overwinters in most of its breeding range, but northern populations may migrate up to 4000 km. Most migration movements, however, are likely shorter distances to areas with lesser snow cover and hence greater prey availability. Populations of this species in the Canary Islands and in Africa are apparently nonmigratory. Winter communal roosts of up to 100 individuals have been reported. Its 4 to 5 eggs (range 2 to 10) are laid in abandoned stick nests built by other species in trees in dense shrub or forest vegetation.

It nests less frequently in tree cavities, cliff crevices or on the ground. It actively hunts at night, taking mainly small rodents. Birds are predated more often in a few of the many areas where it has been studied. Some birds are reportedly captured at night while roosting in the canopy of open forests; it can catch its prey in near complete darkness using sound alone. A courting male's call is a series of 10 or more (sometimes more than 200) low-pitched "hoo" notes regularly spaced 2 to 4 seconds apart. It also gives a barking "ooak ooak ooak" alarm call and a cat-like "wawo" that is sometimes mistaken for a similar fox call. Like the short-eared owl, it also may wing-clap, producing a whip-like sound when the wings are clapped together below the body while in flight. Apparently this display is used both in courtship and in aggressive encounters.

CONSERVATION STATUS: Not Globally Threatened (IUCN). Its numbers are thought to be stable in most of its range, but some North American migration monitoring stations have documented a downward population trend over 3 decades. It is recognized, however, that prey-based nomadism may affect such conclusions. Fluctuating prey populations, high nest predation, habitat changes and nest site availability all regulate population numbers. The use of artificial nest structures could mitigate loss of natural nests where crow populations have been reduced. It is listed as endangered, threatened or of special concern in some of the United States. In some locations it has been demonstrated that long-eared owls died after eating rodents killed with Dieldrin-treated seeds or were poisoned by organophosphate Azodrin.

201 African Long-eared Owl, *Asio abyssinicus*

DESCRIPTION: A medium-sized owl (42 to 44 cm) with a patchy and restricted range. Its yellow eyes, black bill and light gray "eyebrows" are set in a tawny-brown facial disk. The white-edged, long, brown ear tufts are somewhat centrally located on the buff head. The dark brown upperparts

are spotted light and dark. The upper breast is a mix of tawny and dark brown; the pale buff or ivory lower belly is checkered by brown streaks and crossbars.

HABITAT: It uses open grassland and moorlands with patchy oak or cedar forests, and occurs in mountain gorges and valleys up to 3900 m. Plantations may sometimes be occupied.

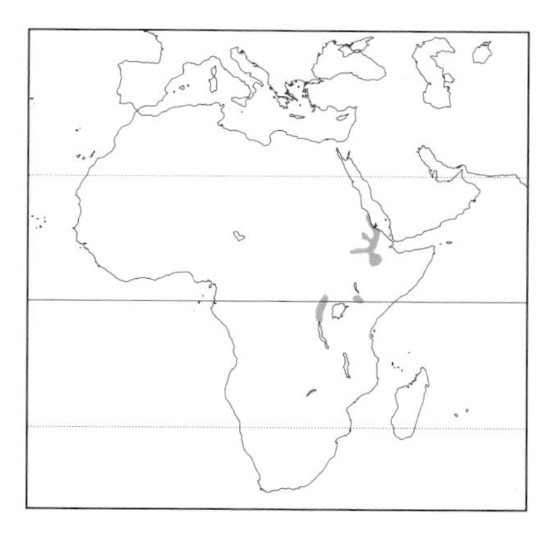

NATURAL HISTORY: There is no information on its movements or breeding biology, but it is thought to use the stick nests of other birds. In places where food is plentiful, it may roost communally with up to 30 others in dense vegetation, e.g., giant heath. Its call is an extended, low-pitched "ooo-oooom" that rises in pitch, but it also utters either a barking sound or a high-pitched squeal when under stress. Rodents, especially exotic species, and shrews are the main prey taken, but it also eats small birds and invertebrates.

CONSERVATION STATUS: Not Globally Threatened (IUCN). Its status is unknown; reported as scarce to rare throughout its range. In some parts of its range it is known from as few as 1 specimen and a couple of sightings. May be threatened by habitat loss and pesticide use.

202 Madagascar Long-eared Owl, *Asio madagascariensis*

Note: downy juvenile shown.

DESCRIPTION: A medium-sized owl (31 to 50 cm). The facial disk varies from dark brown around the dull orange eyes to buff-tan at the edges. The bill is black. The long, thick ear tufts and head are dark brown flecked with tan. The upperparts are black and tan, spotted and vermiculated with brown. Dark brown shaft streaks highlight the light-buff breast and belly.

HABITAT: Found at elevations from sea level to 1600 m, rarely to 1800 m, in a variety of habitats. It occurs in humid evergreen, dry deciduous, gallery, thick secondary and degraded forests.

NATURAL HISTORY: Eats mainly mammals, including lemurs and free-tailed bats, but especially introduced rodents (e.g., rats). Its relatively strong bill and talons suggest it takes larger prey than other related owls. Hunts at night in and adjacent to forests. There is little information on its breeding biology, but it is known to have at least 3 young. It probably lays its eggs in stick nests built by other birds. This resident species roosts in dense foliage during the day. Its call is a long series of barks: "han kan, han kan" varying in loudness and frequency. Also utters another series of "ulooh" notes.

CONSERVATION STATUS: Not Globally Threatened (IUCN). Endemic to Madagascar, its status is uncertain. This species is hard to detect and is possibly rare. It is still perceived as an evil omen and persecuted in some parts of its range. The continued and extensive deforestation on Madagascar is a threat. This species' conservation will depend on much-needed studies of its ecology and biology.

203 Short-eared Owl, *Asio flammeus*

DESCRIPTION: A medium-sized owl (34 to 42 cm) with bright yellow eyes set in black orbits, and a black bill. The well-defined facial disk is grayish-white. It has small, barely visible, ear tufts positioned centrally atop a large round head. Overall mottled brown, resembling dried grass, its underparts are whitish to rusty, with vertical streaks. Females are distinctly darker, browner and more heavily streaked than males. It has relatively long wings and a buoyant and moth-like flight.

HABITAT: Inhabits open country, including tundra, marshes and grasslands throughout much of the world, including distant islands such as the Hawaiian chain and the Galápagos. Also occurs in open areas within treed habitat.

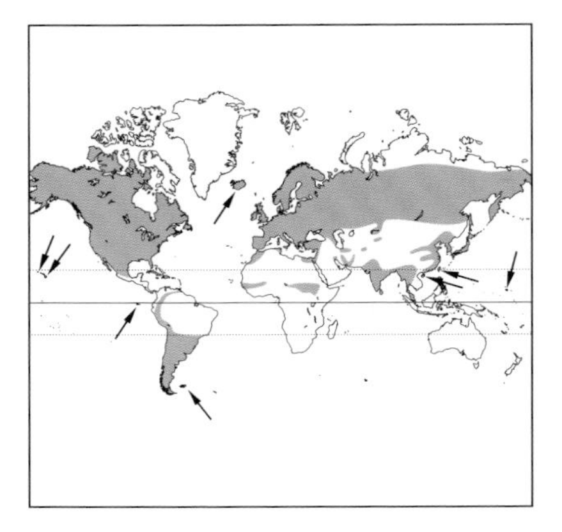

NATURAL HISTORY: It is migratory in the northern parts of its range and generally resident in more southern snow-free areas where prey is more readily available. It is known to move between islands in the Galápagos chain. Population fluctuations of its primary prey (e.g., small mammals such as the meadow vole) greatly influence local population changes, reproductive success and dispersal movements. It less frequently takes birds, even smaller owls (e.g., the burrowing owl). The short-eared owl hunts mostly in flight, occasionally hovering or remaining stationary in the wind on fixed wings, and usually within 3 m of the ground. During breeding periods it often pursues prey by day in open country, thus calling attention to its spectacular nomadic breeding invasions. The male's primary call is described as a long series: "hoo-hoo-hoo-hoo-hoo-hoo," given while displaying to a female during an elaborate and dramatic courtship flight, rising up to 350 m above the ground. With deep and exaggerated wing beats, accompanied by audible wing clapping below the body, the male utters its courtship call. Its 5 to 6 eggs (range 1 to 11) are laid on the ground in a nest bowl dug by the female. Nests have been reported as close together as 55 m, and nesting densities can reach as high as 1 nest per 5.5 ha. The young disperse early, often ranging as far as 50 to 200 m within 4 days. Relatively social, it roosts communally, especially in winter, often as close as 1 m. In arid habitats it is also known to roost by day in large rodent burrows.

CONSERVATION STATUS: Not Globally Threatened (IUCN). Nesting owls and their young are threatened by grassland fires, hay harvesting and increased predation in small remnant patches of prairie grassland habitat. Some may die or be incapacitated as a result of consuming small mammals laden with pesticides.

204 Marsh Owl, *Asio capensis*

DESCRIPTION: A medium-sized (31 to 38 cm) owl with barely visible small ear tufts atop a black-framed facial disk. A dark brown and white face highlights large brown eyes and black bill. It is dark brown above, with a dark-brown chest. Underparts are lighter-colored, with dense barring and distinct shaft streaks. In flight, white wing-tip patches are clearly visible on its long wings. On Madagascar, it is notable larger and more heavily spotted and barred below.

HABITAT: Grassland, marsh and moorland. Found in lowlands and at elevations up to 3000 m. Like the African grass owl, it appears to avoid long grass.

NATURAL HISTORY: This owl can be seen hunting on the wing day or night. Its flight is buoyant, graceful and typically close to the ground. Its most commonly heard call is a coarse, crackling "krrrrrr" sound. A series of croaking "quark" notes is also uttered in flight. In general, its calls are

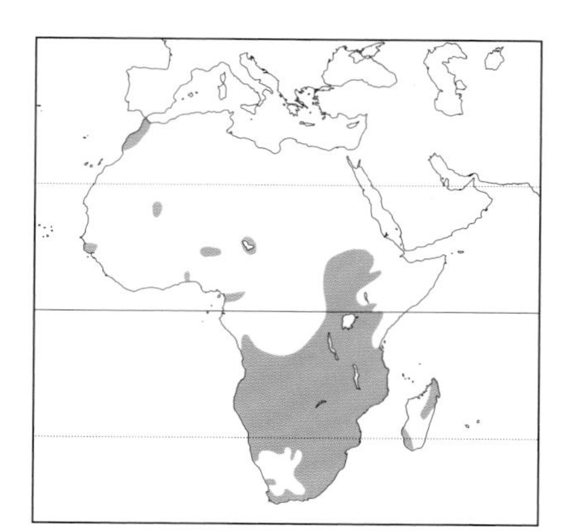

poorly known. Roost sites, which consist of hollows in the grass, are good places to find pellets. The marsh owl eats mainly small rodents, but also takes birds, insects, frogs and lizards. Thirty or more marsh owls may be seen together in areas with concentrated prey, such as during termite emergence, revealing their gregarious behavior. This can also result in breeding concentrations, when nests can be found only 75 m apart. Its nest is a pad of flattened grass in which it lays 2 to 5 eggs, usually 3, which take 4 weeks to hatch. Newly dispersed young roost separately, which may reduce predation. Generally resident, but partly nomadic in dryer areas. The genetic relationship of this species to the short-eared owl needs to be studied.

CONSERVATION STATUS: Not Globally Threatened (IUCN). Perhaps more accurately described as uncertain. Scarce in northwestern Africa and south of the Sahara, although locally common in years with ample prey. Bush fires, floods, overgrazing, pesticides, entanglement in barbed wire and traffic may negatively impact populations. The cumulative affect of human-caused mortality on owl population size is unknown.

205 Fearful Owl, *Nesasio solomonensis*

DESCRIPTION: A medium-sized owl (30 to 38 cm) with strong feet and bill. Its white "eyebrows," chin and throat are offset against a rufous-colored face. The feathers near the inner edges of the facial disk are also white. It is mottled dark brown above, with deep-ochre underparts highlighted by blackish streaks. This owl is sometimes confused with the Solomon hawk owl, which is smaller, more slender and hawk-like. It also bears a striking overall resemblance to the apparently extinct laughing owl, another South Pacific island species. More research is needed to determine if this similarity is the result of a close genetic relationship or simply coincidence.

HABITAT: It occurs mainly in tall lowland and hill forests up to altitudes of 800 m. Nests have been found high in tree holes and cracks, and occasionally on large fig tree epiphytes. The species is found only on Bougainville Island, Papua New Guinea, and Choiseul and Santa Isabel, Solomon Islands, where it is resident, endemic, poorly known and rare.

NATURAL HISTORY: There is very little information published about this species. Perhaps the fearful owl was named for its reportedly mournful, human-like, single note, a protracted call that rises in pitch at

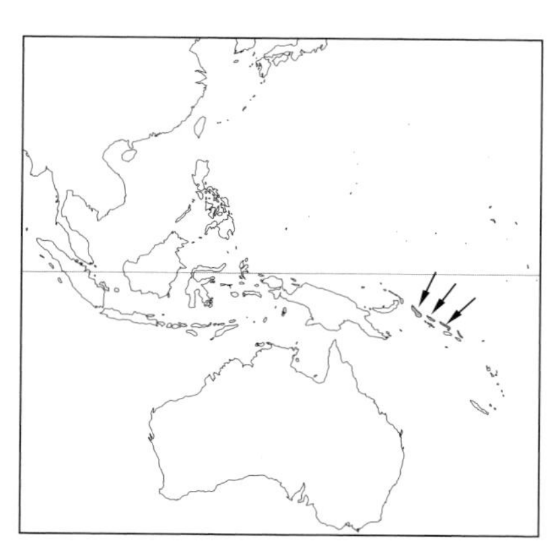

the end. Research has confirmed local knowledge that it feeds mainly on the northern common cuscus, which can be locally abundant. The cuscus is one of a group of New Guinean possums; also known as phalangerids, they are medium to large animals (1 to 6 kg body weight) with a partly naked, prehensile tail. Catching tree-dwelling prey this size requires strong feet and talons.

CONSERVATION STATUS: Vulnerable (IUCN). This enigmatic species is considered at risk because of its small population, which is thought to be declining through habitat loss. Although Solomon Islanders eat the same prey animals as the fearful owl, it is likely to be more threatened by forest destruction. The forests in which it lives are threatened by logging rates estimated at 3 times the sustainable level, with remaining productive forests predicted to be logged within the next 10 years. Remaining populations of this owl are increasingly fragmented into small subpopulations across 3 islands.

List of Contributors

Jerry Batey
4806 Arlene Drive
Corpus Christi, Texas, U.S.A.
78411
jigger@aol.com

Terry Galloway
Department of Entomology
Faculty of Agricultural and Food Sciences
University of Manitoba
Winnipeg, Manitoba, Canada
R3T 2N2
terry_galloway@umanitoba.ca

Greg Hayward
Regional Wildlife Ecologist
USDA Forest Service
Rocky Mountain Region
and
Assistant Professor
University of Wyoming
Laramie, Wyoming, U.S.A.
82071
ghay@uwyo.edu

Patricia Hayward
Wildlife Biologist
786 Roger Canyon Road
Laramie, Wyoming, U.S.A.
82072

Aki Higuchi
Visiting Researcher, Advanced Research Institute for Science and Engineering
Waseda University
Professor Susumu Ishii's Laboratory
Department of Biology, School of Education
Waseda University,
Shinjyuku-ku, Tokyo
169-8050
ahiguchi@mn.waseda.ac.jp

Jeff Hogan
Hogan Films
P.O. Box 10428
Jackson, Wyoming, U.S.A.
83002
hoganfilms@sprynet.com

Geoffrey Holroyd
Canadian Wildlife Service,
Environment Canada
Room 200, 4999-98 Ave.
Edmonton, Alberta, Canada
T6B 2X3
Geoffrey.holroyd@ec.gc.ca

C. Stuart Houston
863 University Drive
Saskatoon, Saskatchewan, Canada
S7N 0J8
Houstons@duke.usask.ca

David Johnson
Northwest Habitat Institute
2344 Summit Lake Shore Rd NW
Olympia, Washington, U.S.A.
98502
djnwhi@aol.com

Ted Leighton and Tyler Stitt
Canadian Cooperative Wildlife Health Centre
Western College of Veterinary Medicine
University of Saskatchewan,
52 Campus Drive
Saskatoon, Saskatchewan, Canada
S7N 5B4
Ted.leighton@usask.ca

Bruce Marcot
USDA Forest Service
620 SW Main St., Suite 400
Portland, Oregon, U.S.A.
97208-3890
brucem@SpiritOne.com

Katherine McKeever
The Owl Foundation
4117, 21st Street, R.R.1
Vineland Station, Ontario,
Canada
L0R 2E0

Irina Menyushina
Prospekt Mira, Bld. 103, Apt. 109
Moscow, Russia
129085,
Ira@nikitaov.msk.ru

Heimo Mikkola
Docent in the Institute of Applied Biotechnology
University of Kuopio, P.O.B 1627
Kuopio, Finland
FIN-70211
fao-gmb@field.fao.org

Robert Nero
Wildlife and Ecosystem
Protection Branch
Manitoba Conservation
Box 24, 200 Saulteaux Crescent
Winnipeg, Manitoba, Canada
R3J 3W3

Jerry Olsen
Applied Ecology Research Group
Faculty of Education
University of Canberra, PO Box 1
Belconnen, Australia
A.C.T. 2616
jerryo@comedu.canberra.edu.au

John M. Penhallurick
Associate Professor
Division of Communication and Education
University of Canberra
Canberra, Australia
A.C.T. 2601
jmp@comedu.canberra.edu.au

Norman Smith
Massachusetts Audubon Society
Blue Hills Trailside Museum
Chickatawbut Hill Education Center
1904 Canton Avenue
Milton, Massachusetts, U.S.A.
02186
nsmith@massaudubon.org

Bernie Solymár
Project Coordinator,
Ontario Barn Owl Recovery
Project R.R. #3, Simcoe, Ontario,
Canada
N3Y 4K2
solymar@nornet.on.ca

Helen Trefry
Canadian Wildlife Service,
Environment Canada
Room 200, 4999-98 Ave.
Edmonton, Alberta, Canada
T6B 2X3
helen.trefry@ec.gc.ca

Susan Trost
Erindale College, McBryde Cres.
Wanniassa, Australia
A.C.T. 2903

Enrique Valdez
Justicia Nº 2570 Circunvalación Vallarta
Guadalajara, Jalisco C.P., Mexico
44680
enriquevaldez@hotmail.com

Selected Bibliography

Alaja, P. 1987. Eläinten värimuunnokset. Metsästys ja Kalastus 9/87:32–33.

Alaja, P. and H. Mikkola 1997. Albinism in the Great Gray Owl (Strix nebulosa) and other owls. Pages 33–37 in Duncan, Johnson, and Nicholls (eds.).

Ali, S. 1987. Indian hill birds. Oxford University Press, Bombay, India.

Ali, S. and S. D. Ripley. 1987. Compact handbook of the birds of India and Pakistan. Oxford University Press, Bombay, India.

Alpers, A. 1964. Maori myths and tribal legends. Longman Paul, Auckland, New Zealand.

Amadon, D., J. Bull, J. T. Marshall and B. F. King, 1988. Hawks and owls of the world: a distributional and taxonomic list. Proceedings of the Western Foundation of Vertebrate Zoology, Volume 3, Number 4, Pages 295–357.

Austin, O. L. 1948. The birds of Korea. Bulletin of the Museum of Comparative Zoology of Harvard College 101:1–301.

Austing, G. R. and J. B. Holt, Jr. 1966. The World of the Great Horned Owl. Lippincott, Philadelphia.

Barringer, J. N. 1980. Unusual plumaged Short-eared Owl. Loon 52:92.

Barss, J. M. 1985. An Analysis of Methodologies for the study of the diets of Long-eared Owls. Portland, OR: Oregon Dept. of Fish and Wildlife, Nongame Wildlife Program.

Batchelor, J. 1901. The Ainu and their folklore. Religious Tract Society, London, U.K.

Bennett, E. B. 1969. Albino Screech Owl. Nebraska Bird Review 37:16.

Bent, A. C. 1961. Life histories of North American birds of prey. Part 2. Orders Falconiformes and Strigiformes. Dover Publications, Inc., New York.

Best, E. 1977. Forest lore of the Maori. E. C. Keating, Government Printer, Wellington, New Zealand.

Best, E. 1982. Maori religion and mythology. Part 2. P.D. Hasselberg, Government Printer, Wellington, New Zealand.

Blakers, M., S. J. J. F. Davies and P. N. Reilly. 1984. The atlas of Australian birds. RAOU and Melbourne University Press, Melbourne.

Bourke, J. G. 1958. An Apache campaign in the Sierra Madre. Charles Scribner's Sons, New York, New York.

Brenner, S. 2002. OPINION: Life Sentences — Ontology recapitulates philology. The Scientist 16[6]:12.

Brewer, E. C. 1898. Dictionary of phrase and fable. Henry Altemius, Philadelphia, Pennsylvania.

Brookes, M. 1998. The species enigma. New Scientist 111:1–4.

Brown, S. C. 1998. Native visions: evolution in Northwest coast art from the eighteenth through the twentieth century. Seattle Art Museum. Douglas and McIntyre Ltd., Vancouver, British Columbia, Canada.

Bull, E. L. and J. R. Duncan. 1993. Great Gray Owl (Strix nebulosa). Birds of North America, No. 41 (Alan Poole and Frank Gill, Eds.). The Academy of Natural Sciences, Philadelphia, and The American Ornithologists' Union, Washington, D.C.

Bull, E. L. and M. G. Henjum. 1990. Ecology of the Great Gray Owl. Portland, OR: U.S. Dept. of Agriculture, Forest Service, Pacific Northwest Research Station, Main Serials SD11 P68 no. 265.

Bunn, D. S. 1982. The Barn Owl. Buteo Books. Vermillion, SD.

Burton, J. A. (ed.). 1973. Owls of the world, their evolution, structure and ecology. E. P. Dutton and Co., Inc., New York.

Cade, T. J. 1983. Hybridization and gene exchange among birds in relation to conservation. Pp. 288–348 in Genetics and conservation. Schonwald-Cox, C. M., Chambers, S. M., MacBryde, B. and Thomas, L. (eds.). Benjamin and Cummings, Menlo Park.

Caduto, M. J. 1994. Keepers of the night: Native American stories and nocturnal activities for children. Fulcrum Pub., Golden, CO.

Canadian Endangered Species Conservation Council (CESCC). 2001. Wild Species 2000: The general status of species in Canada. Ottawa: Minister of Publics Works and Government Services Canada. 48 pp. + CD. (http://www.wild-species.ca).

Cannings, R. J. 1993. Northern Saw-whet Owl (Aegolius acadicus). Birds of North America, No. 42 (Alan Poole and Frank Gill, Eds.). The Academy of Natural Sciences, Philadelphia, and The American Ornithologists' Union, Washington, D.C.

Cenzato, E. and F. Samtopitro. 1990. (English translation copy 1991). Owls: art legend history. Bullfinch Press, Littlebrown and Co., London and Canada.

Christidis, L. and W. E. Boles. 1994. The taxonomy and species of birds of Australia and its territories. Royal Australasian Ornithologists Union Monograph 2:1–112.

Churchfield, S. 1990. The natural history of shrews. Comstock Pub. Associates, Ithaca, NY.

Clark, R. J. and H. Mikkola. 1989. A preliminary revision of threatened and near-threatened nocturnal birds of prey of the world. Pp. 371–388 in Meyburg, B.-U. and Chancellor, R. D. (eds.) Raptors in the modern world. Proceedings of the III World Conference on Birds of Prey and Owls, Eilat, Israel, 22–27, March 1987.

Clark, R. J., D. G. Smith and L. H. Kelso. 1978. Working bibliography of owls of the world. Washington, D.C.: National Wildlife Federation.

Cline, W. 1938. Religion and world view. In: The Sinkaietk or Southern Okanagon of Washington. General Series in Anthropology 6, Menasha, Wisconsin.

Clottes, J. 2001. Chauvet Cave. National Geographic. August 2001, pp. 104–121.

Cocker, M. 2000. African birds in traditional magico-medicinal use—a preliminary survey. Bulletin of the African Bird Club 7(1):60–66.

Cocker, M. Personal communication. On the perception of owls in Western cultures.

Collar, N., M. J. Crosby and A. J. Stattersfield. 1944. Birds to watch 2: the world list of threatened birds. BirdLife International, Cairns.

Cooper, J. E. 2002. Birds of prey: Health and disease. 3rd. Ed. Blackwell Science Ltd., Malden, Massachusetts.

Cramp, S. 1985. Handbook of the birds of Europe, the Middle East and North Africa. Vol. 4. Oxford Univ. Press, Oxford.

de Plancy, J. C. 1965. Dictionary of demonology. Wade Baskin, editor and translator. Philosophical Library, New York, 117 pp. Originally published as Dictionaire infernal, 1863, Paris, France, 7th edition.

Dean, R. 1976. Albinism and melanism among North American birds. Bulletin of the Nuttall Ornithological Club 2:20–24.

del Hoyo, J., A. Elliott and J. Sargatal (eds). 1999. Handbook of the birds of the world. Vol. 5. Barn-owls to Hummingbirds. Lynx Edicons, Barcelona.

Doughty, C., N. Day and A. Plant. 1999. Birds of the Solomons, Vanuatu and New Caledonia. Christopher Helm, London.

Duncan, J. R. and P. A. Lane. 1987. Breeding boreal owls in Roseau County. The Loon 59(4):163–165.

Duncan, J. R. and P. A. Lane. 1988. Great horned owl observed “hawking” insects. J. Raptor Res. 22(3):93.

Duncan, J. R. 1992. Influence of prey abundance and snow cover on Great Gray Owl breeding dispersal. Ph.D. thesis, Zoology Dept., Univ. Manitoba., Winnipeg, MB.

Duncan, J. R. and P. H. Hayward. 1995. Review of Technical Knowledge: Great Gray Owls. pp. 154–175 (Chapter 14) in Hayward, G.D. and J. Verner, eds. Flammulated, Boreal, and Great Gray Owls in the United States: A Technical Conservation Assessment. USDA Forest Service Gen. Tech. Report RM-253.

Duncan, J. R. 1996. Techniques to sex and age Great Gray Owls: A bird in the hand worth two in the bush? Birders Journal 5(5):240–246.

Duncan, J. R. 1997. Great Gray Owls (Strix nebulosa nebulosa) and forest management in North America: A review and recommendations. J. Raptor Res. 31(2):160–166.

Duncan, J. R., D. H. Johnson and T. H. Nicholls (eds). 1997. Biology and Conservation of Owls of the Northern Hemisphere. Second International Symposium: General Technical Report NC – 190, St. Paul, MN, USDA Forest Service.

Duncan, J. R. and R.W. Nero. 1998. Juvenile eastern screech-owl dies eating a red-breasted nuthatch. Blue Jay. 56(4):209–212.

Duncan, J. R. and P. A. Duncan. 1998. Northern Hawk Owl (Surnia ulula). Birds of North America, No. 356 (Alan Poole and Frank Gill, Eds.). The Birds of North America, Inc., Philadelphia, PA.

Elton, C. S. 1942. Voles, mice and lemmings. Problems in population dynamics. Clarendon Press, Oxford.

Enriquez, P. L. and H. Mikkola. 1997. Comparative study of general public owl knowledge in Costa Rica, Central America and Malawi, Africa. Pages 160–166 in Duncan, Johnson and Nicholls (eds).

Evans, D. L. 1978. Partial albinism: Saw-whet Owl and Goshawk. The Loon 50(1):52–53.

Fienup-Riordan, A. 1996. The living tradition of Yup'ik masks. University of Washington Press, Seattle.

Flood, J. 1997. Rock art of the Dreamtime. Harper Collins Publishers, Sydney, Australia.

Ford, N. L. 1967. A systematic study of the owls based on comparative osteology. Ph.D.diss., Univ. of Michigan, Ann Arbor.

Forsyth, A. 1988. The nature of birds. Camden House Publishing, Camden East, ON.

Friederici, P. 2000. Colorless in a world of color. National Wildlife 8-9:1–4.

Gärdenfors, U., J. P. Rodríguez, C. Hilton-Taylor, C. Hyslop, G. Mace, S. Molur and S. Poss. 1999. Draft guidelines for the Application of IUCN Red List Criteria at National and Regional Levels. Species 31-32:58-70.

Gehlbach, F. R. 1994. The Eastern Screech Owl: Life history, ecology, and behavior in the suburbs and countryside. Texas AandM University Press, College Station.

Gill, F. 1990. Ornithology. W. H. Freeman and Co., New York.

Goddard, T. R. 1935. Notes on colour variation and habits of Short-eared Owls. British Birds 28:290–291.

Gore, M. E. J. and P. Won. 1971. The birds of Korea. Royal Asiatic Society, Korea Branch, Seoul, Korea.

Gorman, M. L. 1990. The natural history of moles. Comstock Pub. Associates, Ithaca, NY.

Grant, M. and J. Hazel. 1993. Who's who in classical mythology. Oxford University Press, New York.

Graves, R. 1992. The Greek myths. Penguin Books, London.

Gray, J. 2001. White said Fred (with apologies to Bernard Cribbins). Tyto 6(3):118–119.

Griffiths, A. J. F., J. H. Miller, D. T. Suzuki, R. C. Lewontin and W. M. Gelbart. 1993. An introduction to genetic analysis. 5th edn. W. H. Freeman and Co., New York.

Gromov, I. M. 1992. Voles (Microtinae). Smithsonian Institution Libraries: National Science Foundation, Washington, D.C.

Gross, A. O. 1965. The incidence of albinism in North American birds. Bird-Banding 36:67–71.

Hamer, T. E., E. C. Forsman, A. D. Fuchs, and M. L. Walters. 1994. Hybridization between Barred and Spotted Owls. Auk 111:487–492.

Hands, H. M. 1989. Status of the Common Barn-Owl in the Northcentral United States. The Service, Twin Cities, MN.

Hardy, J. W., B. B. Coffey, Jr. and G. B. Reyard. 1990. Voices of New World Owls. Rev. ed. Florida Mus. Nat. Hist. and Ara Records, Gainesville, FL. [Cassette tape]

Haug, E. A., et al. 1993. Burrowing Owl (Speotyto cunicularia). Birds of North America, No. 61 (Alan Poole and Frank Gill, Eds.). The Academy of Natural Sciences, Philadelphia, and The American Ornithologists' Union, Washington, D.C.

Hawthorn, A. 1979. Kwakiutl art. University of Washington Press, Seattle and London.

Helo, P. 1984. Yön Linnut. Kirja Suomen Pöllöistä. Luontokuva Pekka Helo. Kajaani.

Henry, G. M. 1969. A guide to the birds of Ceylon. Kandy, Ceylon.

Hilton-Taylor, C. (Compiler) 2000. 2000 IUCN Red List of Threatened Species. IUCN, Gland, Switzerland and Cambridge, UK. Xviii + 61 pp. Downloaded Sept. 3, 2001.

Hinam, H. and J. R. Duncan. 2001. The effects of habitat fragmentation and slope on the distribution of three owl species in the Manitoba Escarpment, Manitoba, Canada. Pages 148–161 in Newton, I., R. Kavanagh, J. Olsen and I. Taylor (eds.).

Holland, G. 1991. Second albino Great Gray Owl sighted in Manitoba. Blue Jay 49:32.

Holm, B. 1987. Spirits and ancestors. University of Washington Press, Seattle.

Holmgren, V. C. 1988. Owls in folklore and natural history. Capra Press, Santa Barbara, CA.

Holt, D. W., M. W. Robertson and J. T. Ricks. 1995. Albino Eastern Screech-owl, *Otus asio*. Canadian Field Naturalist 109:121–122.

Hosking, E. and J. Flegg. 1982. Eric Hosking's owls. Mermaid Books, Pelham Books Ltd., London.

Hume R. and T. Boyer. 1991. Owls of the world. Runnig Press, Philadelphia, Pennsylvania.

Ingersoll, E. 1923 (reprinted 1958). Birds in legend, fable and folklore. Longmans, Green and Co., London.

IUCN. 1994. IUCN Red List Categories. IUCN Species Survival Commission, Gland, Switzerland.

Jaeger, E. C. 1955. A source-book of biological names and terms. 3rd Ed. Charles C. Thomas, Publisher, Springfield, IL.

Janossy, D. 1972. Die mittelpleistozäne Vogelfauna der Stránská Skála. Anthropos 20:35–64.

Janossy, D. 1978. Plio-Pleistocene bird remains from the Carpathian Basin, III. Aquila 84:9-36.

Jarvis, K. and D. W. Holt. 1996. Owls: Whoo are they? Mountain Press Publishing Company, Missoula, Montana.

Johnsgard, P. A. 2002. North American Owls: Biology and natural history. 2nd. Ed. Smithsonian Institution Press, Washington, D.C.

Jordaens, J. As the old ones sing, so the young ones pipe. P. 352 ff. in: Peter C. Sutton. The age of Rubens. Museum of Fine Arts. Boston, Massachusetts.

Karalus, K. E. and A. W. Eckert. 1974. The owls of North America. Doubleday, New York.

Kelly, S. 1990. Ornithology: Flocking birds of a feather together. Washington Post. 22-08-90.

Kelsey, E. 1985. Owls. Nature's Children Series. Grolier Limited, Toronto.

Kemp, A. and S. Calburn. 1987. The owls of Southern Africa. Struik Winchester, Foreshore, Cape Town.

Keyser, J. D., C. Pedersen, G. M. Bettis, G. Poetschat and H. Hiczun. 1998. Owl cave. Oregon Archaeological Society Publication No. 11. Portland, Oregon.

Kirk, R. 1978. Exploring Washington archaeology. University of Washington Press, Seattle.

König, C., F. Weick, and J. H. Becking. 1999. Owls: A guide to the owls of the world. Yale University Press, New Haven.

Korpimäki, E. 1986. Gradients in population fluctuations of Tengmalm's Owl *Aegolius funereus* in Europe. Oecologia (Berlin) 69:195–201.

Krahe, R. 2000. Mutation Spectacled Owls. Tyto 5(2):86–87.

Kravitz, D. 1976. Who's who in Greek and Roman mythology. Clarkson N. Potter, Pub. New York.

Lane, P. A. and J. R. Duncan. 1987. Observations of northern hawk-owls nesting in Roseau County. The Loon 59(4):165–174.

Lau, H. Personal communication. On owls in China.

Lawrence, R. D. 2001. Owls: The Silent Fliers. Revised Edition. Key Porter Books, Toronto.

Li, D. 1993. Animals for Chinese medicine in Guizhou Province [in Chinese]. Guizhou Science and Technology Press, Guiyang, Guizhou, China.

Lockwood, W. B. 1993. The Oxford dictionary of British bird names. Oxford University Press, Oxford.

Loyn, R. Personal communication. On aboriginal peoples and owls of North Queensland, Australia.

Lynch, M. Personal communication. On ancient myths and Dutch owl lore.

Marcot, B. G. 1993. Conservation of forests of India: an ecologist's tour. USDA Forest Service, Pacific Northwest Research Station, Portland OR.

Marcot, B. G. 1995. Owls of old forests of the world. USDA Forest Service General Technical Report PNW-GTR-343. Portland, Oregon, U.S.A.

Marshall, J. T. Jr. 1967. Parallel variation in North and Middle American Screech-Owls. Monographs of the Western Foundation of Vertebrate Zoology 1:1–72.

Marshall, J. T., Jr. 1978. Systematics of smaller Asian night birds based on voice. Ornithol. Monograph 25. Amer. Ornith. Union, Washington, D.C. [with cassette tape]

Martin, L. C. 1996. Folklore of birds. Globe Pequot Printers.

Mayr, E. and Paynter R. A. 1964. Checklist of birds of the world, Vol. 10, Museum of Comparative Zoology, Cambridge, Mass.

Mayr, E. and L. Short. 1970. Species taxa of North American birds. A contribution to Comparative systematics. Publ. Nuttall Ornithol. Club 9.

McGaa, E. 1990. Mother Earth spirituality—Native American paths to healing ourselves and our world. Harper Collins Publishers, New York.

McKeever, K. 1983. Care and rehabilitation of injured owls. W. F. Rannie, Publisher. Lincoln, Ontario.

McKeever, L. 1986. A dowry of owls. Lester and Orpen Dennys Ltd., Toronto.

Medlin, F. 1967. (3rd printing 1971) Centuries of owls. Silvermine Publishers, Inc.

Mikkola, H. 1983. Owls of Europe. Buteo Books, Vermillion, S. Dakota.

Mikkola, H. 1992. Wood owls. Pp. 108–140 in Owls of the World (Burton, J. A., ed.). Eurobook/Peter Lowe, London.

Mikkola, H. 1997a. General public owl knowledge in Malawi. Soc. Malawi J. 50(1):13–35.

Mikkola, H. 1997b. Comparative study on general public owl knowledge in Malawi and in eastern and southern Africa. Nyala 20:25–35.

Mikkola, H. 1999. Owls in traditional healing. Tyto 4(3):68–78.

Mikkola, H. 2000a. Albinism in owls. Tyto 4 (5):133–140.

Mikkola, H. 2000b. The Great Gray Owl called Linda. Tyto 5 (4):145–148.

Mikkola, H. 2001. Hybridization between Owls. Tyto 5 (5):202–214.

Miller, A. H. 1935. The vocal apparatus of the Elf Owl and Spotted Screech Owl. Condor 37:288.

Miller, L. 1963. Birds and Indians in the West. Bulletin of Southern California Academy of Sciences 62:178–191.

Mourer-Chauviré, C. 1975. Les oiseaux du Pleistocène moyen et supérieur de France. Doct. Thesis, Univ. Claude Bernard, Lyons.

Mueller, C. D. and F. B. Hutt. 1941. Genetics of the fowl, sex-linked imperfect melanism. Journal of Heredity 32:71–80.

Nengminza, D. S. 1996. The school dictionary Garo to English. Thirteenth edition. Garo Hills Book Emporium, Tura, Meghalaya, India.

Nero, R. W. 1957. Vestigial claws on the wings of a Red-winged Blackbird. Auk 74:262

Nero, R. W. 1980. The Great Gray Owl: Phantom of the Northern Forest. Smithsonian Institution Press, Washington, D.C.

Nero, R.W. and S. L. Loch. 1984. Vestigial wing claws on Great Gray Owls, *Strix nebulosa*. Can. Field-Nat. 98:45–46.

Nero, R. W., R. J. Clark, R. J. Knapton and R. H. Hamre (des). 1987. Biology and conservation of Northern Forest Owls. Sumposium Proceedings General Technical Report RM-142, Fort Collins, CO, USDA Forest Service.

Nero, R. W. 1991. White Great Gray Owl. Blue Jay 49:31.

Nero, R. W. 1994. Lady Grayl: owl with a mission. Natural Heritage/Natural History, Toronto.

Newton, I. 1979. Population ecology of raptors. Buteo Books, Vermillion.

Newton, I., R. Kavanagh, J. Olsen, and I. Taylor (eds.). 2002. Proceedings of Owls the 2000 Conference. Melbourne: CSIRO Publishing.

Norman, J. A., L. Christidis, M. Westerman, and F. A. R. Hill. 1998. Molecular data confirms the species status of the Christmas Island Hawk-Owl *Ninox natalis*. Emu 98:197–208.

Olsen, J., and S. Trost. 1997. Territorial and nesting behavior in Southern Boobook *Ninox novaeseelandiae*. Pages 308–313 in Duncan, Johnson and Nicholls (eds.).

Olsen, J., and S. Taylor. 2001. Winter home range of an adult female Southern Boobook *Ninox novaeseelandiae* in suburban Canberra. Austr. Bird Watcher 19: 109–114.

Olsen, J., S. Trost and G. Hayes. 2002. Vocalisations used by Southern Boobooks (*Ninox novaeseelandiae*) in the Australian Capital Territory. pages 305–319 in Newton, I., R. Kavanagh, J. Olsen and I. Taylor (eds.).

Olsen, J., M. Wink, H. Sauer-Gürth and S. Trost. 2002. A new *Ninox* owl from Sumba, Indonesia. Emu 102:223–231.

Opler, M. 1965. An Apache life-way. Cooper Square, New York.

Payevsky, V. A. and A. P. Shapoval. 2000. Survival rates, life spans, and age structure of bird populations. Annual Reports of the Zoological Institute RAS. Zoological Institute, Russian Academy of Sciences, Universitetskaya nab., 1, St. Petersburg, 199034, Russia. http://www.zin.ru/annrep/2000/17.html

Petty, S. J. 1994. Moult in Tawny Owls Strix aluco in relation to food supply and reproductive success. In: Meyburg, B. U.; Chancellor, R. D., eds. Raptor conservation today. WWGPD. The Pica Press: 521–530.

Rackham, H, translator. 1997. Pliny natural history, Books 8–11. Harvard University Press, Cambridge, Mass.

Ray, V. F. 1939. Cultural relations in the Plateau of Northwestern America. Fredrick Webb Hodge Anniversary Publication Fund 3.

Regan, H. M., M. Colyvan and M. A. Burgman. 2000. A proposal for fuzzy International Union for the Conservation of Nature (IUCN) categories and criteria. Biol. Conserv. 92:101–108.

Ross, C. C. 1973. Some additional records of albinism in North American birds. Cassinia 54:18–19.

Rydberg, V. 1879. The magic of the Middle Ages. Translated from the Swedish by August Hjalmar Edgren. Henry Holt and Company, New York.

Sage, B. L. 1962. Albinism and melanism in Birds. British Birds 55:201–225.

Sage, B. L. 1963. The incidence of albinism and melanism in British Birds. British Birds 56:409–416.

Sage, J. H. 1983. A partial albino Short-eared Owl. Bulletin of the Nuttall Ornithological Club 8:183.

Sangma, D. R. 1993. Jadoreng: the psycho-physical culture of the Garos. M/S Singhania Printing Press, Shillong, Meghalaya, India.

Saunders, N. J. 1995. Animal spirits. MacMillan, London.

Saurola, T. P. 1995. Luomen Pöllöt. Kirjayhtymä Oy, Helsinki.

Saxena, A. Personal communication. On owls in India.

Sayers, B. 1996. A personal view of colour mutations. Tyto 1(1):24–29.

Scherzinger, W. 1983. Beobachtungen an Waldkauz-Habichtskauz-Hybriden. (Strix aluco × Strix uralensis). Zool. Garten, N.F., Jena 53(2):133–148.

Schneider, J. 1969. Albino Screech Owl. Nebraska Bird Review 37:16.

Schodde, R. (1978). The status of endangered Papuasian birds, and Appendix, Pp. 133–145 and 185–206 respectively In: Tyler, M. J. (ed.) The status of endangered Australasian wildlife. Royal Zoological Society of South Australia, Adelaide.

Sclater, P. L. 1858. On the general geographical distribution of the members of the class Aves. Journ. Proc. Linnean Society (London) (Zoology) 2:130–145.

Scriven, R. 1984. A note on albinism in the Great Gray Owl. Blue Jay 42:173–174.

Seebohm, H. 1976. The birds of Siberia. Alan Sutton, Dursley.

Shaw, G. 1990. Barn Owl conservation in Forests. H.M.S.O., London.

Shields, J. and G. King. 1991. Spotted Owls and forestry in the American Northwest. A conflict of interests. Birds International 2:34–45.

Shirihai, H. 1996. The birds of Israel. Academic Press, London.

Sibley, C. G. and B. L. Monroe, Jr. 1990. Distribution and Taxonomy of Birds of the World. Yale University Press, New Haven, CT.

Smythies, B. E. 1953. The birds of Burma. 2nd edition. [Publisher unknown], Edinburgh-London.

Sova, M. Personal communication. On Russian traditions.

Sparks, J. and T. Soper. 1970. Owls: Their natural and unnatural history. Taplinger Publishing Company, New York.

Spofford, W. R. 1952. A partial albino Horned Owl. Kingbird 2:84.

Stefansson, O. 1997. Nordanskogens Vagabond Lappugglan. St. Petersburg.

Steinberg, R. 1999. White mutation Tawny Owls (*Strix aluco*). Tyto 4(3):118–119.

Sterling, R. 1990. Barred, spotted owls hatch hybrid. Mail Tribune, Oregon. 05-08-90.

Steyn, P. 1984. A delight of owls. David Phillip, Publisher (Pty) Ltd., Cape, South Africa.

Sutton, G. M. 1912. An albinistic Burrowing Owl. Bird Lore 14:184.

Sutton, P. and C. Sutton. 1994. How to spot an owl. Chapters Publishing Ltd., Shelburne, VT.

Takats, D. L., C. M. Francis, G. L. Holroyd, J. R. Duncan, K. M. Mazur, R. J. Canings, W. Harris and D. Holt. 2001. Guidelines for nocturnal owl monitoring in North America. Beaverhill Bird Observatory and Bird Studies Canada, Edmonton, AB. 32 pp.

Taylor, I. 1994. Barn Owls: predator-prey relationships and conservation. Cambridge.

Tekenaka, T. Personal communication. On the fish owl ceremony of the Ainu.

Thackway, R. and Cresswell, I. D. (eds.) 1995. An Interim biogeographic regionalisation for Australia: a framework for establishing the national system of reserves, Version 4.0. Australian Nature Conservation Agency, Canberra.

Titani, F. 1997. Albinos want to meet Muluzi. Daily Times, February 13, 1997, Malawi.

Tiwari, B. K., S. K. Barik and R. S. Tripathi. 1999. Sacred forests of Meghalaya. Regional Centre, North Eastern Hill University, Shillong, Meghalaya, India.

Toops, C. The enchanting owl. 1990. Voyageur Press, Stillwater, Minnesota.

Tyler, H. D. and D. Phillips. 1978. Owls by day and night. Naturegraph Publishers, Inc. Happy Camp, CA.

Vallee, J.-L. 1999. La Chouette Effraie. Delachaux et Niestle, Lausanne, Suisse.

VanCamp, L. F. and C. J. Henney. 1975. The Screech Owl: Its life history and population ecology in northern Ohio. U.S. Dept. of Interior, Fish and Wildlife Service, North America Fauna, No. 71.

Voous, K. H. 1988. Owls of the northern hemisphere. The MIT Press, Cambridge, Mass.

Walker, L. W. 1974. The book of owls. Alfred A. Knopf, New York.

Wallace, A. R. 1876. The geographical distribution of animals, with a study of the relations of living and extinct faunas as elucidating the past changes of the earth's surface (2 vols). London.

Wallace, D. R. 1983. The Klamath knot. Sierra Club Books, San Francisco, California.

Weinstein, K. 1989. The owl (in art myth and legend). Crescent Books, New York.

Welty, J. C. 1979. The life of birds. 2nd Ed. Saunders College Publishing, Philadelphia, U.S.A.

Whitfield, M. B., M.E. Maj and J. Kelley. 1995. Incomplete albino Great Gray Owl in Idaho. Blue Jay 53:197–199.

Wink, M. and P. Heidrich. 1999 Molecular Evolution and Systematics of the Owls (Strigiformes) pp. 39–57 in König et al. 1999.

Yolen, J. 1987. Owl moon. Philomel Books, New York.

WORLD WIDE WEB REFERENCES

The Owl Pages (my favorite site)
http://www.owlpages.com

Barn Owl Trust
http://www.barnowltrust.org.uk/

Owl Research Institute
http://www.owlinstitute.org/ori.html

Raptor Research Foundation
http://biology.boisestate.edu/raptor/

World Owl Trust
http://www.owls.org/

Virtual Owl Pellet Dissection
http://www.kidwings.com/
http://www.caosclub.org/nsw/funstuff/mystery.html

Integrated Taxonomic Information System on-line database, 2001
http://www.itis.usda.gov/

Garnett, S. T. and G. M. Crowley. 2000. The Action Plan for Australian Birds 2000. Environment Australia.
http://www.ea.gov.au/biodiversity/threatened/action/birds2000/

North American Bird-Banding Laboratory
http://www.pwrc.usgs.gov/bbl/default.html

The European Union for Bird Ringing
http://www.euring.org/

Raptor Research Foundation
http://biology.boisestate.edu/raptor/

Acknowledgments

Many good things have happened to me since owls became an integral part of my life. One of these things was meeting Patricia Anne Lane, who volunteered to assist me with field studies of the great gray owl. Patsy, a university zoology graduate and volunteer with the Manitoba peregrine falcon recovery project, had the courage to venture out into the Manitoba wilderness with me when she hardly knew me. That first evening, back in September 1986, at close quarters in a derelict camper-trailer, we talked and laughed until 3:00 a.m. Afterwards, Patsy seemed to thrive on the long hours, hard work and strenuous conditions typical of fieldwork. What better preparation for a couple raising children! We have been mutually enjoying our fascination with owls and biology ever since. To Patsy, I extend my deepest gratitude and love.

Retired high school principal Robert McEwin first got me hooked on owls. I was completing a M.Sc. study on American kestrels at McGill University in 1983 under the supervision of Dr. David Bird. Robert, a volunteer at McGill's Raptor Research Centre, helped me with my kestrel research. One day he confessed that he really wanted to work with owls and invited me to help him set up public owl "hoots"—an event where a group is led into the forest at night and either imitates the calls of owls or plays a tape or CD of recorded owl calls. This is done to get owls in the area to respond by calling or flying in close enough to see by flashlight at the nearby Morgan Arboretum. After the first night, I was hooked on owls—it was pretty neat hearing and seeing them at night in the wild, and from then on, all major decisions in my life were partly determined by how the outcome would affect my opportunity to study owls.

I moved to Manitoba in 1985 to work with Dr. Robert W. Nero, one of North America's most gifted natural historians and an expert on the great gray owl. Steeped in the tradition of his professors, Dr. John T. Emlen, Jr., and Aldo Leopold, Bob taught me to appreciate the subtle and nonstatistical aspects of nature study while not compromising scientific integrity. Bob, who had pioneered and willingly shared research techniques, also introduced me to long-time owl enthusiasts Robert R. Taylor, Herbert W. R. Copland and Steven L. Loch. I am, in part, a product of their generosity.

Many folks assisted in the preparation of this book. The photographers who forwarded their amazing images truly make the book a work of art. Colleagues from around the world, many of whom I met at two symposia in Winnipeg, Manitoba, in 1987 and 1997, enthusiastically contributed personal accounts of their work on owls. I am very grateful for their effort, as I know how busy they are with careers and families. As you read their contributions you will sense their strong commitment to owl conservation. Various parts and versions of this text have benefited from the thoughtful and constructive review of my closest colleagues, Bob Nero and Patsy Duncan. Thanks also to the supportive staff at Key Porter Books in Toronto (especially Janice Zawerbny and Michael Mouland), who patiently allowed me to complete the book over several years.

My former supervisors at Manitoba Conservation, namely Dr. Merlin W. Shoesmith, Brian Gillespie and Carol Scott, provided approval and encouragement for my participation on this project. While much of my experience with owls stems from work-related activities for the Manitoba Wildlife and Ecosystem Protection Branch, Manitoba Conservation, it is important to recognize that the views and information presented in this book are not those of the Province of Manitoba.

I thank my wonderful parents, Connie and James Duncan, for their love, patience and support over the years. They encouraged and nurtured the interests of a young boy in love with the natural world. My father instilled in me a love of books, while my mother tolerated all manner of creatures, alive or dead, in our home.

I am indebted to the rest of my family and friends for the continued help and support that enable Patsy and me to continue our efforts to understand these fascinating creatures. Studying owls is not simply a job, but rather a way of life that is both rich and rewarding beyond measure. The thrill of discovery in observing nature is a precious gift that is free for all of us to enjoy.

Photo Credits

Jamie Acker: 112
Pentti Alaja: 140
Ardea: 238 (middle)
Auscape International: 199, 201 (middle), 202 (right), 203, 221, 234 (middle), 243 (left), 248, 272 (left), 276 (middle), 277 (middle)
Robert A. Behrstock/Naturewide Images: 28, 51, 205, 207, 212, 214, 222 (left, right), 223, (right), 224, 225, 226, 227, 228 (middle, right), 229, 240, 250 (left), 251 (left), 253 (right), 254 (left, right), 255, 256 (right), 259, 261 (right), 266 (left), 267, 269 (right), 271, 272 (right), 275, 285 (right)
Dean Berezanski: 114
Hans and Judy Beste/Ardea: 198 (left)
Elizabeth Bomford/Ardea: 106, 244 (left)
Ed Brown: 109
Nancy Camel: 249, 273
Richard Clark/Rudolf Koes: 169 (upper)
John Clarke: 141
Jim Collins: 187
Gordon Court: 260 (left)
Gertrud and Helmut Denzau: 279
Eric Dragesco/Ardea: 258 (middle)
Brooke Duncan: 13
Jim Duncan: 16, 20, 21 (upper), 23, 26, 31, 33, 37,45, 49, 53, 54, 56, 59 (upper), 60, 63, 68, 69, 71 (lower), 74, 75, 107, 108 (upper right), 110, 111 (upper), 112 (right), 113, 115 (lower), 116, 117, 118 (upper), 119, 120, 121,126, 144, 154, 155, 156, 162, 166 (lower), 167, 168 (upper right)
Patsy Duncan: 5, 34, 190, 195
John S. Dunning/Ardea: 262
Göran Ekström: 287 (middle)
Mark Elderkin: 67 (lower)
Environment Canada (with permission): 157
Kenneth W. Fink/Ardea: 245 (right), 260 (middle)
Clifford and Dawn Firth: 241 (left)
Michael and Patricia Fogden: 204
Dick Forsman: 251 (middle)
Frank Lane Picture Agency Ltd. (FLPA): 207 (right), 211 (right), 213 (middle), 237 (middle), 244 (right), 245 (left), 246, 258 (right), 276 (right), 285 (middle)
Terry Galloway: 161, 163
Doug Gilmour: 166 (upper)
Francois Gohier: 32, 233, 250 (middle), 266 (right), 286 (right)
John A Gray: 139
Joanna van Gruisen/Ardea: 269 (left)
Don Hadden/Ardea: 282, 284 (right)
Thomas Hamer: 86
Dan Hartman: 174, 274 (left)
J. Hawkins/FLPA: 58, 247 (left)

Greg Hayward: 181 (left), 183, 184, 185
Patricia Hayward: 181 (right), 186
Aki Higuchi: 177, 178, 179, 180
Jeff Hogan: 25, 175 (lower), 252
Geoffrey Holroyd 118 (lower)
Eric and David Hosking/FLPA: 217, 247 (right)
c/o C. Stuart Houston: 137, 173
Joseph del Hoyo/Lynx Editions: 201 (right)
David H. Johnson: 15, 90, 91, 92, 93, 94, 95, 99, 100, 101, 103
Frank Johnson: 14
Gerry Jones: 17, 19, 24, 27, 44, 58, 59 (lower), 61, 62, 65, 148, 158, 243 (middle), 257, 288 (left)
Steve Kaufman/DRK Photo: 37, 241 (middle)
Rob Leslie: 236 (left)
Alan Lewis: 150, 219
Eric Lindgren/Ardea: 284 (middle)
Manitoba Conservation: 142
Manitoba Conservation/Ted Muir: 22, 35, 168 (lower)
Manitoba Conservation/Robert Taylor: 71, 125
Bruce G Marcot: 102
Luiz Claudio Marigo: 264 (right)
Bill Martz: 111 (lower)
Kurt Mazur: 108 (lower)
Stephen McDaniel: 223
Joe McDonald/Bruce Coleman Inc.: 274 (right)
Katherine McKeever: 191, 192, 193, 194
Irina Menyushina: 127, 128, 129, 130, 131
Pat Morris/Ardea: 238 (right)
Pete Morris: 66, 124, 209, 253 (left), 261 (left), 280 (right), 286 (left), 287 (right)
Jerry Olsen: 81, 122, 123, 124, 277, (right), 278 (middle)
John P. O'Neill: 231, 264 (middle), 270
Jim and Thelma Pepper: 169 (lower)
Tom Reaume: 40
Jim Reimer: 29, 57, 115 (upper), 145 (upper), 164
Ian Ritchie: 21 (lower)
Len Robinson/FLPA: 198 (middle)
S. Robinson: 42
Lior Rubin: 213 (left)
Manuel Ruedi: 207 (left)
Jordi Sargatal/Lynx Editions: 153, 280 (middle)
Wolfgang Scherzinger: 84, 85
Gary Schultz Photography: 196, 239, 265
Kay Shaw: 236 (middle), 288 (right)
Bernt Solymar: 188
Norman Smith: 135, 147
Peter Steyn/Ardea: 235, 278 (right)
Leine Stikkel: 175 (upper)
Morten Strange: 208
Brent Terry: 32, 170, 172
Roy Toft: 202 (left), 220, 242, 256 (left, middle)
Norman Tomalin/Bruce Coleman Inc.: 232
Jon Triffo: 64, 67 (upper), 108 (upper left), 149
Ian Ward: 36, 150
Doug Wechsler/VIREO: 211 (left)
J.T. White: 88
Rod Williams/Bruce Coleman Inc.: 237 (right)
Michael Wink: 78
Anthony Wlock/Reto Zach: 145 (lower)
Günter Ziesler: 234 (left)
K. Zimmer/VIREO: 230

Index

Page citations in **boldface** below refer to **illustrations**.

P

T